T0309713

BEAM PROPAGATION METHOD FOR DESIGN OF OPTICAL WAVEGUIDE DEVICES

BEAM PROPAGATION METHOD FOR DESIGN OF OPTICAL WAVEGUIDE DEVICES

Ginés Lifante Pedrola

Universidad Autónoma de Madrid, Spain

This edition first published 2016
© 2016 John Wiley & Sons, Ltd

Registered Office

John Wiley & Sons, Ltd, The Atrium, Southern Gate, Chichester, West Sussex, PO19 8SQ, United Kingdom

For details of our global editorial offices, for customer services and for information about how to apply for permission to reuse the copyright material in this book please see our website at www.wiley.com.

Library of Congress Cataloging-in-Publication Data

Lifante, Ginés.
   Beam propagation method for design of optical waveguide devices / Ginés Lifante Pedrola.
      pages cm
   Includes bibliographical references and index.
   ISBN 978-1-119-08337-5 (cloth)
1. Beam optics.   2. Light–Transmission.   3. Optical wave guides.   I. Title.
   QC389.L525 2015
   621.36–dc23

                                    2015022455

A catalogue record for this book is available from the British Library.

Cover image: saicle/Getty

Set in 10/12pt Times by SPi Global, Pondicherry, India
Printed and bound in Singapore by Markono Print Media Pte Ltd

1   2016

To my beloved sisters:
Belín, María Cinta, María José, Lidia, Pilar and Zoila.

Contents

Preface

The aim of this book is to provide the fundamentals and the applications of the beam-propagation method (BPM) implemented by finite-difference (FD) techniques, which is a widely used mathematical tool to simulate light wave propagation along axially varying optical waveguide structures. The content covers from the background, variations of the method, numerical implementations and applications of the methodology to many practical examples. Thus, the book gives systematic and comprehensive reviews and tutorials on the analysis and design of integrated photonics devices based on optical waveguides using FD-BPM. It treats almost all aspects of BPM analysis, from fundamentals through the advancements developed by extension and modifications, to the most recent applications to specific integrated optical devices.

The book can be a text for postgraduate courses devoted to numerical simulation of integrated photonic devices. Also, it is suitable for supplementary or background reading in modern curricula graduate courses such as 'Optoelectronics', 'Optical engineering', 'Optical-wave electronics', 'Photonics' or 'Integrated optics'. This book is also of interest for professional researchers and engineers in the area of integrated optics, optoelectronics and optical communications. Although BPM codes are commercially available, or even free, many engineers must develop their own software to suit their particular requirements. This book can serve both the building of home-made codes, as well for use of existing software by understanding the underlying approaches inherent in the BPM and its range of applicability.

Integrated photonics devices are based on optical waveguides with transversal dimensions of the order of microns. This means that the light propagation along these structures cannot be analysed in terms of ray optics; instead the light must be treated as electromagnetic waves. Hence, Chapter 1 presents the basics of the electromagnetic theory of light, starting from Maxwell's equations in inhomogeneous media. Wave equations in terms of the transverse field components in inhomogeneous media are obtained, including the treatment of anisotropic media and second-order non-linear media. Using the slowly varying approximation, full vectorial equations for the electric and magnetic fields are obtained in Chapter 2, which are the basic differential equations for developing BPM algorithms. Finite-difference approximations of the wave equations are then derived for the simplest case of scalar propagation in two-dimensional structures that allows us to study the stability and numerical dissipation of FD-BPM schemes. Chapter 3 develops full vectorial FD-BPM algorithms for the simulation of light propagation in 2D and 3D structures where the numerical implementations of the FD-BPM are detailed.

Extensions and modifications of the BPM approaches based on finite-difference techniques are presented in Chapter 4. These include wide-angle BPM, which relaxes the restriction of the application of BPM to paraxial waves and allows the simulation of light beams with large propagation angles respect of the longitudinal direction. BPM algorithms, which can handle multiple reflections known as bidirectional-BPM, are then discussed. Simulation of light propagation in active media, second-order non-linear media and anisotropic media are also topics covered in Chapter 4. The last sections include the description of time-domain (TD) simulation techniques based on finite differences, which can simulate the propagation of optical pulses and can manage backward waves due to reflections at waveguide discontinuities. Both the time-domain BPM and finite-difference time-domain (FDTD) are explained in detail. The different BPMs supply almost universal numerical tools for describing the performance of a great variety of integrated optical devices. Although particular devices have specific routes to be modelled with their own constraints, the great advantage of the BPM lies in the fact that, as few approximations have been made for its derivation, its applicability is quite wide and almost any integrated photonic device can be modelled by using it. The last chapter, Chapter 5, presents selected examples of integrated optical elements commonly used in practical integrated photonic devices, where their performance and relevant characteristics are analysed by the appropriate BPM approach.

Some appendices have been added at the end of the book. They include material related to BPM algorithms or BPM simulations of some integrated photonics devices, but which is not indispensable to the understanding of the different topics developed along the book chapters. The appendices include mathematical derivations of some formulae, physical phenomena descriptions and even relevant program listing.

Commonly accepted notation and symbols have been utilized throughout this book. However, some of the symbols have multiple meanings and therefore a list of symbols and their meanings is provided at the beginning of the book to clarify symbol usage. Also, a list of acronyms is given to help the reader.

A selection of BPM programs are made available free of charge for the readers at the website of the author (www.uam.es/personal_pdi/ciencias/glifante). Among others, this selection includes the programs 'Vectorial mode solver for planar waveguides', 'Vectorial light propagation in 2D-structures', 'Vectorial light propagation in 3D-structures' and '2D-light propagation in the time domain'.

Ginés Lifante Pedrola
Madrid, February 2015

List of Acronyms

ABC	absorbing boundary conditions
ADI	alternating direction implicit
ASE	amplified spontaneous emission
AWG	arrayed waveguide grating
BC	boundary condition
Bi-BPM	bidirectional beam propagation method
BPM	beam propagation method
CCS	complementary coplanar strip
CFL	Courant–Friedrichs–Levy stability criterion
CN	Crank–Nicolson scheme
CS	coplanar strip
CW	continuous wave
DFR	distributed feedback reflector
EIM	effective index method
EM	electromagnetic
EO	electro-optic
ESW	equivalent straight waveguide
FD	finite difference
FD-BPM	finite-difference beam-propagation method
FDTD	finite-difference time domain
FE	finite element
FE-BPM	finite-element beam-propagation method
FFT	fast Fourier transform
FFT-BPM	fast Fourier transform beam-propagation method
FT	Fourier transform
FPR	free propagation region
FSR	free spectral range
FV-BPM	full vectorial beam propagation method
IFD-NL-BPM	iterative finite difference non-linear beam-propagation method

IL	insertion loss
Im-Dis-BPM	imaginary distance beam-propagation method
IR	infrared
MMI	multimode interference
MPA	modal propagation analysis
MZ	Mach–Zehnder
MZI	Mach–Zehnder interferometer
NL	non-linear
NL-BPM	non-linear beam propagation method
OI	overlap integral
PHASAR	phase-array
PML	perfectly matched layer
QPM	quasi-phase matching
RE	rare earth
SH	second harmonic
SHG	second harmonic generation
SV-BPM	semi-vectorial beam-propagation method
SVE	slowly varying envelope
SVEA	slowly varying envelope approximation
TBC	transparent boundary conditions
TD-BPM	time-domain beam-propagation method
TE	transverse electric field
TF/SF	total field/scattered field
TM	transverse magnetic field
UV	ultraviolet
WDM	wavelength division multiplexing

List of Symbols

Roman Symbols

a_j	tridiagonal system coefficient
a_ν	modal weight
A	attenuation; also, waveguide cross section
A_x	x dependent part of the operator P_{xx} (or Q_{xx})
A_y	y dependent part of the operator P_{xx} (or Q_{xx})
A_{ij}	spontaneous emission probability
b	normalized propagation constant
b_j	tridiagonal system coefficient
B_x	x dependent part of the P_{yy} (or Q_{yy}) operator
B_y	y dependent part of the P_{yy} (or Q_{yy}) operator
$\mathcal{B}(r,t)$	magnetic flux density vector
c	speed of light in free space
c_j	tridiagonal system coefficient
C	P_{xy} (or Q_{xy}) operator (cross-coupling term)
$C_{M,N}(i,j)$	coefficient for the FDTD algorithm
d	thickness, depth
d_{ijk}	second-order non-linear tensor
d_{PML}	PML thickness
D	P_{yx} (or Q_{yx}) operator (cross-coupling term); also, distance, separation
$D_{M,N}(i,j)$	coefficient for the FDTD algorithm
$\mathbf{D}$	(complex amplitude of) displacement vector
$\mathcal{D}$	differential operator (for wide-angle BPM)
$\mathcal{D}(r,t)$	electric displacement vector
E_i	i Cartesian component of E
E_{yx}, E_{yz}	splitting (sub-components) of the magnetic field component E_y in FDTD
$\mathbf{E}(r)$	complex amplitude of $\mathcal{E}(\mathbf{r},t)$ for monochromatic waves
$\mathbf{E}_t(r)$	transverse component of E
$\mathcal{E}(r,t)$	electric field
$f(x)$	mode transversal profile

$f_\nu(x,y)$	eigenmode (eigenvector, transversal field distribution)
F_j	energy flux leaving the j-boundary
$\mathcal{F}$	operator for NL-BPM
G	differential operator (in 3D-scalar wave equation)
G_x, G_y	split of the differential operator G
h	Planck's constant; also, height
$\mathcal{H}$	differential operator for TD-BPM
H_j	i Cartesian component of H
$\mathcal{H}_x, \mathcal{H}_y$	split of the differential operator for TD-BPM
H_{yx}, H_{yz}	splitting (sub-components) of the magnetic field component H_y in FDTD
$H(r)$	complex amplitude of $\mathcal{H}(r,t)$ for monochromatic waves
$H_t(r)$	transverse component of H
$\mathcal{H}(r,t)$	magnetic field
$\mathcal{H}_0(r)$	magnetic field amplitude for monochromatic waves
$\hat{\mathbf{H}}$	matrix differential operator for $\mathit{\Psi}_t$ (or $\mathit{\Phi}_t$)
i	imaginary unity; also, integer
I	intensity (or irradiance)
I_0	intensity of a monochromatic plane wave
I_p	pump intensity
j	integer
$\mathcal{J}(r,t)$	electric current density
k	wavenumber in the medium; also, integer
k_0	wavenumber in free space
K	reference wavenumber
$\widetilde{K}$	complex-valued reference wavenumber
L	window size; also, length
L_x, L_y	transversal grid dimensions
L_z	longitudinal length
L_c	coupling length
$\mathcal{L}_A, \mathcal{L}_B$	pseudo-differential operators
$\mathcal{L}_j$	operators for NL-BPM
m	integer
M_m	polynomial of degree m
$\mathcal{M}$	overall transfer matrix (for Bi-BPM)
$\mathcal{M}(r,t)$	magnetic current density
n	refractive index; also, integer
n_0	reference refractive index
n_c	complex refractive index
n_p	refractive index of the PML medium
N	concentration of active ions
N_e	effective index of the symmetric mode
N_{eff}	effective index of the mode
N_i	population density in the i-th level
N_n	polynomial of degree n
N_o	effective indices of the anti-symmetric mode
N_x, N_y	number of transversal grid points

$\mathcal{O}[..]$	approximation order in FD schemes
p	power exponent of the PML profile
$P(z)$	complex field amplitude correlation function
$P(\xi)$	Fourier transform of $P(z)$
$\mathcal{P}_j$	propagation matrix (for bidirectional BPM)
P_{ij}	differential operator for the transverse SVE field ψ_t
$\mathcal{P}$	differential operator for wide angle BPM; also, for wide-band BPM
$\mathcal{P}(r,t)$	polarization vector
$\mathcal{P}_{NL}$	non-linear polarization
Q_{ij}	differential operators for the transverse SVE field Φ_t
Q_j^m	Von Neumann analysis parameter
Q	operator for wide band BPM
r	reflection coefficient
r_j	tridiagonal system coefficient
r_0	maximum reflection coefficient at the PML region
r	position
R	reflectivity
R_j, R_{ij}	coefficient for FD schemes of BPM
R_{ij}	pump rate (or stimulated emission rate)
$S1_{ij}$–$S4_{ij}$	coefficients for FD schemes of BPM
$\mathcal{S}$	Poynting vector
S	complex Poynting vector
t	time; also, transmission coefficient
T	period
T_j	transmission coefficient
$\mathcal{T}_{AB}$	interface matrix (for bidirectional BPM)
u	SVE-field component Ψ_x (or Φ_x)
$u(x,y,z)$	SVE scalar optical field
u_f	SVE for the fundamental wave
u_s	SVE for the SH wave
u_j^m	discretized SVE optical field
u_j^+	discretized incident field ψ_A^+
u_j^-	discretized reflected field ψ_A^-
$u(r,t)$	temporal envelop of the electric field
u_t	SVE of the transverse electric field
$\mathbf{u}_x$, $\mathbf{u}_y$, $\mathbf{u}_z$	unitary vectors along the x-, y- and z-axis
v	SVE-field component Ψ_y (or Φ_y); also, propagation speed of an EM wave
w	width
$w(k,t)$	spatial frequencies
w_j^+	discretized forward field ψ_B^+
W	width
W_{ij}	stimulated emission rate
W_{ij}^{NR}	non-radiative probability
W_ν	relative power carried by the ν-th mode
x	Cartesian coordinate

X_j	dimensionless operator
y	Cartesian coordinate
z	Cartesian coordinate
Z	total propagation length
$Z1_{ij}$–$Z4_{ij}$	coefficients for FD schemes of BPM

Greek Symbols

α	Crank–Nicolson scheme parameter; also, absorption coefficient
α_{eff}	effective attenuation coefficient (of PML)
$\tilde{\alpha}_s$	intrinsic propagation losses
β	propagation constant
β_ν	propagation constant of the νth order eigenmode
χ_i	polynomial coefficient
χ_L	linear susceptibility
$\chi^{(2)}$	coefficient of second-order non-linear susceptibility
χ_{ijk}	element of the second-order non-linearity susceptibility tensor
δ	delta Kronecker function; also, ABC region thickness (or PML region)
$\Delta x, \Delta y$	grid size
Δz	longitudinal step size
Δk	mismatch parameter
Δt	time step
ε	scalar dielectric permittivity
ε_r	dielectric constant (relative dielectric permittivity)
ε_0	dielectric permittivity of free space
ε_{ij}	element of the permittivity matrix
$\boldsymbol{\varepsilon}$	permittivity tensor
$\phi(i)$	transversal field distribution of a waveguide mode for FDTD
γ	amplification factor (for Von Neumann analysis); also, damping factor (for bi-BPM)
Γ	correlation between optical fields; also, overlap integral
$\boldsymbol{\eta}$	impermeability tensor
η_0	free space impedance
$\varphi(\boldsymbol{r})$	initial phase
φ	incident angle
Φ_x	x-component of the SVE transversal magnetic field
Φ_y	y-component of the SVE transversal magnetic field
$\boldsymbol{\Phi}_t$	SVE field of $\boldsymbol{H}_t(\boldsymbol{r})$
κ	absorption index; also, coupling coefficient
$\kappa_{\max}$	maximum value of $\kappa(x)$
$\kappa(x)$	absorption index profile (in ABC)
λ	wavelength
Λ	grating period
μ	magnetic permeability
μ_0	magnetic permeability of free space
ν	frequency
θ_i	angle of reflection (or transmission)

ρ	parameters for ABC region
ρ_i	magnetic conductivity of the PML
$\rho(\boldsymbol{r},t)$	charge density
$\sigma(\boldsymbol{r})$	electrical conductivity
$\sigma(\rho)$, σ_i	electrical conductivity profile of the Bérenger layer
σ_{ij}	absorption (or emission) cross-section
$\sigma_{\max}$	maximum conductivity of the PML
τ	pulse temporal width; also, lifetime
ω	angular frequency
$\omega(k)$	relation dispersion
ω_s	angular frequency for the SH wave
ω_f	angular frequency for the fundamental wave
ω_0	carrier frequency
ξ_i	polynomial coefficient
ξ_ν	eigenvalue (relative propagation constant)
$\psi(x, y, z)$	generic scalar field
ψ^+	forward field
ψ^-	backward field
ψ_A^+	incident field in region A
ψ_A^-	reflected field in region A
ψ_B^+	transmitted field in region B
$\boldsymbol{\Psi}$	slowly varying electric field
$\boldsymbol{\psi}_t$	SVE field of $\boldsymbol{E}_t(\boldsymbol{r})$
Ψ_x	x component of the SVE transverse electric field $\boldsymbol{\psi}_t$
Ψ_y	y component of the SVE transverse electric field $\boldsymbol{\psi}_t$

Mathematical Symbols

∂	partial differential
∇	gradient operator
∇	divergence operator
$\nabla\times$	curl operator

1

Electromagnetic Theory of Light

Introduction

Integrated photonics devices are based on optical waveguides with transversal dimensions of the order of microns, comparable to the wavelength of the optical radiation used in the integrated devices (visible and near infrared). This fact implies that the performance of the optical chips cannot be analysed in terms of ray optics, but instead the light must be treated as vectorial waves. Thus, to describe adequately the light propagation along the waveguide structures that define an integrated photonics device, the electromagnetic theory of light is required, which deals the light as optical waves in terms of their electric and magnetic fields. This treatment retains the vectorial character of the waves. Nevertheless, in some cases the vectorial nature of the electromagnetic waves can be simplified and a scalar treatment of the optical waves is enough for an accurate description of the light propagation through the optical waveguides.

Along this chapter the basics of the electromagnetic theory of light is described, which is the start point to derive the beam propagation equations to model the light propagation in optical waveguides and integrated photonic devices. First, the Maxwell's equations for light propagation in free space are presented in terms of the electric and magnetic field. Then, the electric displacement vector and the magnetic flux density vector are introduced to describe the optical propagation in material media. The constitutive relations allow then to establish a set of equations in terms of the electric and magnetic fields. Using these equations, the wave equations in inhomogeneous media are derived where the refractive index can be then defined. Then the wave equation for monochromatic waves in inhomogeneous media is obtained, where the temporal dependence of the fields is in the form of harmonic function. The especial cases of light propagation in absorbing media, anisotropic media and in second-order non-linear media are discussed, and wave equations for each case are derived. Finally, the wave equations in inhomogeneous isotropic and linear media in terms of the transverse field components are obtained, for both electric and magnetic fields. Also, wave equation for anisotropic media

Beam Propagation Method for Design of Optical Waveguide Devices, First Edition. Ginés Lifante Pedrola.
© 2016 John Wiley & Sons, Ltd. Published 2016 by John Wiley & Sons, Ltd.

and second-order non-linear media are established in terms of the electric transverse field components. These equations will serve to derive in the subsequent chapters the beam propagation formalism.

1.1 Electromagnetic Waves

1.1.1 Maxwell's Equations

Light is, in terms of classical theory, the flow of electromagnetic (EM) radiation through free space or through a material medium in the form of oscillating electric and magnetic fields. Although electromagnetic radiation occurs over an extremely wide range from gamma rays to long radio waves, the term 'light' is restricted to the part of the electromagnetic spectrum that covers from the vacuum ultraviolet (UV) to the far infrared. This part of the spectrum is often also called optical range. The EM radiation propagates in the form of two mutually perpendicular and coupled vectorial waves: the electric field $\mathcal{E}(\mathbf{r},t)$ and the magnetic field $\mathcal{H}(\mathbf{r},t)$. These two vectorial magnitudes are dependent on the position ($\mathbf{r}$) and time (t). Therefore, to describe properly the light propagation in a medium, be it the vacuum or a material medium, it is necessary in general to know six scalar functions with their dependence of the position and the time. These functions are not independent but linked through Maxwell's equations.

Maxwell's equations form a set of four coupled equations involving the electric field vector and the magnetic field vector of the light and are based on experimental evidence, two of them being scalar equations and the other two vectorial equations. In their differential form, Maxwell's equations for light propagating in the free space are:

$$\nabla \cdot \mathcal{E} = 0; \tag{1.1a}$$

$$\nabla \cdot \mathcal{H} = 0; \tag{1.1b}$$

$$\nabla \times \mathcal{E} = -\mu_0 \frac{\partial \mathcal{H}}{\partial t}; \tag{1.1c}$$

$$\nabla \times \mathcal{H} = \varepsilon_0 \frac{\partial \mathcal{E}}{\partial t}, \tag{1.1d}$$

where the constants $\varepsilon_0 = 8.85 \times 10^{-12} \, \text{m}^{-3} \, \text{kg}^{-1} \, \text{s}^4 \, \text{A}^2$ and $\mu_0 = 4\pi \times 10^{-7} \, \text{m} \, \text{kg} \, \text{s}^{-2} \, \text{A}^{-2}$ represent respectively the dielectric permittivity and the magnetic permeability of free space and the ∇ and $\nabla \times$ denote the divergence and curl operators, respectively.

The differential operator ∇ is defined as:

$$\nabla \equiv \left(\frac{\partial}{\partial x} \mathbf{u}_x + \frac{\partial}{\partial y} \mathbf{u}_y + \frac{\partial}{\partial z} \mathbf{u}_z \right), \tag{1.2}$$

where $\mathbf{u}_x$, $\mathbf{u}_y$ and $\mathbf{u}_z$ represent the unitary vectors along the x-, y- and z-axis, respectively. This differential operator acting to a scalar field gives rise to a vector (gradient). In particular, if $\xi(x,y,z)$ represents a scalar field, we have:

$$\nabla \xi(x,y,z) \equiv \left(\frac{\partial \xi}{\partial x} \mathbf{u}_x + \frac{\partial \xi}{\partial y} \mathbf{u}_y + \frac{\partial \xi}{\partial z} \mathbf{u}_z \right). \tag{1.3}$$

On the other hand, if $\mathbf{A}(x,y,z) = A_x(x,y,z)\mathbf{u}_x + A_y(x,y,z)\mathbf{u}_y + A_z(x,y,z)\mathbf{u}_z$ is a vector field, the divergence operator $(\nabla \cdot)$ acts as follows:

$$\nabla \cdot \mathbf{A} \equiv \frac{\partial A_x}{\partial x} + \frac{\partial A_y}{\partial y} + \frac{\partial A_z}{\partial z}, \tag{1.4}$$

which is a scalar magnitude. Finally, the curl differential operator $(\nabla \times)$ acting on the vector field $\mathbf{A}$ gives another vector with the following components:

$$\nabla \times \mathbf{A} \equiv \begin{vmatrix} \mathbf{u}_x & \mathbf{u}_y & \mathbf{u}_z \\ \frac{\partial}{\partial x} & \frac{\partial}{\partial y} & \frac{\partial}{\partial z} \\ A_x & A_y & A_z \end{vmatrix} = \left(\frac{\partial A_z}{\partial y} - \frac{\partial A_y}{\partial z} \right) \mathbf{u}_x + \left(\frac{\partial A_x}{\partial z} - \frac{\partial A_z}{\partial x} \right) \mathbf{u}_y + \left(\frac{\partial A_y}{\partial x} - \frac{\partial A_x}{\partial y} \right) \mathbf{u}_z. \tag{1.5}$$

For the description of the electromagnetic field in a material medium it is necessary to define two additional vectorial magnitudes: the electric displacement vector $\mathcal{D}(\mathbf{r},t)$ and the magnetic flux density vector $\mathcal{B}(\mathbf{r},t)$. Maxwell's equations in a material medium, involving these two magnitudes and the electric and magnetic fields, are expressed as:

$$\nabla \cdot \mathcal{D} = \rho; \tag{1.6a}$$

$$\nabla \cdot \mathcal{B} = 0; \tag{1.6b}$$

$$\nabla \times \mathcal{E} = -\frac{\partial \mathcal{B}}{\partial t}; \tag{1.6c}$$

$$\nabla \times \mathcal{H} = \mathcal{J} + \frac{\partial \mathcal{D}}{\partial t}, \tag{1.6d}$$

where $\rho(\mathbf{r},t)$ and $\mathcal{J}(\mathbf{r},t)$ denote the charge density and the current density vector, respectively. If the medium if free of charges, which is the most common situation in optics, Maxwell's equations simplify to the form:

$$\nabla \cdot \mathcal{D} = 0; \tag{1.7a}$$

$$\nabla \cdot \mathcal{B} = 0; \tag{1.7b}$$

$$\nabla \times \mathcal{E} = -\frac{\partial \mathcal{B}}{\partial t}; \tag{1.7c}$$

$$\nabla \times \mathcal{H} = \mathcal{J} + \frac{\partial \mathcal{D}}{\partial t}. \tag{1.7d}$$

Now, in order to solve these differential coupled equations it is necessary to establish additional relations between the vectors $\mathcal{D}$ and $\mathcal{E}$, $\mathcal{J}$ and $\mathcal{E}$ as well as the vectors $\mathcal{H}$ and $\mathcal{B}$. These relations are called constitutive relations and depend on the electric and magnetic properties of the considered medium. In the most simple case of linear and isotropic media, the constitutive relations are given by:

$$\mathcal{D} = \varepsilon \mathcal{E}; \tag{1.8a}$$

$$\mathcal{B} = \mu \mathcal{H}; \tag{1.8b}$$

$$\mathcal{J} = \sigma \mathcal{E}, \tag{1.8c}$$

where $\varepsilon = \varepsilon(r)$ is the dielectric permittivity, $\mu = \mu(r)$ is the magnetic permeability and $\sigma = \sigma(r)$ is the electrical conductivity of the medium. Here, their dependence on the position vector r has been explicitly indicated. If the medium is not linear, it is necessary to include additional terms involving power expansion of the electric and magnetic fields. Besides, in an isotropic medium (glasses for instance) these optical constants are scalar magnitudes and independent of the direction of the vectors $\mathcal{E}$ and $\mathcal{H}$, implying that the vectors $\mathcal{D}$ and $\mathcal{J}$ are parallel to the electric field $\mathcal{E}$ and the vector $\mathcal{B}$ is parallel to the magnetic field $\mathcal{H}$. By contrast, in an anisotropic medium (for instance, most of the dielectric crystals) the optical constants must be treated as tensorial magnitudes.

By using the constitutive relations for a linear and isotropic medium, Maxwell's equations can be written in terms of the electric field $\mathcal{E}$ and magnetic field $\mathcal{H}$ only:

$$\nabla \cdot (\varepsilon \mathcal{E}) = 0; \tag{1.9a}$$

$$\nabla \cdot \mathcal{H} = 0; \tag{1.9b}$$

$$\nabla \times \mathcal{E} = -\mu \frac{\partial \mathcal{H}}{\partial t}; \tag{1.9c}$$

$$\nabla \times \mathcal{H} = \sigma \mathcal{E} + \varepsilon \frac{\partial \mathcal{E}}{\partial t}. \tag{1.9d}$$

A perfect dielectric medium is defined as a material in which the conductivity is very low and can be neglected ($\sigma \approx 0$). In this category fall most of the materials used for integrated optical devices, such as glasses, ferroelectric crystals, polymers or even semiconductors, while metals do not belong to this category because of their high conductivity. In addition, in most of materials (non-magnetic materials) and in particular, dielectric media, the magnetic permeability is very close to that of free space and the approximation $\mu \approx \mu_0$ holds. Then, in dielectric and non-magnetic media, Maxwell's equations simplify in the form:

$$\nabla \cdot (\varepsilon \mathcal{E}) = 0; \tag{1.10a}$$

$$\nabla \cdot \mathcal{H} = 0; \tag{1.10b}$$

$$\nabla \times \mathcal{E} = -\mu_0 \frac{\partial \mathcal{H}}{\partial t}; \tag{1.10c}$$

$$\nabla \times \mathcal{H} = \varepsilon \frac{\partial \mathcal{E}}{\partial t}. \tag{1.10d}$$

In what follows, we will restrict ourselves to non-magnetic and low conductivity materials, where Maxwell's equations (1.10a)–(1.10d) apply.

1.1.2 Wave Equations in Inhomogeneous Media

Combining the four Maxwell's equations (1.10a)–(1.10d) it is possible to obtain an equation involving the electric field alone and another equation that involves only the magnetic field.

Taking the curl operation over the Eqs. (1.10c) and (1.10d) we obtain:

$$\nabla \times (\nabla \times \mathcal{E}) = -\nabla \times \left(\mu_0 \frac{\partial \mathcal{H}}{\partial t} \right); \tag{1.11a}$$

$$\nabla \times (\nabla \times \mathcal{H}) = \nabla \times \left(\varepsilon \frac{\partial \mathcal{E}}{\partial t} \right) = (\nabla \varepsilon) \times \frac{\partial \mathcal{E}}{\partial t} + \varepsilon \left(\nabla \times \frac{\partial \mathcal{E}}{\partial t} \right). \tag{1.11b}$$

Having in mind the vectorial identity $\nabla \times \nabla \times \equiv \nabla (\nabla \cdot) - \nabla^2$, these equations transform to:

$$\nabla (\nabla \cdot \mathcal{E}) - \nabla^2 \mathcal{E} = -\mu_0 \frac{\partial}{\partial t} (\nabla \times \mathcal{H}); \tag{1.12a}$$

$$\nabla (\nabla \cdot \mathcal{H}) - \nabla^2 \mathcal{H} = (\nabla \varepsilon) \times \frac{\partial \mathcal{E}}{\partial t} + \frac{\partial}{\partial t} [\varepsilon \nabla \times \mathcal{E}], \tag{1.12b}$$

where we have used the fact that the temporal and spatial derivatives commute and it is assumed that permittivity, ε, is time independent. On the other hand, expanding the first Maxwell's equation (1.10a):

$$\nabla \cdot (\varepsilon \mathcal{E}) = (\nabla \varepsilon) \cdot \mathcal{E} + \varepsilon \nabla \cdot \mathcal{E} = 0 \quad \Rightarrow \quad \nabla \cdot \mathcal{E} = -\mathcal{E} \cdot \nabla \ln \varepsilon, \tag{1.13}$$

where we have used the relationship: $\nabla \ln \varepsilon \equiv \dfrac{\nabla \varepsilon}{\varepsilon}$.

Introducing Eqs. (1.13) and (1.10d) into Eq. (1.12a), we have:

$$\nabla^2 \mathcal{E} + \nabla (\mathcal{E} \cdot \nabla \ln \varepsilon) - \mu_0 \varepsilon \frac{\partial^2 \mathcal{E}}{\partial t^2} = 0, \tag{1.14}$$

and similarly using Eqs. (1.10b) and (1.10c) into Eq. (1.12b), we obtain:

$$\nabla^2 \mathcal{H} + (\nabla \ln \varepsilon) \times (\nabla \times \mathcal{H}) - \mu_0 \varepsilon \frac{\partial^2 \mathcal{H}}{\partial t^2} = 0. \tag{1.15}$$

These last two differential equations are known as wave equations in inhomogeneous media, which are valid for linear, non-magnetic and isotropic material media. It is worth noting that, although we have obtained a wave equation for the electric field $\mathcal{E}$ and another for the magnetic field $\mathcal{H}$, the solution of both equations are not independent, because the electric and magnetic fields are related through the Maxwell's equations (1.10c) and (1.10d). The solutions of the wave equations are known as electromagnetic waves.

The electromagnetic waves transport energy and the flux of energy (measured in units of W/m^2) carried by the EM wave is given by the Poynting vector $\mathcal{S}$, defined as:

$$\mathcal{S} \equiv \mathcal{E} \times \mathcal{H}. \tag{1.16}$$

On the other hand, the intensity (or irradiance) I of an EM wave, defined as the amount of energy passing through the unit area in the unit of time, is given by the time average of the Poynting vector modulus:

$$I = \langle |\mathcal{S}| \rangle. \tag{1.17}$$

The reason for using an averaged value instead of an instant value for defining the intensity of an EM wave is because the electric and magnetic fields associated to the EM wave oscillate at very high frequency and the apparatus used to detect that intensity (light detectors) cannot follow the instant values of the Poynting vector modulus.

1.1.3   Wave Equations in Homogeneous Media: Refractive Index

An optically homogeneous medium is defined as a material in which its optical properties are independent on the position. Then, for homogeneous dielectric media the second terms in Eqs. (1.14) and (1.15) vanish:

$$\nabla \ln \varepsilon = \frac{\nabla \varepsilon}{\varepsilon} = 0, \tag{1.18}$$

and the wave equations simplify on the forms:

$$\nabla^2 \mathcal{E} = \mu_0 \varepsilon \frac{\partial^2 \mathcal{E}}{\partial t^2}; \tag{1.19a}$$

$$\nabla^2 \mathcal{H} = \mu_0 \varepsilon \frac{\partial^2 \mathcal{H}}{\partial t^2}. \tag{1.19b}$$

Each of these two vectorial wave equations can be split onto three scalar wave equations, expressed as:

$$\nabla^2 \xi = \mu_0 \varepsilon \frac{\partial^2 \xi}{\partial t^2}, \tag{1.20}$$

where the scalar variable $\xi(\mathbf{r},t)$ may represent each of the six Cartesian components of either the electric and magnetic fields. The solution of this scalar equation represents a wave that propagates with a speed v (phase velocity) given by:

$$v = \frac{1}{\sqrt{\varepsilon \mu_0}}. \tag{1.21}$$

Therefore, the complete solution of the vectorial wave equations (1.19a) and (1.19b) represents an electromagnetic wave, where each of the Cartesian components of the electric and magnetic fields propagate in the form of waves of equal speed v in the homogeneous medium.

For propagation in free space (vacuum) and using the values for ε_0 and μ_0 we obtain:

$$c = \frac{1}{\sqrt{\varepsilon_0 \mu_0}} \approx 3.00 \times 10^8 \, \text{m/s}, \tag{1.22}$$

which corresponds to the speed of light in free space measured experimentally. It is worth noting that here the speed of light is obtained by using only values of electric and magnetic constants.

Usually, it is convenient to express the propagation speed v of the electromagnetic waves in a medium as a function of the speed of light in free space, c, through the relation:

$$v \equiv \frac{c}{n}, \tag{1.23}$$

where n represents the refractive index of the dielectric medium. Taking into account the relations (1.21) and (1.22), the refractive index can be related to the dielectric permittivity of the medium and that of the free space by:

$$n = \sqrt{\frac{\varepsilon}{\varepsilon_0}} = \sqrt{\varepsilon_r}, \tag{1.24}$$

where we have introduced the magnitude relative dielectric permittivity, ε_r, also called the dielectric constant, defined as the relation between the dielectric permittivity of the material medium and that of the free space. As we will see in the following chapters, the refractive index of a medium is the most important parameter for defining optical waveguide structures used in integrated photonic devices. As well as the refractive index of 1 corresponding to propagation through free space, the refractive index ranges from values close to 1.5 for glasses and some dielectric crystals (for instance, $n(SiO_2) = 1.55$) to values close to 4 for semiconductor materials (for instance, $n(Si) = 3.75$).

1.2 Monochromatic Waves

The temporal dependence of the electric and magnetic fields within the wave equations admits solutions on the form of harmonic functions. The electromagnetic waves with such sinusoidal dependence on the time variable are called monochromatic waves, which are characterized by their angular frequency ω (in units of rad/s). In a general form, the electric and magnetic fields associated to a monochromatic wave can be expressed as:

$$\mathcal{E}(r,t) = \mathcal{E}_0(r) \cos[\omega t + \phi(r)]; \tag{1.25a}$$

$$\mathcal{H}(r,t) = \mathcal{H}_0(r)\cos[\omega t + \phi(r)], \tag{1.25b}$$

where the field amplitudes $\mathcal{E}_0(r)$ and $\mathcal{H}_0(r)$, and the initial phase $\varphi(r)$ have dependence on the position r, and the time dependence of the fields is only in the cosine argument through ωt.

Usually, when dealing with monochromatic waves it is convenient to express the monochromatic fields using complex notation. Using this notation, the electric and magnetic fields are expressed as:

$$\mathcal{E}(r,t) = \text{Re}\left[E(r)e^{+i\omega t}\right]; \tag{1.26a}$$

$$\mathcal{H}(r,t) = \text{Re}\left[H(r)e^{+i\omega t}\right], \tag{1.26b}$$

where $E(r)$ and $H(r)$ denote the complex amplitudes of the electric and magnetic fields, respectively, i is the imaginary unity and Re stands for the real part. The angular frequency, ω, which characterizes the monochromatic wave, is related with the frequency, v, and the period, T, by:

$$\omega = 2\pi v = 2\pi/T. \tag{1.27}$$

The electromagnetic spectrum covered by light (optical spectrum) ranges from frequencies of 3×10^5 Hz corresponding to the far infrared (IR), to 6×10^{15} Hz corresponding to vacuum UV, being the frequency of visible light in the range of 430–770 THz.

The average of the Poynting vector for monochromatic waves as a function of the complex fields amplitudes takes the form:

$$\langle \mathcal{S} \rangle = \left\langle Re\{Ee^{+i\omega t}\} \times Re\{He^{+i\omega t}\} \right\rangle = Re\{S\}, \tag{1.28}$$

where S is defined here as:

$$S = 1/2 E \times H^*, \tag{1.29}$$

which is called the complex Poynting vector. Using this definition, the intensity carried by a monochromatic EM wave can be expressed in a compact form as:

$$I = |Re\{S\}|. \tag{1.30}$$

Maxwell's equations (1.10a)–(1.10d) using the complex fields amplitudes E and H simplify notably in the case of monochromatic waves, because the partial derivatives with respect to the time can be directly obtained by multiplying by the factor $i\omega$, resulting in:

$$\nabla \cdot (\varepsilon E) = 0; \tag{1.31a}$$

$$\nabla \cdot H = 0; \tag{1.31b}$$

$$\nabla \times E = -i\mu_0 \omega H; \tag{1.31c}$$

$$\nabla \times H = i\varepsilon \omega E, \tag{1.31d}$$

where a dielectric and non-magnetic medium has been assumed in which $\sigma = 0$ and $\mu = \mu_0$. The corresponding wave equations for the electric and magnetic fields Eqs. (1.14) and (1.15) are given by:

$$\nabla^2 \boldsymbol{E} + \nabla(\boldsymbol{E} \cdot \nabla \ln \varepsilon) + \omega^2 \mu_0 \varepsilon \boldsymbol{E} = 0; \tag{1.32a}$$

$$\nabla^2 \boldsymbol{H} + (\nabla \ln \varepsilon) \times (\nabla \times \boldsymbol{H}) + \omega^2 \mu_0 \varepsilon \boldsymbol{H} = 0. \tag{1.32b}$$

Often, it is more convenient to rewrite these equations as a function of the refractive index $n(\boldsymbol{r})$ of the medium as follows:

$$\nabla^2 \boldsymbol{E} + \nabla\left(\frac{1}{n^2}\nabla n^2 \cdot \boldsymbol{E}\right) + n^2 k_0^2 \boldsymbol{E} = 0; \tag{1.33a}$$

$$\nabla^2 \boldsymbol{H} + \frac{1}{n^2}(\nabla n^2) \times (\nabla \times \boldsymbol{H}) + n^2 k_0^2 \boldsymbol{H} = 0, \tag{1.33b}$$

where we have defined the wavenumber k_0 as:

$$k_0 \equiv \omega/c \equiv 2\pi/\lambda, \tag{1.34}$$

λ being the wavelength of light in free space.

1.2.1 Homogeneous Media: Helmholtz's Equation

For light propagation in homogeneous media the wave equations are substantially simplified. If we substitute the solutions on the form of monochromatic waves ((1.26a) and (1.26b)) in the wave equations ((1.19a) and (1.19b)), we obtain a new wave equation, valid only for monochromatic waves, known as Helmholtz equation:

$$\nabla^2 \xi(\boldsymbol{r}) + k^2 \xi(\boldsymbol{r}) = 0, \tag{1.35}$$

where $\xi(\boldsymbol{r})$ now represents each of the six Cartesian components of the $\boldsymbol{E}(\boldsymbol{r})$ and $\boldsymbol{H}(\boldsymbol{r})$ vectors defined in Eqs. (1.26a) and (1.26b), and where we have defined the wavenumber in the medium k as:

$$k \equiv \omega(\varepsilon \mu_0)^{1/2} = n k_0, \tag{1.36}$$

n being the refractive index of the homogeneous medium.

1.2.2 Light Propagation in Absorbing Media

An absorbing medium is characterized by the fact that the energy of the EM radiation is dissipated in it. This would imply that the amplitude of a plane EM wave will decrease in exponential form as the wave propagates along the absorbing medium. The mathematical description of light propagation in absorbing media can be treated by considering that the dielectric permittivity is no longer a real number, but a complex quantity ε_c. In terms of the fields' descriptions, this implies that the electric displacement will not be generally in phase with the electric field. As the refractive index is defined as the function of the dielectric permittivity, in general it will be a complex number, now defined by:

$$n_c = \sqrt{\frac{\varepsilon_c}{\varepsilon_0}}, \tag{1.37}$$

where n_c is called the complex refractive index. It is useful to work separately with the real and imaginary part and for doing this the complex refractive index is put in the form:

$$n_c = n - i\kappa, \tag{1.38}$$

where n is now the real refractive index and κ is called the absorption index (both dimensionless quantities).

The most important aspect concerning light propagation in absorbing media is the intensity variation suffered by the electromagnetic radiation as it propagates. For the case of a monochromatic plane of the form:

$$E(r) = E_0 e^{-ik \cdot r}, \tag{1.39}$$

and assuming that the propagation is along the z-axis without loss of generality, the intensity associated to that planar wave can be calculated by using Eq. (1.30), and takes the form:

$$I(z) = \frac{1}{2c\mu_0} |E_0|^2 e^{-2\kappa k_0 z}. \tag{1.40}$$

If we define now I_0 as the intensity associated to the wave at the plane $z = 0$, it follows:

$$I_0 = \frac{1}{2c\mu_0} |E_0|^2, \tag{1.41}$$

and the expression for $I(z)$ becomes more compact as:

$$I(z) = I_0 e^{-2\kappa k_0 z}. \tag{1.42}$$

This formula indicates that the intensity of a monochromatic plane wave propagating in an absorbing medium decreases exponentially as a function of the propagation distance.

In some applications it is convenient to deal with the absorption by using the absorption coefficient, α, defined as:

$$\alpha \equiv 2\kappa k_0 = 2\kappa\omega/c, \tag{1.43}$$

which has dimensions of m^{-1}. In this way, the attenuation of a light beam passing an absorbing medium is expressed in a compact manner by:

$$I(z) = I_0 e^{-\alpha z}. \tag{1.44}$$

When working with optical waveguides or optical fibres, light attenuation A is often referred to in decibels, whose relation with the absorption coefficient is:

$$A(\mathrm{dB}) \equiv 10 \log_{10}(I_0/I) = 4.3\alpha d, \tag{1.45}$$

where I/I_0 represents the fraction of light intensity after a distance d.

1.2.3 Light Propagation in Anisotropic Media

In an isotropic medium, whose properties do not depend of the direction considered, we have seen that the electric displacement vector D and its associated electric field E are related through $D = \varepsilon E$, where ε is the scalar electric permittivity. This relation implies that the vectors D and E are always parallel, and the refractive index is given by Eq. (1.24). Nevertheless, in the most general case of light propagation in anisotropic media, whose optical properties depend on the polarization and propagation direction of the light, the displacement vector D and the electric field E are no longer necessary parallel, and they are related by:

$$D = \varepsilon E, \tag{1.46}$$

where ε is the permittivity tensor, which can be expressed as a 3×3 matrix referring to three arbitrary orthogonal axes:

$$\varepsilon = \begin{pmatrix} \varepsilon_{xx} & \varepsilon_{xy} & \varepsilon_{xz} \\ \varepsilon_{yx} & \varepsilon_{yy} & \varepsilon_{yz} \\ \varepsilon_{zx} & \varepsilon_{zy} & \varepsilon_{zz} \end{pmatrix}. \tag{1.47}$$

The elements of the permittivity tensor fulfil the following relation:

$$\varepsilon_{ij} = \varepsilon_{ji}^*. \tag{1.48}$$

This implies that for materials that are neither absorbing nor amplifying, where the elements of the tensor are real, the permittivity tensor is symmetrical.

For some practical calculations concerning light propagation in anisotropic materials, it is often useful to work with the impermeability tensor η, defined as:

$$\eta \equiv \varepsilon_0 \varepsilon^{-1}, \tag{1.49}$$

and then $E = \varepsilon^{-1} D \equiv \frac{1}{\varepsilon_0} \eta D$. Taking advantage of the symmetry of the permittivity tensor for non-absorbing media, the impermeability matrix can be expressed as:

$$\eta = \begin{pmatrix} \eta_1 & \eta_6 & \eta_5 \\ \eta_6 & \eta_2 & \eta_4 \\ \eta_5 & \eta_4 & \eta_3 \end{pmatrix}, \tag{1.50}$$

where we have used the following contracted notation for the sub-indices:

$$xx \rightarrow 1 \qquad yy \rightarrow 2 \qquad zz \rightarrow 3;$$
$$yz, zy \rightarrow 4 \quad xz, zx \rightarrow 5 \quad xy, yx \rightarrow 6.$$

Using the impermeability, the optical properties of an anisotropic medium can be conveniently described through its index ellipsoid or optical indicatrix (Figure 1.1) defined by [1]:

$$\eta_1 x^2 + \eta_2 y^2 + \eta_3 z^2 + 2\eta_4 zy + 2\eta_5 xz + 2\eta_6 yx = 1. \tag{1.51}$$

Often, the components of the impermeability tensor are written as:

$$\eta_i \equiv \left(\frac{1}{n_i^2} \right), \tag{1.52}$$

but let us remember that the quantity n_i in expression (1.52) does not represent, in general, a true refractive index. Using this nomenclature, the index ellipsoid takes the form:

$$\frac{x^2}{n_1^2} + \frac{y^2}{n_2^2} + \frac{z^2}{n_3^2} + \frac{2zy}{n_4^2} + \frac{2xz}{n_5^2} + \frac{2yx}{n_6^2} = 1. \tag{1.53}$$

By making an appropriate choice of axes (x', y' and z'), called the principal axes of the material, the permittivity tensor (Eq. (1.47)) can be diagonalized and it becomes:

$$\boldsymbol{\varepsilon} = \begin{pmatrix} \varepsilon_{x'} & 0 & 0 \\ 0 & \varepsilon_{y'} & 0 \\ 0 & 0 & \varepsilon_{z'} \end{pmatrix}, \tag{1.54}$$

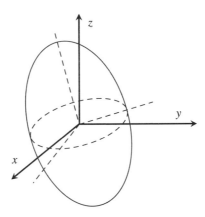

Figure 1.1　Index ellipsoid of an anisotropic medium. Its principal axes (dashed lines) do not coincide in general with the x-, y- and z-axes

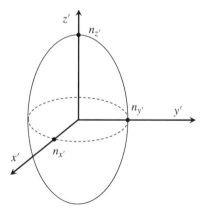

Figure 1.2 Index ellipsoid of an anisotropic medium, where the principal axes are coincident to the x'-, y'- and z'-axes

where all the elements are zero except to the diagonal. The $\varepsilon_{x'}$, $\varepsilon_{y'}$ and $\varepsilon_{z'}$ elements are called the principal permittivities along the x', y' and z' directions, respectively. In this reference system, the index ellipsoid (Figure 1.2) simplifies to the form:

$$\frac{x'^2}{n_{x'}^2} + \frac{y'^2}{n_{y'}^2} + \frac{z'^2}{n_{z'}^2} = 1, \tag{1.55}$$

where $n_{x'}$, $n_{y'}$, $n_{z'}$ now represent the true refractive indices that are related to the diagonal elements of the permittivity tensor by:

$$n_{x'} = \sqrt{\frac{\varepsilon_{x'}}{\varepsilon_0}}, n_{y'} = \sqrt{\frac{\varepsilon_{y'}}{\varepsilon_0}}, n_{z'} = \sqrt{\frac{\varepsilon_{z'}}{\varepsilon_0}}. \tag{1.56}$$

1.2.4 Light Propagation in Second-Order Non-Linear Media

For a non-linear medium, the constitutive relations given by Eq. (1.8) are no longer valid and it becomes necessary to include additional terms in serial powers of the fields. The non-linearity can be introduced by means of the polarization vector $\mathcal{P}$. In this way, Eq. (1.8a) may be rewritten as:

$$\mathcal{D} = \varepsilon_0 \mathcal{E} + \mathcal{P}. \tag{1.57}$$

For a non-linear medium, the polarization is split in a linear and a non-linear part through:

$$\mathcal{P} = \varepsilon_0 \chi_L \mathcal{E} + \mathcal{P}_{NL}, \tag{1.58}$$

where χ_L is the linear susceptibility and $\mathcal{P}_{NL}$ incorporates the non-linear polarization. In the case of second-order non-linear media, non-linear polarization is given by [2]:

$$(\mathcal{P}_{NL})_i = \sum_{j,k} 2 d_{ijk} \mathcal{E}_j \mathcal{E}_k, \tag{1.59}$$

where d_{ijk} is the second-order non-linear tensor.

In this way, the displacement vector is expressed by:

$$\mathcal{D} = \varepsilon_0 \mathcal{E} + \mathcal{P} = \varepsilon_0 \mathcal{E} + \varepsilon_0 \chi_L \mathcal{E} + \mathcal{P}_{NL} = \varepsilon \mathcal{E} + \mathcal{P}_{NL}, \tag{1.60}$$

where the linear permittivity is given by:

$$\varepsilon = \varepsilon_0 (1 + \chi_L). \tag{1.61}$$

Taking into account the non-linearity of the medium through relation (1.60), Maxwell's equations remain as:

$$\nabla \cdot (\varepsilon \mathcal{E} + \mathcal{P}_{NL}) = 0; \tag{1.62a}$$

$$\nabla \cdot \mathcal{H} = 0; \tag{1.62b}$$

$$\nabla \times \mathcal{E} = -\mu_0 \frac{\partial \mathcal{H}}{\partial t}; \tag{1.62c}$$

$$\nabla \times \mathcal{H} = \frac{\partial}{\partial t}(\varepsilon \mathcal{E} + \mathcal{P}_{NL}). \tag{1.62d}$$

Combining the four Maxwell's equations it is possible to obtain an equation involving the electric field alone. Taking the curl operation over Eq. (1.62c) and having in mind the vectorial identity, $\nabla \times \nabla \times \equiv \nabla(\nabla \cdot) - \nabla^2$, that equation transforms to:

$$\nabla(\nabla \cdot \mathcal{E}) - \nabla^2 \mathcal{E} = -\mu_0 \frac{\partial}{\partial t}(\nabla \times \mathcal{H}), \tag{1.63}$$

where we have used the fact that the temporal and spatial derivatives commute. Inserting Eq. (1.62d) into Eq. (1.63) we have:

$$\nabla^2 \mathcal{E} - \nabla(\nabla \cdot \mathcal{E}) - \mu_0 \frac{\partial^2}{\partial t^2}(\varepsilon \mathcal{E} + \mathcal{P}_{NL}) = 0, \tag{1.64}$$

and assuming that the permittivity, ε, is time independent:

$$\nabla^2 \mathcal{E} - \nabla(\nabla \cdot \mathcal{E}) - \mu_0 \varepsilon \frac{\partial^2 \mathcal{E}}{\partial t^2} - \mu_0 \frac{\partial^2 \mathcal{P}_{NL}}{\partial t^2} = 0. \tag{1.65}$$

This equation is known as the non-linear wave equation for light propagating in a non-linear medium.

Here, we will limit the analysis to monochromatic waves in a second-order non-linear medium, and we consider that only three frequencies ω_1, ω_2 and ω_3 are involved. The corresponding monochromatic fields are in the form:

$$\mathcal{E}^{\omega_1}(r,t) = \frac{1}{2}\left[E_1(r)e^{i\omega_1 t} + \text{c.c.}\right]; \tag{1.66a}$$

$$\mathcal{E}^{\omega_2}(r,t) = \frac{1}{2}\left[E_2(r)e^{i\omega_2 t} + \text{c.c.}\right]; \tag{1.66b}$$

$$\mathcal{E}^{\omega_3}(r,t) = \frac{1}{2}\left[E_3(r)e^{i\omega_3 t} + \text{c.c.}\right], \tag{1.66c}$$

where E_1, E_2 and E_3 stand for the position dependent complex amplitudes of the fields at frequencies ω_1, ω_2 and ω_3, respectively, and c.c. denotes the complex conjugate. The electric field that appears in Eq. (1.65) is simply given by the sum of the fields at the three frequencies:

$$\mathcal{E}(r,t) = \mathcal{E}^{\omega_1}(r,t) + \mathcal{E}^{\omega_2}(r,t) + \mathcal{E}^{\omega_3}(r,t). \tag{1.67}$$

On the other hand, the i components of the non-linear polarization at frequencies ω_1, ω_2 and ω_3 are given by [3]:

$$\left[\mathcal{P}_{NL}^{\omega_1}(r,t)\right]_i = \sum_{j,k} d_{ijk} E_{3j}(r) E_{2k}^*(r) e^{i(\omega_3 - \omega_2)t} + \text{c.c.}; \tag{1.68a}$$

$$\left[\mathcal{P}_{NL}^{\omega_2}(r,t)\right]_i = \sum_{j,k} d_{ijk} E_{3j}(r) E_{1k}^*(r) e^{i(\omega_3 - \omega_1)t} + \text{c.c.}; \tag{1.68b}$$

$$\left[\mathcal{P}_{NL}^{\omega_3}(r,t)\right]_i = \sum_{j,k} d_{ijk} E_{2j}(r) E_{1k}(r) e^{i(\omega_2 + \omega_1)t} + \text{c.c.}. \tag{1.68c}$$

Introducing the expression of the electric fields (Eqs. (1.66a)–(1.66c)) and the non-linear polarizations (Eqs. (1.68a)–(1.68c)) into Eq. (1.65), and after equating terms with the same frequency, the i-component of the wave equation corresponding to the frequency ω_1 yields:

$$\frac{1}{2}\left\{\nabla^2 E_{1i}(r)e^{i\omega_1 t} + \text{c.c.}\right\} - \frac{1}{2}\left\{\frac{\partial}{\partial x_i}[\nabla \cdot E_1(r)]e^{i\omega_1 t} + \text{c.c.}\right\}$$

$$+ \frac{1}{2}\mu_0 \varepsilon_1 \omega_1^2 \left\{E_{1i}(r)e^{i\omega_1 t} + \text{c.c.}\right\} + \mu_0 \omega_1^2 \left\{\sum_{j,k} d_{ijk} E_{3j}(r) E_{2k}^*(r) e^{i(\omega_3 - \omega_2)t} + \text{c.c.}\right\} = 0, \tag{1.69a}$$

where x_i indicates either x, y or z coordinates. Clearly, this equation couples the three Cartesian components of the field at frequency ω_1 through the second term and also depends on the components of the fields at frequencies ω_2 and ω_3 through the non-linear term. In addition to

Eq. (1.69a), two similar equations are obtained which involve oscillations at frequencies ω_2 and ω_3:

$$\frac{1}{2}\left\{\nabla^2 E_{2i}(\boldsymbol{r})e^{i\omega_2 t} + \text{c.c.}\right\} - \frac{1}{2}\left\{\frac{\partial}{\partial x_i}[\nabla \cdot \boldsymbol{E}_2(\boldsymbol{r})]e^{i\omega_2 t} + \text{c.c.}\right\}$$

$$+ \frac{1}{2}\mu_0\varepsilon_2\omega_2^2\left\{E_{2i}(\boldsymbol{r})e^{i\omega_2 t} + \text{c.c.}\right\} + \mu_0\omega_2^2\left\{\sum_{j,k}d_{ijk}E_{3j}(\boldsymbol{r})E_{1k}^*(\boldsymbol{r})e^{i(\omega_3-\omega_1)t} + \text{c.c.}\right\} = 0; \tag{1.69b}$$

$$\frac{1}{2}\left\{\nabla^2 E_{3i}(\boldsymbol{r})e^{i\omega_3 t} + \text{c.c.}\right\} - \frac{1}{2}\left\{\frac{\partial}{\partial x_i}[\nabla \cdot \boldsymbol{E}_3(\boldsymbol{r})]e^{i\omega_3 t} + \text{c.c.}\right\}$$

$$+ \frac{1}{2}\mu_0\varepsilon_3\omega_3^2\left\{E_{3i}(\boldsymbol{r})e^{i\omega_3 t} + \text{c.c.}\right\} + \mu_0\omega_3^2\left\{\sum_{j,k}d_{ijk}E_{1j}(\boldsymbol{r})E_{2k}(\boldsymbol{r})e^{i(\omega_1+\omega_2)t} + \text{c.c.}\right\} = 0. \tag{1.69c}$$

From Eqs. (1.69a)–(1.69c) we finally obtain:

$$\nabla^2 E_{1i}(\boldsymbol{r}) - \frac{\partial}{\partial x_i}[\nabla \cdot \boldsymbol{E}_1(\boldsymbol{r})] + n_1^2 k_{01}^2 E_{1i}(\boldsymbol{r}) + k_{01}^2 \sum_{j,k}\chi_{ijk}E_{3j}(\boldsymbol{r})E_{2k}^*(\boldsymbol{r}) = 0; \tag{1.70a}$$

$$\nabla^2 E_{2i}(\boldsymbol{r}) - \frac{\partial}{\partial x_i}[\nabla \cdot \boldsymbol{E}_2(\boldsymbol{r})] + n_2^2 k_{02}^2 E_{2i}(\boldsymbol{r}) + k_{02}^2 \sum_{j,k}\chi_{ijk}E_{3j}(\boldsymbol{r})E_{1k}^*(\boldsymbol{r}) = 0; \tag{1.70b}$$

$$\nabla^2 E_{3i}(\boldsymbol{r}) - \frac{\partial}{\partial x_i}[\nabla \cdot \boldsymbol{E}_3(\boldsymbol{r})] + n_3^2 k_{03}^2 E_{3i}(\boldsymbol{r}) + k_{03}^2 \sum_{j,k}\chi_{ijk}E_{1j}(\boldsymbol{r})E_{2k}(\boldsymbol{r}) = 0, \tag{1.70c}$$

where we have used the definition of the non-linear susceptibility given by $\chi_{ijk} \equiv \frac{2d_{ijk}}{\varepsilon_0}$ and the wavenumber corresponding to the field at frequency ω_i ($i = 1$–3) has been denoted by k_{0i}.

1.3 Wave Equation Formulation in Terms of the Transverse Field Components

For the description of light propagation in inhomogeneous media, such as optical waveguides, it is useful to rewrite the wave equations in terms of the transversal components of the electric or magnetic fields. If the refractive index profile of a given structure varies slowly along the direction of wave propagation, which is a common situation in integrated photonic devices and optical fibres, the transverse and longitudinal components of the electric and magnetic fields are decoupled [4]. Here we assume that the light is in the form of monochromatic waves and that the propagation is mainly along the z-direction. In the following we will develop the wave equations for the transverse components, in terms of the electric field and in terms of the magnetic field. These equations will be the starting point to obtain the paraxial wave equation for developing 'beam propagation' algorithms.

1.3.1 Electric Field Formulation

To obtain the wave equation in terms of the transversal electric field components, we start with the wave equation for monochromatic fields in terms of the electric field given by Eq. (1.32a):

$$\nabla^2 E - \nabla(\nabla \cdot E) + n^2 k_0^2 E = 0, \tag{1.71}$$

where we have used the relationship $E \cdot \nabla \ln \varepsilon = -\nabla \cdot E$, derived from the first Maxwell equation. Next, we split the electric field in its transversal and longitudinal components in the following way:

$$E(r) = E_t(r) + E_z(r)\mathbf{u}_z, \tag{1.72}$$

being:

$$E_t(r) = E_x(r)\mathbf{u}_x + E_y(r)\mathbf{u}_y. \tag{1.73}$$

Also, the differential operator ∇ is split in its transversal and longitudinal parts as:

$$\nabla = \nabla_t + \frac{\partial}{\partial z}\mathbf{u}_z, \tag{1.74}$$

the transversal operator being defined by:

$$\nabla_t = \frac{\partial}{\partial x}\mathbf{u}_x + \frac{\partial}{\partial y}\mathbf{u}_y. \tag{1.75}$$

In this way, Eq. (1.71) expands as:

$$\nabla^2 E_t - \nabla^2 E_z \mathbf{u}_z - \nabla\left(\nabla_t \cdot E_t + \frac{\partial E_z}{\partial z}\right) + n^2 k_0^2 E = 0. \tag{1.76}$$

Taking the transversal components of this equation, it yields:

$$\nabla^2 E_t - \nabla_t\left(\nabla_t \cdot E_t + \frac{\partial E_z}{\partial z}\right) + n^2 k_0^2 E_t = 0. \tag{1.77}$$

On the other hand, the first Maxwell equation (1.9a) (valid for isotropic and non-linear media) can be expanded as:

$$\nabla \cdot (n^2 E) = \nabla_t \cdot (n^2 E_t) + \frac{\partial}{\partial z}(n^2 E_z) = 0, \tag{1.78}$$

or:

$$\nabla_t \cdot (n^2 E_t) + n^2\frac{\partial E_z}{\partial z} + E_z\frac{\partial n^2}{\partial z} = 0. \tag{1.79}$$

Now assuming that the refractive index change in the propagation direction is negligible, that is, there are not abrupt discontinuities in the z direction, it holds that:

$$\frac{\partial n^2}{\partial z} \approx 0. \tag{1.80}$$

Using this approximation, the partial derivative of the z component of the electric field respective to the z coordinate equation that appears in Eq. (1.79) can be expressed as:

$$\frac{\partial E_z}{\partial z} \approx -\frac{1}{n^2} \nabla_t \cdot \left(n^2 E_t\right) = -\frac{\nabla_t n^2}{n^2} \cdot E_t - \nabla_t \cdot E_t, \tag{1.81}$$

which is exact for uniform (z invariant) structures. Substituting this partial derivative on Eq. (1.77) now, it becomes:

$$\nabla^2 E_t + \nabla_t \left[\frac{\nabla_t n^2}{n^2} \cdot E_t\right] + n^2 k_0^2 E_t = 0. \tag{1.82}$$

Expressed explicitly in terms of the x and y components, it finally yields:

$$\nabla^2 E_x + \frac{\partial}{\partial x}\left[\frac{1}{n^2}\frac{\partial n^2}{\partial x} E_x\right] + \frac{\partial}{\partial x}\left[\frac{1}{n^2}\frac{\partial n^2}{\partial y} E_y\right] + n^2 k_0^2 E_x = 0; \tag{1.83a}$$

$$\nabla^2 E_y + \frac{\partial}{\partial y}\left[\frac{1}{n^2}\frac{\partial n^2}{\partial x} E_x\right] + \frac{\partial}{\partial y}\left[\frac{1}{n^2}\frac{\partial n^2}{\partial y} E_y\right] + n^2 k_0^2 E_y = 0. \tag{1.83b}$$

1.3.2   Magnetic Field Formulation

In an analogous way, we can proceed to obtain a wave equation in inhomogeneous media in terms of the transversal components of the magnetic field. Starting from Eq. (1.33b), and considering a structure in which the refractive index varies slowly along the direction of the wave propagation ($\partial n^2/\partial z \approx 0$), we obtain:

$$\nabla^2 H_t + \frac{1}{n^2}\left(\nabla_t n^2\right) \times \left(\nabla_t \times H_t\right) + n^2 k_0^2 H_t = 0. \tag{1.84}$$

And by separating this vectorial equation in two scalar equations, we finally get:

$$\nabla^2 H_x - \frac{1}{n^2}\frac{\partial n^2}{\partial y}\left(\frac{\partial H_x}{\partial y} - \frac{\partial H_y}{\partial x}\right) + n^2 k_0^2 H_x = 0; \tag{1.85a}$$

$$\nabla^2 H_y - \frac{1}{n^2}\frac{\partial n^2}{\partial x}\left(\frac{\partial H_y}{\partial x} - \frac{\partial H_x}{\partial y}\right) + n^2 k_0^2 H_y = 0. \tag{1.85b}$$

It is worth mentioning that the Eqs. (1.82) and (1.84) are exact for z-invariant structures, for example, in straight optical waveguides. For structures where their refractive indices change along the propagation direction (z-coordinate), for example, in Y-junctions these equations are only approximated. Nevertheless, most of the optical waveguides used in integrated optical devices are designed with small angles in their geometry, typically in the order of 2–10°, where this approximation is very successfully fulfilled. Also, when sinusoidal index gratings are written in waveguides or fibres (where the refractive index along the propagation directions is periodically modulated), their index variation in a wavelength distance is very small and therefore the condition $\partial n^2/\partial z \approx 0$ is also fulfilled.

1.3.3   Wave Equation in Anisotropic Media

As in the case of isotropic media, if the transverse components of an electromagnetic field are known, then the longitudinal component can be obtained by using the first Maxwell equation $\nabla(\varepsilon E) = 0$, but now considering that the permittivity is a tensor and not a scalar. This will allow us to describe the vectorial properties of the electromagnetic field in anisotropic media using the transverse components of the field [5]. For convenience, let us assume that the permittivity tensor (Eq. (1.47)) has the following not null components:

$$\boldsymbol{\varepsilon} = \begin{pmatrix} \varepsilon_{xx} & \varepsilon_{xy} & 0 \\ \varepsilon_{yx} & \varepsilon_{yy} & 0 \\ 0 & 0 & \varepsilon_{zz} \end{pmatrix}. \tag{1.86}$$

To obtain a wave equation for the transversal components, we will follow a similar procedure to that described in the previous section. To start with, we take the transverse component of Eq. (1.71) now assuming the tensorial character of the medium permittivity, which is written as:

$$\nabla^2 \boldsymbol{E}_t - \nabla_t \left(\nabla_t \cdot \boldsymbol{E}_t + \frac{\partial E_z}{\partial z} \right) + (\boldsymbol{\varepsilon}_{tt}/\varepsilon_0)k_0^2 \boldsymbol{E}_t = 0, \tag{1.87}$$

where the sub-matrix $\boldsymbol{\varepsilon}_{tt}$, which includes the transverse components of the permittivity tensor, is defined as:

$$\boldsymbol{\varepsilon}_{tt} \equiv \begin{pmatrix} \varepsilon_{xx} & \varepsilon_{xy} \\ \varepsilon_{yx} & \varepsilon_{yy} \end{pmatrix}. \tag{1.88}$$

On the other hand, using Gauss' law:

$$\nabla \cdot (\boldsymbol{\varepsilon} E) = 0, \tag{1.89}$$

we obtain:

$$\nabla_t \cdot (\boldsymbol{\varepsilon}_{tt} \boldsymbol{E}_t) + \varepsilon_{zz}\frac{\partial E_z}{\partial z} + E_z\frac{\partial \varepsilon_{zz}}{\partial z} = 0. \tag{1.90}$$

If the permittivity component $\varepsilon_{zz}(x,y,z)$ varies slowly along the propagation direction z, which is valid for most photonic guided-wave devices, then the last term can be neglected and Eq. (1.90) simplifies to:

$$\frac{\partial E_z}{\partial z} \approx -\frac{1}{\varepsilon_{zz}} \nabla_t \cdot (\boldsymbol{\varepsilon}_{tt} \boldsymbol{E}_t). \tag{1.91}$$

This relation is exact for z-invariant structures, where $\dfrac{\partial \varepsilon_{zz}}{\partial z} = 0$.

By substituting Eq. (1.91) into Eq. (1.87) we finally derive the wave equation for the transverse electric field:

$$\nabla^2 \boldsymbol{E}_t + (\boldsymbol{\varepsilon}_{tt}/\varepsilon_0)k_0^2 \boldsymbol{E}_t = \nabla_t \left[\nabla_t \cdot \boldsymbol{E}_t - \frac{1}{\varepsilon_{zz}} \nabla_t \cdot (\boldsymbol{\varepsilon}_{tt} \boldsymbol{E}_t) \right]. \tag{1.92}$$

1.3.4   Second Order Non-Linear Media

In general, it is not possible to obtain a wave equation in terms of the transverse components for second-order non-linear media, as non-linear polarization can induce transverse and longitudinal components of the fields. Nevertheless, we can derive a general wave equation that will serve as the starting point to develop 'beam propagation' equations for non-linear media, where some extra approximation will be needed.

The first Maxwell equation (1.62a) considering a non-linear medium can be expanded as:

$$(\nabla \varepsilon) \cdot \boldsymbol{\mathcal{E}} + \varepsilon(\nabla \cdot \boldsymbol{\mathcal{E}}) + \nabla \cdot \boldsymbol{\mathcal{P}}_{NL} = (\nabla_t \varepsilon) \cdot \boldsymbol{\mathcal{E}}_t + \frac{\partial \varepsilon}{\partial z} \boldsymbol{\mathcal{E}}_z + \varepsilon(\nabla \cdot \boldsymbol{\mathcal{E}}) + \nabla \cdot \boldsymbol{\mathcal{P}}_{NL} = 0. \tag{1.93}$$

Assuming that the structure is slowly changing with the propagation direction, $\dfrac{\partial \varepsilon}{\partial z} \approx 0$, and neglecting the last term in expression (1.93) $\nabla \cdot \boldsymbol{\mathcal{P}}_{NL} \approx 0$ (small non-linearity), we get:

$$\nabla \cdot \boldsymbol{\mathcal{E}} \approx -\frac{1}{\varepsilon}(\nabla_t \varepsilon) \cdot \boldsymbol{\mathcal{E}}_t. \tag{1.94}$$

Introducing this approximation in the wave equations ((1.70a)–(1.70c)) we obtain finally:

$$\nabla^2 \boldsymbol{E}_{1t} + \nabla_t \left[\frac{\nabla_t n_1^2}{n_1^2} \cdot \boldsymbol{E}_{1t} \right] + n_1^2 k_{10}^2 \boldsymbol{E}_{1t} + \left[k_{10}^2 \chi_{ijk} E_{3j} E_{2k}^* \right]_t = 0; \tag{1.95a}$$

$$\nabla^2 \boldsymbol{E}_{2t} + \nabla_t \left[\frac{\nabla_t n_2^2}{n_2^2} \cdot \boldsymbol{E}_{2t} \right] + n_2^2 k_{20}^2 \boldsymbol{E}_{2t} + \left[k_{20}^2 \chi_{ijk} E_{3j} E_{1k}^* \right]_t = 0; \tag{1.95b}$$

$$\nabla^2 \boldsymbol{E}_{3t} + \nabla_t \left[\frac{\nabla_t n_3^2}{n_3^2} \cdot \boldsymbol{E}_{3t} \right] + n_3^2 k_{30}^2 \boldsymbol{E}_{3t} + \left[k_{30}^2 \chi_{ijk} E_{1j} E_{2k} \right]_t = 0. \tag{1.95c}$$

Note that these equations involve the transversal components of the fields, but also may contain longitudinal components as well through the non-linear terms, depending on the non-null components of the non-linear tensor χ_{ijk}.

References

[1] Davis, C.C. (1996) *Laser and Electro-Optics*. Cambridge University Press.

[2] Shen, Y.R. (2003) *Principles of Nonlinear Optics*. John Wiley & Sons, Inc., Hoboken, NJ.

[3] Yariv, A. and Yeh, P. (2003) in *Optical Waves in Crystals: Propagation and Control of Laser Radiation*. John Wiley & Sons, Inc., Hoboken, NJ.

[4] Huang, W.P. and C.L. Xu, (1993) Simulation of three-dimensional optical waveguides by a full-vector beam propagation method. *Journal of Quantum Electronics* **29**, 2639–2649.

[5] Xu, C.L., Huang, W.P., Chrostowski, J. and Chaudhuri, S.K. (1994) A full-vectorial beam propagation method for anisotropic waveguides. *Journal of Lightwave Technology* **12**, 1926–1931.

2

The Beam-Propagation Method

Introduction

Integrated photonic devices are based on light propagation along optical waveguides. The design and optimization of such devices requires numerical wave propagation techniques for the accurate simulation of light along inhomogeneous optical structures. In this respect, a wide range of powerful numerical techniques has been developed, which can solve the Maxwell's equations to calculate the electromagnetic field distribution in the considered domain subject to a distribution of sources and imposed boundary conditions, being the finite difference time domain (FDTD), one of the most popular algorithms used so far [1]. Nevertheless, when applied to the optical waveguiding structures that are used in photonics devices they suffer from low computational efficiency. This is because, although optical waveguides typically have transversal dimensions of the order of the wavelength of the guided light, their longitudinal dimension is usually many orders of magnitude larger than the operating wavelength. Therefore, when dealing with optical waveguides, special numerical techniques are required, which take advantage of the fact that in integrated photonic devices a preferred direction of optical propagation can be identified (usually taken as the z-axis) and this is referred as paraxial propagation. This direction is defined as the one along which the major optical power is transported, while the power flow in the perpendicular direction is small, which is a direct consequence of the fact that optical integrated circuits vary slowly along the z direction. In paraxial propagation it is assumed that the dielectric constant distribution $\varepsilon(x,y,z)$ (or refractive index) varies slowly along z; that is, $\partial \varepsilon / \partial z \approx 0$, and under this assumption the transverse components of the electromagnetic field decouple from the longitudinal ones. Furthermore, with light propagation along these optical structures it is very convenient to introduce the so called slowly varying envelope (SVE) field, where most of the EM (electromagnetic) field variation is extracted by defining a suitable selected reference propagation constant. The wave equations that govern the electromagnetic field in the slowly varying envelope approximation (SVEA) are the starting point of the beam propagation methods

Beam Propagation Method for Design of Optical Waveguide Devices, First Edition. Ginés Lifante Pedrola.
© 2016 John Wiley & Sons, Ltd. Published 2016 by John Wiley & Sons, Ltd.

(BPMs), in which the algorithms permit the propagation of an initial monochromatic EM-field distribution along the axial direction by longitudinal steps of sufficient small length.

The first beam propagation algorithm proposed was based on fast Fourier transformation (FFT-BPM) and it was successfully applied to simulate optical propagation in optical fibres [2]. Nevertheless, it can only be applied to scalar propagation and it is only suitable for weakly guiding waveguides. Also, as the propagation step needs to be small, it requires a long computation time. These disadvantages were overcome after the development of other kind of BPM analysis based on either finite differences (FD-BPM) [3] or finite elements (FE-BPM) [4]. These BPM algorithms can be implemented to treat full vectorial propagation in large index contrast optical waveguides and are far superior to FFT analysis in terms of computer efficiency. In the FD-BPM the transversal plane at each propagation step along the longitudinal direction is divided up with a rectangular grid of points that may be of constant or variable spacing. The first and second order derivatives that appear in the wave equation are represented by their finite difference approaches, where the field at each grid point is related to the fields at adjacent points, both in the transversal and longitudinal directions. On the other hand, in the FE-BPM the transverse plane is divided into hybrid edge/nodal elements and the transverse and longitudinal field components are expressed as a function of the edge and nodal variables for each element, shape function vectors for edge elements and shape function vector for nodal elements. The order and number of elements can be arbitrarily selected, depending on the required computation accuracy, where the non-uniform finite element meshes can be used, which in turn can be adaptively updated in the direction of propagation depending on the optical distribution, so that computational efficiency can be improved without degrading numerical accuracy. Thus, while FE-BPM is superior in terms of a better discretization of a given structure, FD-BPM using rectangular grids is much easier to implement numerically. Using both finite differences and finite-element approaches, BPM algorithms have been developed to simulate light propagation in anisotropic media and non-linear media [5–8], where wide-angle BPMs have also been implemented [9, 10]. Finally, bidirectional algorithms have been successfully applied to treat bidirectional propagation using FD-BPM and FE-BPM numerical schemes [11, 12].

Here we focus on the simulation of light propagation along waveguide structures by using the finite difference approach. First, the SVEA is introduced as a convenient tool to deal paraxial propagation. Using the slowly varying field, full vectorial equations for the electric and magnetic fields are obtained, which are the basic differential equations for developing BPM algorithms. Particular BPM equations for semi-vectorial and scalar propagation can then be derived. The following sections introduce the finite difference schemes for BPM in the simplest case of scalar propagation in two-dimensional structures, which allows us to study the stability and numerical dissipation of the FD-BPM scheme. The suppression of reflected waves from the computation edges are then studied, as this is an important issue for accurate light propagation modelling. Finally, the use of BPM techniques for obtaining the eigenmodes supported by a z invariant waveguide is presented.

2.1 Paraxial Propagation: The Slowly Varying Envelope Approximation (SVEA). Full Vectorial BPM Equations

Let us consider now the case of propagation mainly along the z direction, also called paraxial propagation (Figure 2.1). In typical guided-wave problems the most rapid variation in the fields is the phase variation due to propagation along the guiding axis. Assuming that this

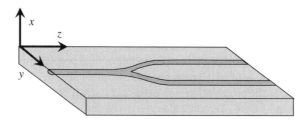

Figure 2.1   Three-dimensional optical circuit, where the refractive index of the structure varies slowly along z, which permits paraxial propagation when the angle of the Y-junction is small

axis is predominantly along the z direction, it is beneficial to factor out the problem by the introducing a so-called SVE field $\boldsymbol{\Psi}_t$ associated to the transverse electric field $\boldsymbol{E}_t$, defined through:

$$E_t = \boldsymbol{\Psi}_t e^{-in_0 k_0 z}, \tag{2.1}$$

where n_0 represents the refractive index of a reference medium, called the reference refractive index, and the SVE field is a two-component vector:

$$\boldsymbol{\Psi}_t \equiv \begin{bmatrix} \boldsymbol{\Psi}_x \\ \boldsymbol{\Psi}_y \end{bmatrix}. \tag{2.2}$$

Using this proposed solution, the second derivative of the x and y components of the electric field with respect to the z coordinate can be expressed as:

$$\frac{\partial^2 E_x}{\partial z^2} = \left[\frac{\partial^2 \boldsymbol{\Psi}_x}{\partial z^2} - 2in_0 k_0 \frac{\partial \boldsymbol{\Psi}_x}{\partial z} - n_0^2 k_0^2 \boldsymbol{\Psi}_x \right] e^{-in_0 k_0 z}; \tag{2.3a}$$

$$\frac{\partial^2 E_y}{\partial z^2} = \left[\frac{\partial^2 \boldsymbol{\Psi}_y}{\partial z^2} - 2in_0 k_0 \frac{\partial \boldsymbol{\Psi}_y}{\partial z} - n_0^2 k_0^2 \boldsymbol{\Psi}_y \right] e^{-in_0 k_0 z}. \tag{2.3b}$$

Introducing these expansions into Eqs. (1.83a) and (1.83b), the following equations are yielded for the slowly varying fields $\boldsymbol{\Psi}_x$ and $\boldsymbol{\Psi}_y$:

$$\frac{\partial^2 \boldsymbol{\Psi}_x}{\partial x^2} + \frac{\partial^2 \boldsymbol{\Psi}_x}{\partial y^2} + \frac{\partial^2 \boldsymbol{\Psi}_x}{\partial z^2} - 2in_0 k_0 \frac{\partial \boldsymbol{\Psi}_x}{\partial z} + \frac{\partial}{\partial x} \left[\frac{1}{n^2} \frac{\partial n^2}{\partial x} \boldsymbol{\Psi}_x + \frac{1}{n^2} \frac{\partial n^2}{\partial y} \boldsymbol{\Psi}_y \right] + k_0^2 \left(n^2 - n_0^2 \right) \boldsymbol{\Psi}_x = 0 \tag{2.4a}$$

$$\frac{\partial^2 \boldsymbol{\Psi}_y}{\partial x^2} + \frac{\partial^2 \boldsymbol{\Psi}_y}{\partial y^2} + \frac{\partial^2 \boldsymbol{\Psi}_y}{\partial z^2} - 2in_0 k_0 \frac{\partial \boldsymbol{\Psi}_y}{\partial z} + \frac{\partial}{\partial y} \left[\frac{1}{n^2} \frac{\partial n^2}{\partial x} \boldsymbol{\Psi}_x + \frac{1}{n^2} \frac{\partial n^2}{\partial y} \boldsymbol{\Psi}_y \right] + k_0^2 \left(n^2 - n_0^2 \right) \boldsymbol{\Psi}_y = 0 \tag{2.4b}$$

where we have factorized out the term $\exp(-in_0 k_0 z)$. At this point, equations (2.4) are exact (assuming that $\partial n^2 / \partial z = 0$), and completely equivalent to the Eqs. (1.83a) and (1.83b), except that they are expressed in terms of the slowly varying fields $\boldsymbol{\Psi}_x$ and $\boldsymbol{\Psi}_y$ instead of E_x and E_y.

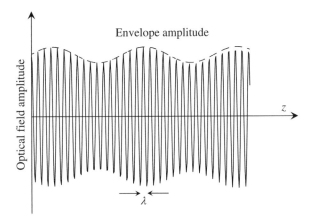

Figure 2.2 Fast optical field amplitude variation along the propagation direction and the slowly varying envelope amplitude (dashed line)

The SVEA consists of neglecting the second derivative on Ψ_x and Ψ_y for the z coordinate, assuming that:

$$\left|\frac{\partial^2 \Psi_x}{\partial z^2}\right| \ll 2n_0 k_0 \left|\frac{\partial \Psi_x}{\partial z}\right|; \tag{2.5a}$$

$$\left|\frac{\partial^2 \Psi_y}{\partial z^2}\right| \ll 2n_0 k_0 \left|\frac{\partial \Psi_y}{\partial z}\right|. \tag{2.5b}$$

Figure 2.2 shows the physical meaning of this approximation. While the optical field (for instance, one electric field transverse component) oscillates very fast as a function of the propagation direction, its envelope amplitude varies very slowly, with a period of many wavelengths.

With the SVEA, Eqs. (2.4a) and (2.4b) simplify to [3]:

$$2in_0 k_0 \frac{\partial \Psi_x}{\partial z} = \frac{\partial^2 \Psi_x}{\partial x^2} + \frac{\partial^2 \Psi_x}{\partial y^2} + \frac{\partial}{\partial x}\left(\frac{1}{n^2}\frac{\partial n^2}{\partial x}\Psi_x + \frac{1}{n^2}\frac{\partial n^2}{\partial y}\Psi_y\right) + k_0^2\left(n^2 - n_0^2\right)\Psi_x; \tag{2.6a}$$

$$2in_0 k_0 \frac{\partial \Psi_y}{\partial z} = \frac{\partial^2 \Psi_y}{\partial x^2} + \frac{\partial^2 \Psi_y}{\partial y^2} + \frac{\partial}{\partial y}\left(\frac{1}{n^2}\frac{\partial n^2}{\partial x}\Psi_x + \frac{1}{n^2}\frac{\partial n^2}{\partial y}\Psi_y\right) + k_0^2\left(n^2 - n_0^2\right)\Psi_y. \tag{2.6b}$$

These are the basic of the full vectorial BPM equations (also referred as Fresnel or paraxial vector wave equations), in terms of the electric field, for three-dimensional structures. Given an input field $\Psi_t(x,y,z=0)$, these equations can be used to determine the evolution of the fields in the space $z > 0$. To have completely determined the electric field, its longitudinal component E_z can be obtained easily just by using Eq. (1.81) by direct derivation.

The SVEA can be also applied to the wave equation for the magnetic field in a similar way [13]. By putting the transversal magnetic field in the form:

$$H_t = \Phi_t e^{-in_0 k_0 z},\tag{2.7}$$

and assuming slow variation of the second derivative of the envelope amplitudes:

$$\frac{\partial^2 \Phi_x}{\partial z^2} \ll 2n_0 k_0 \left| \frac{\partial \Phi_x}{\partial z} \right|;\tag{2.8a}$$

$$\frac{\partial^2 \Phi_y}{\partial z^2} \ll 2n_0 k_0 \left| \frac{\partial \Phi_y}{\partial z} \right|,\tag{2.8b}$$

the SVE transversal magnetic field components Φ_x and Φ_y are governed by:

$$2in_0 k_0 \frac{\partial \Phi_x}{\partial z} = \frac{\partial^2 \Phi_x}{\partial x^2} + \frac{\partial^2 \Phi_x}{\partial y^2} - \frac{1}{n^2}\frac{\partial n^2}{\partial y}\left(\frac{\partial \Phi_x}{\partial y} - \frac{\partial \Phi_y}{\partial x}\right) + k_0^2\left(n^2 - n_0^2\right)\Phi_x;\tag{2.9a}$$

$$2in_0 k_0 \frac{\partial \Phi_y}{\partial z} = \frac{\partial^2 \Phi_y}{\partial x^2} + \frac{\partial^2 \Phi_y}{\partial y^2} - \frac{1}{n^2}\frac{\partial n^2}{\partial x}\left(\frac{\partial \Phi_y}{\partial x} - \frac{\partial \Phi_x}{\partial y}\right) + k_0^2\left(n^2 - n_0^2\right)\Phi_y,\tag{2.9b}$$

which form the basic full vectorial BPM equations in terms of the magnetic field for three-dimensional structures. Equations (2.9a) and (2.9b) are known as paraxial vector wave equations (based on the magnetic field representation).

The advantage of using the slowly varying fields instead of the electric or magnetic fields amplitudes come from the fact that by factorizing of the rapid phase variation allows the slowly field to be represented numerically on a longitudinal grid (i.e. along z) that can be much coarser than the wavelength for many problems, contributing in part to the efficiency of the BPM technique [14].

Equations (2.6) and (2.9) constitute the full vectorial wave equation in terms of the transverse field components that full describe the optical propagation in inhomogeneous media. Each set of equations are differential coupled equations and therefore they must be solved simultaneously to obtain the correct fields. To put emphasis in the coupled character of the equations, it is convenient to rewrite them in terms of four differential operators, which in the case of electric field formulation are expressed as follows [3]:

$$2in_0 k_0 \frac{\partial \Psi_x}{\partial z} = P_{xx}\Psi_x + P_{xy}\Psi_y;\tag{2.10a}$$

$$2in_0 k_0 \frac{\partial \Psi_y}{\partial z} = P_{yy}\Psi_y + P_{yx}\Psi_x,\tag{2.10b}$$

where the differential operators P_{ij} are defined as:

$$P_{xx}\Psi_x \equiv \frac{\partial}{\partial x}\left(\frac{1}{n^2}\frac{\partial}{\partial x}\left(n^2\Psi_x\right)\right) + \frac{\partial^2 \Psi_x}{\partial y^2} + k_0^2\left(n^2 - n_0^2\right)\Psi_x;\tag{2.11a}$$

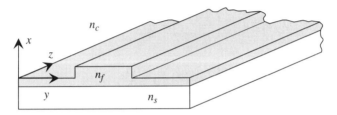

Figure 2.3   Three-dimensional rib-type waveguide, showing abrupt index changes along the x and y directions at the boundaries. Therefore, coupling between polarizations should be present

$$P_{yy}\Psi_y \equiv \frac{\partial^2 \Psi_y}{\partial x^2} + \frac{\partial}{\partial y}\left(\frac{1}{n^2}\frac{\partial}{\partial y}\left(n^2\Psi_y\right)\right) + k_0^2\left(n^2 - n_0^2\right)\Psi_y; \tag{2.11b}$$

$$P_{xy}\Psi_y \equiv \frac{\partial}{\partial x}\left(\frac{1}{n^2}\frac{\partial}{\partial y}\left(n^2\Psi_y\right)\right) - \frac{\partial^2 \Psi_y}{\partial x\partial y}; \tag{2.11c}$$

$$P_{yx}\Psi_x \equiv \frac{\partial}{\partial y}\left(\frac{1}{n^2}\frac{\partial}{\partial x}\left(n^2\Psi_x\right)\right) - \frac{\partial^2 \Psi_x}{\partial y\partial x}. \tag{2.11d}$$

Following Eq. (2.10), it is clear that the vector nature of the electromagnetic waves dictates that the transverse electric field components are polarization-dependent and coupled with each other. The operators P_{ij} in Eq. (2.11) show that the vectorial properties become important when the variation of the refractive index over the transverse cross section of the waveguide structure is abrupt and/or the index differences at the discontinuities are large, such as in rib-type waveguides (Figure 2.3). The vector properties will break the degeneracy between the x-and y-polarized waves due to $P_{xx} \neq P_{yy}$ and cause coupling between the two polarizations through P_{xy} and P_{yx}. For two-dimensional optical structures the polarization coupling terms associated with P_{xy} and P_{yx} vanish. In this case the vector waves may be decomposed into the TE (y-polarized transverse electric field) and TM (x-polarized transverse magnetic field) waves, which can be treated and solved separately [15].

In a similar way, the set of the two coupled differential equations for the SVE transversal magnetic field components can be rewritten as:

$$2in_0k_0\frac{\partial \Phi_x}{\partial z} = Q_{xx}\Phi_x + Q_{xy}\Phi_y; \tag{2.12a}$$

$$2in_0k_0\frac{\partial \Phi_y}{\partial z} = Q_{yy}\Phi_y + Q_{yx}\Phi_x, \tag{2.12b}$$

where the differential operators Q_{ij} have been defined as:

$$Q_{xx}\Phi_x \equiv \frac{\partial^2 \Phi_x}{\partial x^2} + n^2\frac{\partial}{\partial y}\left(\frac{1}{n^2}\frac{\partial \Phi_x}{\partial y}\right) + k_0^2\left(n^2 - n_0^2\right)\Phi_x; \tag{2.13a}$$

$$Q_{yy}\Phi_y \equiv n^2 \frac{\partial}{\partial x}\left(\frac{1}{n^2}\frac{\partial \Phi_y}{\partial x}\right) + \frac{\partial^2 \Phi_y}{\partial y^2} + k_0^2\left(n^2 - n_0^2\right)\Phi_y; \qquad (2.13b)$$

$$Q_{xy}\Phi_y \equiv -n^2 \frac{\partial}{\partial y}\left(\frac{1}{n^2}\frac{\partial \Phi_y}{\partial x}\right) + \frac{\partial^2 \Phi_y}{\partial y \partial x}; \qquad (2.13c)$$

$$Q_{yx}\Phi_x \equiv -n^2 \frac{\partial}{\partial x}\left(\frac{1}{n^2}\frac{\partial \Phi_x}{\partial y}\right) + \frac{\partial^2 \Phi_x}{\partial x \partial y}. \qquad (2.13d)$$

The discontinuities of $\partial \Phi_x/\partial y$ and $\partial \Phi_y/\partial x$ across the index interfaces along the y and x directions are responsible for the polarization dependence (i.e. $Q_{xx} \neq Q_{yy}$) and coupling (i.e. Q_{xy} and $Q_{yx} \neq 0$) [13]. For a two-dimensional structure (planar waveguides), there is no coupling between the two polarized waves, and Eqs. (2.12a) and (2.12b) can be solved for the TE (x-polarization) and the TM (y-polarization) separately. For three-dimensional structures, such as the waveguide shown in Figure 2.3, the coupling between the two polarizations occurs, so that the fields are in general hybrid.

The main advantage of using the wave equations for the electric field and magnetic fields in the forms of Eqs. (2.10) and (2.12), respectively, is that in these formulations the boundary conditions for the transverse electric and magnetic fields are contained in the operators defined by Eqs. (2.11) and (2.13) [13, 15].

Sometimes it is useful to put the full vectorial BPM Eqs. (2.10) and (2.12) in a more compact form:

$$\frac{\partial \boldsymbol{\Psi}_t}{\partial z} = -i\hat{\mathbf{H}}\boldsymbol{\Psi}_t; \qquad (2.14a)$$

$$\frac{\partial \boldsymbol{\Phi}_t}{\partial z} = -i\hat{\mathbf{H}}\boldsymbol{\Phi}_t, \qquad (2.14b)$$

where $\boldsymbol{\Psi}_t$ and $\boldsymbol{\Phi}_t$ refer to the column vectors of the SVE transversal electric and magnetic fields, respectively, and $\hat{\mathbf{H}}$ refers to a matrix differential operator defined by:

$$\hat{\mathbf{H}}\boldsymbol{\Psi}_t \equiv \frac{1}{2n_0 k_0}\begin{bmatrix} P_{xx} & P_{xy} \\ P_{yx} & P_{yy} \end{bmatrix}\boldsymbol{\Psi}_t; \qquad (2.15a)$$

$$\hat{\mathbf{H}}\boldsymbol{\Phi}_t \equiv \frac{1}{2n_0 k_0}\begin{bmatrix} Q_{xx} & Q_{xy} \\ Q_{yx} & Q_{yy} \end{bmatrix}\boldsymbol{\Phi}_t. \qquad (2.15b)$$

It is important to note that the equations for the transverse fields E_t and H_t are decoupled, so they can be solved independently. Any of the two formulations (based either on the electric or magnetic fields) can be chosen to simulate the optical propagation of vector waves in inhomogeneous media. A comparison of the results using both formalisms indicates excellent agreement between them [16].

2.2 Semi-Vectorial and Scalar Beam Propagation Equations

For three-dimensional structures, polarization coupling is always present, so the propagation waves are hybrid. However, the coupling between the two polarizations is usually weak and may not have appreciable effect unless the two polarizations are synchronized. Therefore, under certain circumstances, we may still neglect the coupling terms and treat the two polarizations independently by using a semi-vectorial beam propagation method (SV-BPM) to solve Eqs. (2.10a) and (2.10b).

For a structure with a very small index gradient for one direction (let's say in the y direction), for example, channel waveguides with very low lateral index contrast (see Figure 2.4), the gradient of the refractive index along that particular direction can be neglected, that is:

$$\partial n^2/\partial y \approx 0. \tag{2.16}$$

Under this assumption, Eq. (2.6a) simplifies to:

$$2in_0k_0\frac{\partial \Psi_x}{\partial z} = \frac{\partial^2 \Psi_x}{\partial x^2} + \frac{\partial^2 \Psi_x}{\partial y^2} + \frac{\partial}{\partial x}\left(\frac{1}{n^2}\frac{\partial n^2}{\partial x}\Psi_x\right) + k_0^2\left(n^2 - n_0^2\right)\Psi_x, \tag{2.17}$$

which involves only the x component of the transverse electric field amplitude. This equation is referred as the wave equation for quasi-TM propagation.

For quasi-TE propagation, the semi-vectorial wave equation can be solved in terms of the magnetic field component Φ_x. From Eq. (2.9a) and taking $\partial n^2/\partial y \approx 0$, it follows:

$$2in_0k_0\frac{\partial \Phi_x}{\partial z} = \frac{\partial^2 \Phi_x}{\partial x^2} + \frac{\partial^2 \Phi_x}{\partial y^2} + k_0^2\left(n^2 - n_0^2\right)\Phi_x. \tag{2.18}$$

Another semi-vectorial formulation can also be obtained by assuming that the electric field propagating in the waveguide is linearly polarized and parallel to one of the transverse coordinates. Assuming that $\Psi_y = 0$ in Eq. (2.6a), quasi-TM propagation is governed by:

$$2in_0k_0\frac{\partial \Psi_x}{\partial z} = \frac{\partial^2 \Psi_x}{\partial x^2} + \frac{\partial^2 \Psi_x}{\partial y^2} + \frac{\partial}{\partial x}\left(\frac{1}{n^2}\frac{\partial n^2}{\partial x}\Psi_x\right) + k_0^2\left(n^2 - n_0^2\right)\Psi_x, \tag{2.19}$$

which is coincident with Eq. (2.17) obtained by assuming $\partial n^2/\partial y \approx 0$.

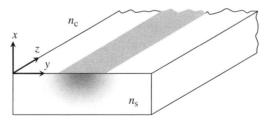

Figure 2.4 Three-dimensional diffused waveguide, showing smooth index changes along the x and y directions. Therefore, coupling between the two orthogonal polarizations can be neglected

If the predominant transversal component of the electric field is the y component, the quasi-TE propagation is governed by:

$$2in_0k_0\frac{\partial\Psi_y}{\partial z} = \frac{\partial^2\Psi_y}{\partial x^2} + \frac{\partial^2\Psi_y}{\partial y^2} + \frac{\partial}{\partial y}\left(\frac{1}{n^2}\frac{\partial n^2}{\partial y}\Psi_y\right) + k_0^2\left(n^2 - n_0^2\right)\Psi_y. \tag{2.20}$$

A last case can be contemplated for a semi-vectorial formulation and this is when the coupling terms in the vector equations (2.10) are neglected. For this case, the formulation is based on considering the following decoupled equations for the SVE of the electric field:

$$2in_0k_0\frac{\partial\Psi_x}{\partial z} = P_{xx}\Psi_x; \tag{2.21a}$$

$$2in_0k_0\frac{\partial\Psi_y}{\partial z} = P_{yy}\Psi_y. \tag{2.21b}$$

Similarly, the wave equations (2.12) for the magnetic field envelope are reduced to:

$$2in_0k_0\frac{\partial\Phi_x}{\partial z} = Q_{xx}\Phi_x; \tag{2.22a}$$

$$2in_0k_0\frac{\partial\Phi_y}{\partial z} = Q_{yy}\Phi_y. \tag{2.22b}$$

These are the approaches that we will consider here. As the transverse field components are completely decoupled, the problem of beam propagation is considerably simplified, while retaining what are usually the most significant polarization effects. Unless a structure is specifically designed to induce coupling, such as polarization converters, the effect of the off-diagonal terms (coupling terms) is extremely weak and the semi-vectorial approximation is an excellent one, even in high index contrast structures such as semiconductor waveguides.

2.2.1   Scalar Beam Propagation Equation

For waveguide structures with very small index contrast in both transverse directions, the whole gradient of refractive index term can be neglected:

$$\partial n^2/\partial x \approx 0; \tag{2.23a}$$

$$\partial n^2/\partial y \approx 0, \tag{2.23b}$$

which gives the following scalar wave equation:

$$2in_0k_0\frac{\partial u}{\partial z} = \frac{\partial^2 u}{\partial x^2} + \frac{\partial^2 u}{\partial y^2} + k_0^2\left(n^2 - n_0^2\right)u, \tag{2.24}$$

where u may represent any of the transverse SVE fields, either electric or magnetic, and it is frequently called optical field in a generic manner. In the scalar approach, the optical field $u(x,y,z)$ and its derivative are simply assumed to be continuous across all boundaries. Scalar approximation is well suited for weakly guiding problems, which are a wide range of typical waveguide structures.

2.3 BPM Based on the Finite Difference Approach

In the finite-difference method, the continuous space is replaced by a discrete lattice structure defined in the computational region. The field in the transverse plane (xy) is represented only at discrete points on a grid (Figure 2.5) and at discrete planes along the longitudinal or main propagation direction (z). The spatial domain of interest, of dimensions $L_x \times L_y$ in the transversal direction, is divided in a mesh of $N_x \times N_y$ discrete points, with grid spacing Δx and Δy. Therefore, each cell has dimensions of $\Delta x \times \Delta y$, where $\Delta x = L_x/(N_x - 1)$ and $\Delta y = L_y/(N_y - 1)$. The coordinates x_i and y_j of a generic point (i,j) in the transversal plane are related to the integer numbers i and j by:

$$x_i = i\Delta x = i L_x/(N_x - 1), y_j = j\Delta y = j L_y/(N_y - 1). \tag{2.25}$$

The computational region must include not only the relevant waveguide structure, but also must be wide enough to describe at least the evanescent waves corresponding to the propagation modes.

On the other hand, the longitudinal dimension is also discretized, with spacing Δz. The different fields (scalar or vectorial) at the lattice point of $x = i\Delta x$, $y = j\Delta y$ and $z = m\Delta z$ are represented by their discrete versions and can be conveniently written according to the following nomenclature:

$$\psi(x,y,z) \equiv \psi(i\Delta x, j\Delta y, m\Delta z) \equiv \psi_{i,j}^m. \tag{2.26}$$

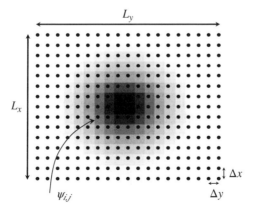

Figure 2.5 In the finite difference beam propagation method, the field in the transverse (xy) plane, for a fixed longitudinal coordinate z, is represented at discrete points on a rectangular grid

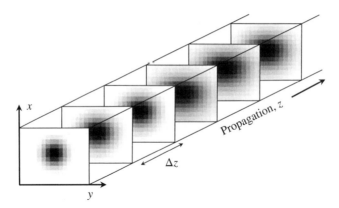

Figure 2.6   The BPM algorithm for the optical propagation allows us to determine the field at the $(z + \Delta z)$ plane knowing the discretized field at the transversal plane at the position z

Also, the operators in the wave equations are approximated by the appropriate finite differences formulae (see Appendix A). The partial differentiations with respect to the transverse coordinates are replaced by their finite-difference equations and here special care must be taken at index discontinuities. Given the discretized field at one z plane, the goal is to derive numerical equations that determine the field at the next z plane ($z + \Delta z$ plane). This elementary propagation step is then repeated to calculate the field throughout the structure. Thus, the BPM algorithm for the propagation along an arbitrary distance z is realized through several discrete steps of length Δz (Figure 2.6).

2.4   FD-Two-Dimensional Scalar BPM

In order to develop the different numerical schemes based on finite differences corresponding to the wave equation at each approximation level, let us start with the simplest case, which is the scalar wave equation in two-dimensional structures (one-dimensional transversal section). This equation allows us to model planar waveguides, where the physical problem depends only on two coordinates (x and z) and where only one field component is enough to describe the light propagation along the structure.

In the case of 2D structures, the computational domain at the transverse plane is reduced to a grid of N points and the field at a generic longitudinal position z is defined by a vector of N elements (Figure 2.7). In the scalar approach, the optical field $u(x,z)$ and its derivative are assumed to be continuous across all boundaries. The scalar wave equation (2.24) for 2D structures, removing any dependence on the y coordinate, simplifies to:

$$2in_0 k_0 \frac{\partial u}{\partial z} = \frac{\partial^2 u}{\partial x^2} + k_0^2 \left(n^2 - n_0^2\right) u. \tag{2.27}$$

The partial derivative with respect to the z coordinate is represented by its finite difference as (see Appendix A):

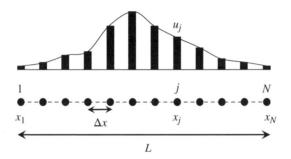

Figure 2.7 In the 2D-FD-BPM, the field in the transverse direction (x), for a fixed longitudinal coordinate z, is represented at discrete points on a linear grid

$$\frac{\partial u}{\partial z} \Rightarrow \frac{u_j^{m+1} - u_j^m}{\Delta z}, \tag{2.28}$$

where u_j^m denotes the optical field at the position ($j\Delta x$, $m\Delta z$) with $j = 1, 2,..., N$. On the other hand, the second-order derivative with respect to the transversal coordinate x is represented by the standard second-order finite-difference operator (see Appendix A):

$$\frac{\partial^2 u}{\partial x^2} \Rightarrow \frac{u_{j-1}^m - 2u_j^m + u_{j+1}^m}{(\Delta x)^2}, \tag{2.29}$$

which is exact to the second order on Δx.

Using these representations, the paraxial scalar wave equation (2.27) for two-dimensional structures can be expressed as:

$$2ik_0 n_0 \frac{u_j^{m+1} - u_j^m}{\Delta z} = \frac{u_{j-1}^m - 2u_j^m + u_{j+1}^m}{(\Delta x)^2} + k_0^2 \left[\left(n_j^m \right)^2 - n_0^2 \right] u_j^m. \tag{2.30}$$

This numerical scheme in finite differences, known as 'forward-difference', allows us to calculate the optical field u_j^{m+1} at the longitudinal position ($z + \Delta z$) knowing the complete field u_j^m at the position z [17]. The calculation of u_j^{m+1} is straightforward from Eq. (2.30) and indicates that the optical field u_j^{m+1} is computed from the known field values u_{j-1}^m, u_j^m, and u_{j+1}^m at the position z. This explicit method based on finite differences is accurate to first order on the longitudinal step Δz. On the other hand, from the point of view of numerical stability, the method is conditionally stable, where the stability condition is given by:

$$\Delta z \leq (\Delta x)^2 n_0 \pi / \lambda. \tag{2.31}$$

For practical applications, unfortunately the value of Δz necessary to assure numerical stability is too small. As an example, Figure 2.8 shows the light propagation of the fundamental mode

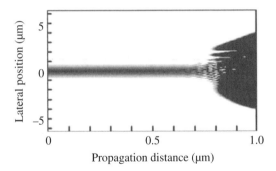

Figure 2.8 Light propagation through a graded-index 2D-waveguide using the forward scheme with a propagation step of $\Delta z = 0.03$ µm

along a uniform graded-index planar waveguide, simulated by using the forward difference numerical scheme performed with a longitudinal step of $\Delta z = 0.03$ µm. The refractive index of the guiding structure follows a Gaussian profile given by $n(x) = n_s + \Delta n \, e^{-(x/d)^2}$, with a substrate index of $n_s = 1.5$, maximum index change of $\Delta n = 0.03$ and a width of $d = 2$ µm. The simulation uses a window of 256 grid points, with grid spacing of $\Delta x = 0.05$ µm and a working wavelength of $\lambda = 1.55$ µm. Even using such a small propagation step, after a distance of ~ 0.6 µm the propagation becomes highly unstable. Indeed, to obtain a stable propagation, the propagation step must be less than $\Delta z < 0.008$ µm (according to formula (2.31)). This indicates that for modelling a device of, let's say, 5 mm length it will be necessary $\sim 7 \times 10^5$ propagation steps! Therefore, this numerical scheme is of little practical use.

One way to overcome this problem of instability consists of using a finite difference scheme somehow similar to the former one, known as 'backward-difference'. This consists of evaluating the finite difference scheme on the right hand by using the field at the $(z + \Delta z)$ position. The paraxial scalar wave equation expressed using this scheme takes the following form:

$$2ik_0n_0\frac{u_j^{m+1}-u_j^m}{\Delta z} = \frac{u_{j-1}^{m+1}-2u_j^{m+1}+u_{j+1}^{m+1}}{(\Delta x)^2} + k_0^2\left[\left(n_j^{m+1}\right)^2 - n_0^2\right]u_j^{m+1}. \qquad (2.32)$$

This method, at variance from the former one, has the advantage of being unconditionally stable, thus allowing arbitrarily long propagation steps (Figure 2.9). Nevertheless, the approximated solution obtained in the simulation is equivalent with respect to the forward-difference method (first-order error in Δz) and, therefore, no more accuracy is gained. In addition, numerical dissipation occurs, which gives rise to an apparent loss of energy (non-physical losses).

Fortunately, there is a method, also based on finite difference schemes, which is not only unconditionally stable but also provides more accurate solutions than the two previous methods. This method, called Crank–Nicolson scheme, is a linear combination of the forward-difference method and the backward-difference method (Figure 2.10).

The finite difference method following a Crank–Nicolson scheme is obtained by multiplying Eq. (2.30) by $(1 - \alpha)$ and Eq. (2.32) by α and adding the two results, yielding:

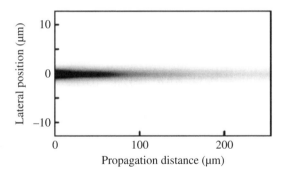

Figure 2.9 Light propagation through a graded-index 2D-waveguide using the backward scheme. The propagation is stable, even by using a large propagation step ($\Delta z = 1$ µm), but it presents a clear loss of power

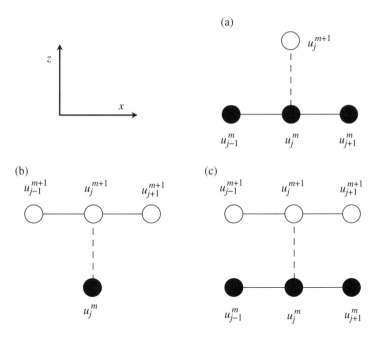

Figure 2.10 Diagram showing the three schemes used in the simulation of light propagation based on finite differences: (a) 'forward' or 'fully explicit'; (b) 'backward' or 'fully implicit' and (c) 'Crank–Nicolson' [17]

$$\frac{2ik_0n_0}{\Delta z}\left[u_j^{m+1} - u_j^m\right] = (1-\alpha)\frac{u_{j-1}^m - 2u_j^m + u_{j+1}^m}{(\Delta x)^2} + (1-\alpha)\,k_0^2\left[\left(n_j^m\right)^2 - n_0^2\right]u_j^m$$

$$+\alpha\frac{u_{j-1}^{m+1} - 2u_j^{m+1} + u_{j+1}^{m+1}}{(\Delta x)^2} + \alpha k_0^2\left[\left(n_j^{m+1}\right)^2 - n_0^2\right]u_j^{m+1}. \tag{2.33}$$

The parameter, α, which indicates the contribution of the forward and backward schemes ($0 \leq \alpha \leq 1$), is known as the scheme parameter. Equation (2.33) relates the optical field at $z + \Delta z$; that is, u^{m+1}, with the field at z; that is, u^m. Rearranging terms in this equation, we obtain:

$$a_j u_{j-1}^{m+1} + b_j u_j^{m+1} + c_j u_{j+1}^{m+1} = r_j, \tag{2.34}$$

where the a_j, b_j, c_j and r_j coefficients are given by:

$$a_j = -\frac{\alpha}{(\Delta x)^2}; \tag{2.35a}$$

$$b_j = \frac{2\alpha}{(\Delta x)^2} - \alpha \left[\left(n_j^{m+1} \right)^2 - n_0^2 \right] k_0^2 + \frac{2ik_0 n_0}{\Delta z}; \tag{2.35b}$$

$$c_j = -\frac{\alpha}{(\Delta x)^2}; \tag{2.35c}$$

$$r_j = \frac{(1-\alpha)}{(\Delta x)^2} \left[u_{j-1}^m + u_{j+1}^m \right] + \left\{ (1-\alpha) \left[\left(n_j^m \right)^2 - n_0^2 \right] k_0^2 - \frac{2(1-\alpha)}{(\Delta x)^2} + \frac{2ik_0 n_0}{\Delta z} \right\} u_j^m. \tag{2.35d}$$

Equation (2.34), besides the coefficients defined by the expressions in Eq. (2.35), in fact form a tridiagonal system of N linear equations ($j = 1, 2, \ldots, N$), which can be solved very efficiently [18]. One of the algorithms used for solving this tridiagonal system is the Thomas method, which allows rapid solution in order $O(N)$ operations (see Appendix B). In addition, it can be demonstrated that the solution of this equation system shows excellent numerical stability for $\alpha > 0.5$, although some numerical dissipation can be generated. Strictly speaking, the Crank–Nicolson scheme is unconditionally stable for $\alpha > 0.5$ if the refractive index is independent of x and z. Nevertheless, if the refractive index varies slowly, or if it is uniform with small discontinuities, the Crank–Nicolson method leads to valid solutions, even for the most adverse situations.

Figure 2.11 presents the simulation of light propagation along the waveguide structure previously examined, but using the 'standard' Crank–Nicolson scheme ($\alpha = 0.5$). Even with

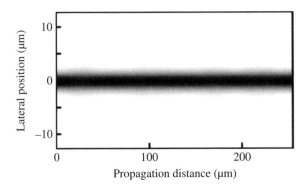

Figure 2.11 Light propagation through a graded-index 2D-waveguide using the 'standard' Crank–Nicolson scheme ($\alpha = 0.5$). The propagation is stable and without any sign of energy dissipation

a propagation step of $\Delta z = 1$ μm, the propagation is stable and the energy is conserved. In the following section the numerical dissipation, numerical dispersion and stability in the Crank–Nicolson scheme is analysed in more detail, putting special attention to the role played by the scheme parameter α.

2.5   Von Neumann Analysis of FD-BPM

Here, we will proceed to analyse the stability, numerical dissipation and numerical dispersion of the numerical schemes based on the finite difference scheme for TE propagation in 2D structures (planar waveguides) in terms of the electric field component following a Von Neumann stability analysis [3].

The Von Neumann method [17] starts by assuming that the optical field has solution of the form:

$$u_j^m = \gamma^m e^{-ik_x j\Delta x},\tag{2.36}$$

where u_j^m denotes the optical field at the nodal position $x = j\Delta x$ and the longitudinal position $z = m\Delta z$. The key fact is that the advance of the optical field from u_j^m to u_j^{m+1} by one longitudinal step corresponds to multiplication by the complex number γ, that is, the z-dependence of a single eigenmode is nothing more than successive integer powers of the number γ. Therefore, the difference equation is unstable (has exponentially growing modes) if $|\gamma| > 1$. The number γ is called the amplification factor. This amplification factor contains the information about the stability, numerical dissipation and dispersion of the finite-difference scheme.

To find γ, as the scalar wave equation for 2D structures is equivalent to that the TE-wave equation (as we will see in the next chapter), the proposed solution (2.36) is substituted back into the finite difference Eq. (2.33), yielding:

$$\frac{2ik_0n_0}{\Delta z}\left[\gamma^{m+1}e^{-ik_x j\Delta x} - \gamma^m e^{-ik_x j\Delta x}\right] = (1-\alpha)\frac{\gamma^m e^{-ik_x(j-1)\Delta x} - 2\gamma^m e^{-ik_x j\Delta x} + \gamma^m e^{-ik_x(j+1)\Delta x}}{(\Delta x)^2}$$

$$+ (1-\alpha)k_0^2\left[\left(n_j^m\right)^2 - n_0^2\right]\gamma^m e^{-ik_x j\Delta x} + \alpha\frac{\gamma^{m+1}e^{-ik_x(j-1)\Delta x} - 2\gamma^{m+1}e^{-ik_x j\Delta x} + \gamma^{m+1}e^{-ik_x(j+1)\Delta x}}{(\Delta x)^2}$$

$$+ \alpha k_0^2\left[\left(n_j^{m+1}\right)^2 - n_0^2\right]\gamma^{m+1}e^{-ik_x j\Delta x}$$

$$\tag{2.37}$$

Dividing by $\gamma^m e^{-ik_x j\Delta x}$, we obtain:

$$\gamma = \frac{1 - \dfrac{i(1-\alpha)\Delta z}{2k_0n_0}\left\{k_0^2\left[\left(n_j^m\right)^2 - n_0^2\right] - \dfrac{4}{(\Delta x)^2}\sin^2\left(\dfrac{k_x\Delta x}{2}\right)\right\}}{1 + \dfrac{i\alpha\Delta z}{2k_0n_0}\left\{k_0^2\left[\left(n_j^{m+1}\right)^2 - n_0^2\right] - \dfrac{4}{(\Delta x)^2}\sin^2\left(\dfrac{k_x\Delta x}{2}\right)\right\}}.\tag{2.38}$$

This can be expressed in a more compact form by:

$$\gamma = \frac{1 - i(1-\alpha)Q_j^m}{1 + i\alpha Q_j^{m+1}}, \tag{2.39}$$

where the quantity Q_j^m has been defined as:

$$Q_j^m \equiv \frac{\Delta z}{2k_0 n_0}\left\{ k_0^2\left[\left(n_j^m\right)^2 - n_0^2\right] - \frac{4}{(\Delta x)^2}\sin^2\left(\frac{k_x \Delta x}{2}\right)\right\}. \tag{2.40}$$

Strictly speaking, the Von Neumann analysis presented here is valid only when the refractive index is independent of x and z [3]. If the refractive index slowly varies in the region of interest or piecewise uniform with small index discontinuities, then the Von Neumann analysis may be applied locally. Under these circumstances, the analysis usually leads to conclusions valid for the worst possible situations.

2.5.1 Stability

Following the proposed solution (2.36), it is clear that the stability of the numerical scheme is assured for values of the amplification factor satisfying the stability criterion:

$$|\gamma| \le 1, \tag{2.41}$$

which implies the following condition:

$$(1-\alpha)\left\{ k_0^2\left[\left(n_j^m\right)^2 - n_0^2\right] - \frac{4}{(\Delta x)^2}\sin^2\left(\frac{k_x \Delta x}{2}\right)\right\} \le \alpha\left\{ k_0^2\left[\left(n_j^{m+1}\right)^2 - n_0^2\right] - \frac{4}{(\Delta x)^2}\sin^2\left(\frac{k_x \Delta x}{2}\right)\right\}, \tag{2.42}$$

or in a more compact form:

$$(1-\alpha)Q_j^m \le \alpha Q_j^{m+1}. \tag{2.43}$$

If the structure is z invariant, that is, the refractive index of the optical structure does not depend on the z coordinate, the quantities inside the brackets are equal (i.e. $Q_j^m = Q_j^{m+1}$) and therefore the condition (2.43) is simply given by:

$$\alpha \ge 0.5. \tag{2.44}$$

If the refractive index is not uniform along the propagation direction z, this condition may no longer be true. The numerical boundary conditions used, which will be treated in the subsequent sections, may also have some effects on the stability of the numerical scheme. Under these circumstances, the standard Crank–Nicolson scheme with $\alpha = 0.5$ may give rise to some instabilities, which can be avoided by using a scheme parameter slightly greater that 0.5 ($\alpha =$ $0.5 + \varepsilon$, $\varepsilon > 0$). Nevertheless, using a too high value of ε can have negative consequences regarding a correct simulation, as we will show in the following.

2.5.2 Numerical Dissipation

A stable numerical scheme may not conserve the power. The numerical dissipation introduces non-physical power loss and thus can limit the usefulness of the BPM for prediction of the guided power. Theoretically, a non-dissipative finite-difference scheme requires that $|\gamma| = 1$. For TE waves propagating in uniform (z invariant) waveguide structures (where $Q_j^m = Q_j^{m+1}$), the formula (2.39) shows that the non-dissipative scheme condition is given by:

$$|\gamma|^2 = 1 \quad \Rightarrow \quad Q_j^m(1-2\alpha) = 0 \tag{2.45}$$

indicating that the standard Crank–Nicolson scheme ($\alpha = 0.5$) is non-dissipative for straight waveguides. For z-dependent waveguide structures, although this conclusion is no longer valid, generally speaking the Crank–Nicolson scheme is the least dissipative scheme, while the standard implicit scheme ($\alpha = 1$) is the most. Also, in general, as is evident from Eq. (2.45), the numerical dissipation can always be reduced by having small values of Q_j^m, which implies shortening the longitudinal propagation step Δz, as can be deduced from the definition (2.40).

Figure 2.12 shows the attenuation suffered by TE and TM light propagation using FD-BPM (the implementation of these methods will be given in next chapter) in a straight 2D step-index waveguide as a function of the reference index for three different scheme parameters α. The 2D structure is an asymmetric step-index planar waveguide, which is monomode at the working

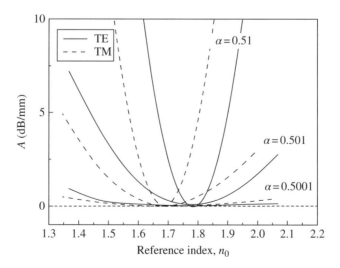

Figure 2.12 Attenuation due to numerical dissipation in FD-BPM as a function of the reference index used for performing TE and TM propagation of the fundamental modes for three different control parameters

wavelength of 1.55 μm. The cover, film and substrate indices are $n_c = 1$, $n_f = 1.95$ and $n_s = 1.45$, respectively, the film thickness being $d = 0.6$ μm. The simulation is performed by using a grid spacing of $\Delta x = 0.005$ μm, with 1000 mesh points in the transversal direction and a longitudinal step of $\Delta z = 0.2$ μm. The input beams for the simulations are the fundamental guided-modes for TE and TM polarization. Each curve in Figure 2.12 corresponding to a particular scheme parameter shows a minimum as a function of the reference index n_0 used in the simulation. These values coincide with the effective indices of the guided modes for TE and TM propagation ($N_{eff} = 1.774255$ for TE and $N_{eff} = 1.684615$ for TM). On the other hand, for a given reference index, the attenuation increases strongly as the scheme parameter deviates from 0.5, indicating that the standard Crank–Nicolson scheme ($\alpha = 0.5$) is the least dissipative numerical implementation. Also, the attenuation for $\alpha = 0.5001$ is maintained at very low values for both TE and TM propagation within a wide range of reference indices. Therefore, regarding power analysis in waveguides, the right choice is to set the scheme parameter as close to 0.5 as possible, while maintaining stability.

The positions of the minima for TE and TM propagations found on the attenuation are not a coincidence. In fact, the reference index should be adequately chosen so that the SVEA is satisfied. To fulfil the SVEA, n_0 should be chosen so that the variation of transverse fields along the longitudinal direction are minimized and the optimum refractive index of reference should be the average of the effective indices of all the propagating modes involved in the propagation. If the difference of the refractive index over the transverse cross section is small, n_0 can be chosen to be the refractive index of the substrate. In the case of single-mode waveguides the reference index can be set to the effective index of the guided-mode, as has been evidenced in Figure 2.12. In the particular case of bi-modal waveguides, such as directional couplers, the reference index can be chosen as:

$$n_0 = \frac{N_e + N_o}{2},\qquad(2.46)$$

where N_e and N_o are the effective indices of the symmetric and anti-symmetric modes of the coupler.

The use of numerical boundary conditions may also affect the numerical dissipation. In practice, one should carefully choose the scheme parameter, α, as the standard Crank–Nicolson scheme ($\alpha = 0.5$) may generate some high frequency oscillations in the field distribution. When the numerical boundary conditions are correctly used, the high frequency components are absorbed at the edge of the computational domain, leading to non-physical loss.

2.5.3 Numerical Dispersion

Due to the discretization, some phase errors will be introduced in a finite-difference scheme. As a result, the numerical dispersion will develop and degrade the accuracy of the propagation constants of the waveguide modes calculated using BPM, degrading, for example, the correct coupling length of modes between parallel waveguides. By assuming that the refractive index of the structure slowly varies, we may still apply the Von Neumann method to analyse this numerical dispersion. The solution proposed in Eq. (2.36) may be rewritten as:

$$u_j^m = |\gamma^m| e^{-ik_x j \Delta x} e^{-ik_z \Delta z}.\qquad(2.47)$$

Assuming z invariant structures $(Q_j^m = Q_j^{m+1})$, the amplification factor can be expressed as:

$$\gamma = \frac{1 - i(1-\alpha)Q_j^m}{1 + i\alpha Q_j^m} = \frac{1 - \alpha(1-\alpha)\left(Q_j^m\right)^2 - iQ_j^m}{1 + \left(\alpha Q_j^m\right)^2} = |\gamma^m| e^{-ik_z \Delta z}, \qquad (2.48)$$

and the phase factor $k_z \Delta z$ is therefore:

$$k_z \Delta z = \text{atn}\left[\frac{Q_j^m}{1 - \alpha(1-\alpha)\left(Q_j^m\right)^2}\right]. \qquad (2.49)$$

Now using the approximation $Q_j^m \ll 1$, which is fulfilled for small longitudinal steps, it follows that:

$$\text{atn}\left[\frac{-Q_j^m}{1 - \alpha(1-\alpha)\left(Q_j^m\right)^2}\right] \approx \text{atn}\left(-Q_j^m\right) \approx -Q_j^m. \qquad (2.50)$$

Now, from Eq. (2.40), the propagation constant for the envelope of the transverse electric field along z can be approximated by:

$$k_z \approx \frac{1}{2k_0 n_0}\left\{ k_0^2\left[\left(n_j^m\right)^2 - n_0^2\right] - \frac{4}{(\Delta x)^2}\sin\left(\frac{k_x \Delta x}{2}\right) \right\}. \qquad (2.51)$$

Equation (2.51) indicates that, as far as the approximation (2.50) holds, the scheme parameter α and the longitudinal step Δz do not affect the numerical dispersion.

As has been mentioned, the coupling length of directional couplers is very sensitive to the propagation constant of the guided modes, so it is an excellent example to test the correct description of light propagation under different BPM numerical schemes [19]. To show the effect of the propagation step and the reference index on the numerical dispersion, let us calculate the coupling length for a directional coupler made of two identical parallel waveguides (Figure 2.13) for TE propagation at a wavelength of $\lambda = 1.55\,\mu\text{m}$. The proposed dual-waveguide structure supports two guided modes: a symmetric mode, with an effective index of $N_e = 1.470727$ and an anti-symmetric mode with an effective index of $N_o = 1.469551$. Following the coupled-mode theory [20], the coupling length, L_c, defined as the minimum distance required to a maximum power exchange between waveguides, is given by:

$$L_c = \frac{\lambda}{N_e - N_o}, \qquad (2.52)$$

which results in a coupling length of $L_c = 659.2\,\mu\text{m}$.

Figure 2.14 shows the coupling length calculated by 2D-TE-BPM simulation, for three different reference indices, as a function of the propagation step, and using a scheme parameter of

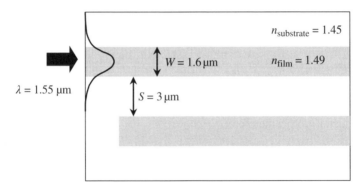

Figure 2.13   Directional coupler made of two identical planar waveguides

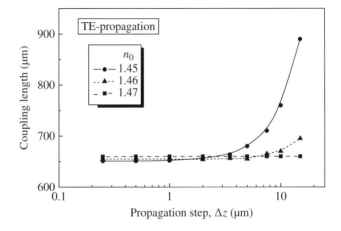

Figure 2.14   Coupling length calculated by TE-propagation, as a function of the propagation step for three different reference indices

$\alpha = 0.5001$ to assure stability and power conservation. The numerical simulation uses a calculation transversal window of 25.5 μm with 512 mesh points (grid spacing $\Delta x = 0.05$ μm). To perform the simulation, the fundamental mode of a single (independent) waveguide is launched to the upper waveguide and the coupling length is determined from where the power remaining in the original waveguide is at the first minimum. It is observed that, for a given reference index, the coupling length is independent of the propagation step for small Δz, as was deduced from Eq. (2.51). For larger values of Δz the coupling length deviates from the saturation value found for small Δz and this deviation is less pronounced as the reference index approaches to 1.47. Indeed, for this reference index the coupling length is almost independent of the propagation step length, at least up to values as large as $\Delta z \sim 20$ μm. Once again, the reference index 1.47 is very close of the one obtained from Eq. (2.46) ($n_0 = 1.470139$). Also, the values of the coupling length for small Δz are very close to that calculated by formula (2.52), being coincident when using a reference index of $n_0 = 1.47$ ($L_c = 660 \pm 0.5$ μm).

Very similar behaviour is found for the coupling length calculated by using TM-BPM propagation, as indicated in Figure 2.15. In this case, the coupling length calculated by the relation (2.52) gives $L_c = 626.0\,\mu m$ ($N_e = 1.470144$ and $N_o = 1.468906$). The coupling length using a reference index of $n_0 = 1.47$ (close to that provided from Eq. (2.46)) yields a value of $626.0 \pm 0.5\,\mu m$, coincident to that obtained from coupled wave theory. As the reference index used to perform the BPM simulation deviates from 1.47, the coupling length calculated deviated from the exact value and a smaller propagation step are needed to reach correct values.

Even when using reference indices different to 1.47, the coupling length calculated for small values of propagation step differs only a few percent from the correct value. This circumstance is examined in Figure 2.16 where the relative error in the coupling length obtained by BPM

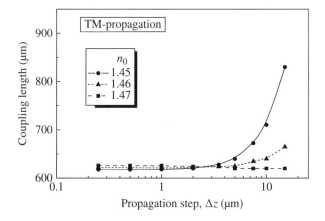

Figure 2.15 Coupling length calculated by TM-propagation, as a function of the propagation step for three different reference indices

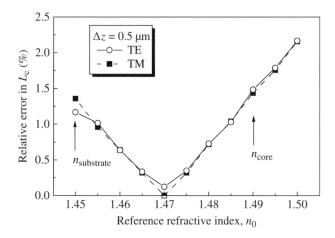

Figure 2.16 Influence of the reference index on the accuracy of the coupling length calculated by FD-BPM numerical simulations

simulation using $\Delta z = 0.5$ is plotted against the reference index for both TE and TM propagation. Relative error values lower than 1.5% are obtained for reference indices in the range 1.45–1.49, corresponding to the refractive indices of the substrate and core, respectively. Again, the minimum error is found for a reference index of ~ 1.47, which corresponds with the value supplied by relation (2.46).

Obviously, another source of numerical dispersion comes from the fact of transversal discretization of the waveguide structure. To minimize this, we need to use a sufficiently fine mesh in the transverse cross section, that is, reduce the grid spacing Δx.

Apart from the numerical stability and numerical dissipation control, the greatest advantage of the Crank–Nicolson method comes from the fact that it provides a better approximation to the exact solution of the problem. While the simple finite difference method (implicit or explicit standard schemes) allows an approximation of first order in the propagation step O $[\Delta z + (\Delta x)^2]$, the Crank–Nicolson method establishes an approximation of second order in the propagation $O[(\Delta z)^2 + (\Delta x)^2]$. In this way the finite difference method is a powerful numerical method that allows the use of large propagation steps, with a consequential saving in computational time.

2.6   Boundary Conditions

As the computational domain in BPM is finite, the simulation of the optical propagation needs to specify boundary conditions for the optical field at the limits of the computational window. In FD-BPM calculations, when the Crank–Nicolson method is applied at the boundary points $j = 1$ and $j = N$, we find that it refers to unknown quantities outside the domain (u_0 and u_{N+1}). Thus, as the fields at the points $j = 0$ and $j = N + 1$ are not defined, but they are necessary for calculating the field in the interior points ($j = 1$ and N), two extra equations are needed. This implies that for the points at the computational domain limits, Eq. (2.34) must be replaced by appropriate boundary conditions that complete the system of equations. These boundary conditions must be adequately chosen in such a way that the effect of the boundaries does not introduce errors in the propagation description of the optical field. If these conditions are not well specified, the radiation could be reflected on the limits of the computational window, returning to enter the region of interest and avoiding a correct propagation description of the optical field.

Dirichlet boundary conditions provide the simplest possibility by specifying the boundary values u_1 and u_N directly, for example, by setting their values to zero. Other possibility is the Neumann boundary condition, which imposes conditions on the field gradients (for instance, $u_1 = u_2$ and $u_N = u_{N-1}$). Unfortunately, none of these boundary conditions give satisfactory results, as their implementation provokes optical field reflections at the window limits. In particular, the condition of zero fields at the boundaries is not realistic when the optical perturbation reaches the limits of the computational window, because simply requiring the field to vanish at the boundaries is equivalent to placing perfectly reflecting walls at the edge of the domain. This effect can be appreciated in Figure 2.17 where the propagation of a 20° tilted Gaussian beam in a homogeneous medium is simulated by FD-BPM, with Dirichlet boundary conditions implemented. The simulation of the beam propagation proceeds correctly as long as the optical beam does not reach the computational window limits. But once the optical field reaches the upper window limit, it suffers unphysical reflection from the boundary. The observed fringe pattern is in fact due to the interference between the incident and reflected

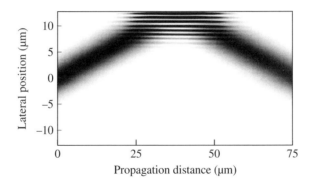

Figure 2.17 FD-BPM simulation for a tilted Gaussian beam propagating through a homogeneous medium. When the radiation reaches the upper limit, it suffers reflection and re-enters the domain

fields. The final result is that the optical beam re-enters the computational region. This unphysical behaviour is a consequence of the perfectly reflecting walls derived from the implementation of the Dirichlet boundary condition. Thus, it is necessary to impose realistic and reasonably physical boundary conditions for adequately describing and simulating the optical propagation by BPM.

An evident solution for avoiding that the window limits interfere in the correct optical propagation is to build a wide enough computational region. In this way, the optical field at the window limits can be almost negligible, thus suppressing such unwanted behaviour. Nevertheless, this procedure implies an unnecessary increment in the window size and to compute the optical field in regions out of interest for the solution of the problem, increasing computer memory requirements and computing time. An evident way to prevent reflections from the boundaries is to build artificial absorbing regions adjacent to the computation edges (absorbing boundary conditions, or ABC method), allowing the refractive index to be complex where its imaginary part controls the absorption [21]. This method implies that the computational region must be increased to include these extra absorption regions. One alternative approach to ABC consists of implementing realistic boundary conditions from the physical point of view, known as transparent boundary conditions (TBC), which prevent radiation flux back into the computational domain [22, 23]. The TBC algorithm allows the radiation to freely pass the domain limits without appreciable reflection, thus simulating a non-existent boundary. Another method to avoid reflections at the window edges is to build a layer with a sort of anisotropic conductivity (known as a perfectly matched layer, PML) [24, 25]. Except for grazing incidence, the reflection coefficient for a plane wave incident in such PML region can be practically zero.

In this section, after examining the energy conservation in the 2D-FD equation, the basics concepts and implementations of these different boundary condition algorithms will be described and some examples applied to 2D-FD-BPM will be presented.

2.6.1 Energy Conservation in the Difference Equations

In order to understand the origin of the reflection of radiation at the boundaries when using FD-BPM, let us examine the energy conservation during propagation, where, for the sake of simplicity, the discussion will be focused on two-dimensional propagation [23].

Since only the boundary region is of interest, the energy conservation analysis is restricted to the diffraction term in the paraxial scalar beam-propagation equation (2.27):

$$\frac{\partial u}{\partial z} = \frac{1}{2iK} \frac{\partial^2 u}{\partial x^2}, \tag{2.53}$$

which represents the optical propagation in an homogeneous medium of refractive index $n_0 = K/k_0$. By simple manipulations, Eq. (2.53) may be rewritten in an energy conservation equation:

$$\frac{\partial}{\partial z} \int_a^b |u|^2 dx = \frac{1}{2iK} \left(u^* \frac{\partial u}{\partial x} - u \frac{\partial u^*}{\partial z} \right) \Big|_a^b \equiv -F_b + F_a, \tag{2.54}$$

where F_b represents the energy flux leaving the upper boundary and F_a represents that entering through the bottom boundary. Since the treatment of the two boundaries is essentially identical, only the upper boundary is considered.

Applying the standard Crank–Nicolson scheme ($\alpha = 0.5$) to Eq. (2.53), the following finite difference equation is obtained:

$$u_j^{m+1} - u_j^m = \frac{\Delta z}{4iK(\Delta x)^2} \left[u_{j-1}^{m+1} - 2u_j^{m+1} + u_{j+1}^{m+1} + u_{j-1}^m - 2u_j^m + u_{j+1}^m \right]. \tag{2.55}$$

Multiplying this equation by $\left(u_j^{m+1^*} + u_j^{m^*} \right)/2$, its complex conjugate by $\left(u_j^{m+1} + u_j^m \right)/2$ and adding the results, it yields:

$$\left| u_j^{m+1} \right|^2 - \left| u_j^m \right|^2 = \frac{\Delta z}{8iK(\Delta x)^2} \left\{ \begin{array}{l} \left[\left(u_j^m \right)^* + \left(u_j^{m+1} \right)^* \right] \left[u_{j+1}^m + u_{+1}^{m+1} \right] + \\ \left[\left(u_j^m \right)^* + \left(u_j^{m+1} \right)^* \right] \left[u_{j-1}^m + u_{j-1}^{m+1} \right] - \text{c.c.} \end{array} \right\}. \tag{2.56}$$

This equation can also be derived from the equation of energy conservation (2.54) by setting $a = x_{j-(1/2)}$ and $b = x_{j+(1/2)}$, provided that all terms are well centred on both the x- and z-axis prior to being multiplied together. Taking into account this expression, the energy flux passing out of the control volume centred at j and into that centred at $j+1$ is identified as:

$$F_{j+(1+2)} = \frac{1}{8iK(\Delta x)^2} \left\{ \left[\left(u_j^m \right)^* + \left(u_j^{m+1} \right)^* \right] \left[u_{j+1}^m + u_{j+1}^{m+1} \right] \right.$$
$$\left. - \left[u_j^m + u_j^{m+1} \right] \left[\left(u_{j+1}^m \right)^* + \left(u_{j+1}^{m+1} \right)^* \right] \right\}. \tag{2.57}$$

With this definition, Eq. (2.56) may be compactly written as:

$$\left| u_j^{m+1} \right|^2 - \left| u_j^m \right|^2 = \Delta z \left(F_{j-(1/2)} - F_{j+(1/2)} \right). \tag{2.58}$$

The form of Eqs. (2.57) and (2.58) shows that the flux entering a control volume through a given interface is identical to that leaving the adjacent control volume. We may thus sum (2.58) over all interior mesh points and, due to multiple cancellations, we obtain the following formula for the net change in energy between propagation steps:

$$\sum_{j=2}^{N-1} \left(\left| u_j^{m+1} \right|^2 - \left| u_j^m \right|^2 \right) = \Delta z \left(F_{3/2} - F_{N-(1/2)} \right). \tag{2.59}$$

This expression is the direct finite difference analogue of Eq. (2.54) and shows that the total energy only changes if there is a nonzero flux of energy through one or both boundaries. Taking only the upper boundary into account now, we obtain:

$$F_{N-(1/2)} = \frac{1}{8iK(\Delta x)^2} \left\{ \left[\left(u_{N-1}^m \right)^* + \left(u_{N-1}^{m+1} \right) \right] \left[u_N^m + u_N^{m+1} \right] - \left[u_{N-1}^m + u_{N-1}^{m+1} \right] \left[\left(u_N^m \right)^* + \left(u_N^{m+1} \right)^* \right] \right\}.$$

$$\tag{2.60}$$

It is now evident from this equation that both standard Dirichlet ($u_N = 0$) and Neumann ($u_N = u_{N-1}$) boundary conditions will result in zero flux through this boundary and therefore the radiation cannot escape from the computational region, being totally reflected at the boundaries, as we observed in Figure 2.17.

2.6.2 *Absorbing Boundary Conditions (ABCs)*

A method to prevent reflection at the boundaries is the insertion of artificial absorption regions adjacent to the pertinent boundaries [21]. For such a purpose, the refractive index in the regions near the boundaries is allowed to be complex, with an imaginary part responsible to induce absorption of the radiation ($\tilde{n} = n + i\kappa$). The absorbing region must be carefully tailored, as the absorbing region itself provokes reflections. Thus, a gradual absorption coefficient must be designed to induce strong enough attenuation of the optical waves but with a small gradient to prevent high reflection [26]. In practice, the absorption coefficient is ramped up from zero at the region's leading edge to some maximum value at the boundary node. Thus, the thickness of the region, the maximum absorption coefficient and functional shape must all be carefully chosen for the method to work properly. The implementation in the FD-BPM is straightforward, because the algorithm is still tridiagonal, but the computational region must be increased to include these extra absorption regions, thus increasing the computational effort.

Figure 2.18 shows the geometry of the absorbing regions adjacent to the two boundaries. There δ is the extension of the absorbing regions and L is the total window size. These regions should be adequately designed to avoid reflections from them and one possible functional form of the absorption index profiles $\kappa(x)$ follows a parabolic relationship on the form of:

$$\kappa(x) = \kappa_{max} \left(\frac{x - L/2 + \delta}{\delta} \right)^2, \quad x \in \left[\frac{L}{2}, \frac{L}{2} - \delta \right]; \tag{2.61a}$$

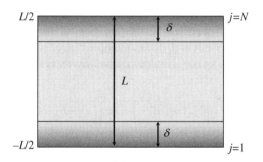

Figure 2.18   Geometry of the absorption layers built at the two boundaries of the computational domain, which allows absorption of the radiation entering these two regions, thus avoiding reflection from the window edges

$$\kappa(x) = \kappa_{max} \left(\frac{L/2 + \delta - x}{\delta} \right)^2 , \quad x \in \left[-\frac{L}{2} + \delta, -\frac{L}{2} \right]. \tag{2.61b}$$

As the implementation of the ABC still involves a tridiagonal matrix, the code is similar to conventional BPM but allows the refractive index to be complex. The complex refractive index at the position j is split in its real part n_j and its imaginary part κ_j. Therefore, the coefficients for the Thomas algorithm are the following:

$$a_j = -\frac{\alpha}{(\Delta x)^2}; \tag{2.62a}$$

$$b_j = \frac{2\alpha}{(\Delta x)^2} - \alpha \left[\left(n_j^{m+1} \right)^2 - \left(\kappa_j^{m+1} \right)^2 + 2in_j^{m+1} \kappa_j^{m+1} - n_0^2 \right] k_0^2 + \frac{2iK}{\Delta z}; \tag{2.62b}$$

$$c_j = -\frac{\alpha}{(\Delta x)^2}; \tag{2.62c}$$

$$r_j = \frac{(1-\alpha)}{(\Delta x)^2} \left[u_{j-1}^m + u_{j+1}^m \right] + \left\{ (1-\alpha) \left[\left(n_j^m \right)^2 - \left(\kappa_j^m \right)^2 + 2in_j^m \kappa_j^m - n_0^2 \right] k_0^2 - \frac{2(1-\alpha)}{(\Delta x)^2} + \frac{2iK}{\Delta z} \right\} u_j^m \tag{2.62d}$$

Figure 2.19 shows the paraxial propagation of a 5 µm wide Gaussian beam tilted at 20° through a homogeneous medium of $n = 1.45$, using a wavelength of $\lambda = 0.81$ µm and a propagation step of $\Delta z = 0.25$ µm (300 points). The computational domain has 512 transversal points with a grid spacing of $\Delta x = 0.05$ µm. At the edges of this computational domain, two absorbing layers have been added with $\delta = 5$ µm and $\kappa_{max} = 0.2$. It can be seen that when the radiation reaches the upper PML region, it attenuates and virtually disappears from the computational domain, with almost no sign of reflected radiation.

   The insertion of artificial absorbing regions adjacent to the pertinent boundaries is quite an accurate procedure, providing that the absorbing region is carefully tailored; that is, by using a small enough absorption gradient so that the absorber itself does not generate reflections and a

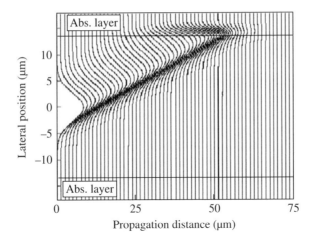

Figure 2.19 Simulation of the propagation of a Gaussian beam tilted 20° in a homogeneous medium, when absorbing boundary conditions are implemented. The two ABC regions are 5 μm wide and the maximum imaginary refractive index (absorption index) was set at 0.2, the wavelength being 0.81 μm

thickness sufficient to absorb all radiation impinging upon the region. The practical problem of using the ABC for avoiding reflections is that the task for ensuring that these conditions are properly met for each new problem is often a cumbersome and time-consuming process. Even when one is successful, the addition of extra problem zones results in computational penalties of run time and storage space.

2.6.3 Transparent Boundary Conditions (TBC)

One elegant way to overcome the problem of radiation coming back to the domain after reaching the boundaries is to try to implement realistic boundary conditions from the physical point of view; that is, an algorithm that allows the wave to leave the computational region when it reaches the window limits. This algorithm is known as the transparent boundary condition because it simulates a non-existent boundary [22]. Radiation is allowed to freely escape the problem without appreciable reflection, whereas radiation flux back into the problem region is prevented. The TBC employs no adjustable parameters and so is problem independent and can be directly applied to any waveguide structure. In addition, it is easily incorporated into the standard Crank–Nicolson differencing scheme in both two and three dimensions, and is applicable to longitudinally varying structures.

Here, the basic principles of the TBC algorithm are presented, starting with the analysis of the fluxes passing the boundaries [23]. Since the treatment of the two boundaries is essentially identical, we consider only the upper boundary. We start with the important assumption that at this boundary, the field is of the form:

$$u = u_0 e^{ik_x x}, \tag{2.63}$$

where u_0 and k_x are, in general, complex and k_x is unknown for the moment. With this assumption, the flux in the upper boundary F_b becomes:

$$F_b = \frac{\mathrm{Re}(k_x)|u(b)|^2}{K}, \tag{2.64}$$

where Re indicates the real part. Therefore, as long as the real part of k_x is positive, the contribution to the overall change in energy from this boundary will always be negative; that is, radiative energy can only flow out of the problem region.

Within the Crank–Nicolson scheme based on finite differences, and assuming the same exponential dependence described previously, the optical field in the limit of the window u_N^m prior to the start of the $(m+1)$th propagation step should fulfil the following relation:

$$\frac{u_N^m}{u_{N-1}^m} = \frac{u_{N-1}^m}{u_{N-2}^m} = e^{ik_x \Delta x}. \tag{2.65}$$

This expression allows the determination of k_x after completion of the mth step, using two interior points close to the boundary. Then, the boundary condition for the next propagation step $(m+1)$ is:

$$u_N^{m+1} = u_{N-1}^{m+1} \cdot e^{ik_x \Delta x}, \tag{2.66}$$

where the k_x value is that previously calculated using Eq. (2.65) in the mth step. However, prior to the application of Eq. (2.66), the real part of k_x must be restricted to be positive to ensure only radiation outflow. Therefore, if the real part of k_x from Eq. (2.66) is negative, it is reset to zero and the field at the boundary is recalculated using the corrected value of k_x.

Since the boundary condition itself is linear and only involves the two mesh points nearest the boundary, the ratio between the interior points previously used to determine k_x is allowed to change. Thus, the value of k_x computed for the next propagation step will be different in general. Such an adaptive procedure is required for an accurate algorithm that reflects the minimum amount of energy back into the problem region.

The great advantage of the TBC method lies in its convenience and utility. While the use of artificial absorbers to remove scattered radiation is clumsy, it imposes some penalties of computer runtime and storage and must be re-tailored for each new problem, the TBC algorithm utilizes no adjustable parameters, thus being problem independent and imposes no storage penalty because it requires no extra computational zones to be considered.

The TBC algorithm shows, besides its accuracy and efficiency, a high degree of robustness. The implementation is easily performed for two-dimensional problems as well as propagation through 3D structures. In each case, the paraxial propagation leads to the solution of a tridiagonal system and thus the inclusion of TBC does not complicate the resolution of the problem.

In order to show the accuracy and robustness of the TBC, Figure 2.20 shows the propagation of the $20°$ tilted Gaussian beam propagation through a homogeneous medium simulated by the FD-BPM technique where the TBC algorithm has been implemented in both upper and lower boundaries. It is observed that when the radiation reaches the upper window limit it virtually disappears into the boundary without any sign of reflection or distortion in the beam shape as it passes through the boundary. Indeed, this is the expected result from a physical point of view: the total disappearance of the beam as it crosses the limit of the computational region.

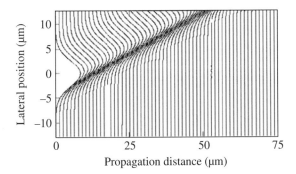

Figure 2.20 Tilted Gaussian beam propagation by means of FD-BPM, using the same parameters as in Figures 2.17 and 2.19, when TBC is implemented. When the radiation reaches the upper window limit, it passes through the boundary and virtually disappears, leaving the computational region

The effective reflection coefficient, defined as the ratio between the initial energy in the computational region and the energy in it after a long enough propagation distance, in this case is lower than 10^{-5}, indicating the efficiency and accuracy of the TBC method for simulating optical propagation based on FD-BPM.

2.6.4 Perfectly Matched Layers (PMLs)

A third technique to avoid reflections at the boundaries is the PML boundary condition, initially introduced by Bérenger for the FDTD method for Maxwell's equations [24]. In this technique, a layer with anisotropic conductivity (Bérenger layer) is placed at the edge of the computational domain. In this layer, the propagation equation is mapped via an anisotropic complex transformation. This mapping produces a wave that is perfectly matched with the outgoing waves and, in principle, the reflection coefficient of the Bérenger layer can be practically zero for a plane wave impinging this layer with an arbitrary propagation angle (except for grazing incidence).

In the PML the corresponding wave equation of the scalar field $\psi(x,z)$ is written as [27]:

$$
\frac{1}{1-i\sigma_x/\omega\varepsilon_0 n_p^2}\frac{\partial}{\partial x}\left(\frac{1}{1-i\sigma_x/\omega\varepsilon_0 n_p^2}\frac{\partial\psi}{\partial x}\right) + \frac{1}{1-i\sigma_z/\omega\varepsilon_0 n_p^2}\frac{\partial}{\partial z}\left(\frac{1}{1-i\sigma_z/\omega\varepsilon_0 n_p^2}\frac{\partial\psi}{\partial z}\right)
$$
$$
+ \frac{n_p^2\omega^2}{c^2}\psi = 0
$$

(2.67)

where n_p is a constant refractive index of the PML medium, and can be chosen to be equal to that of the medium adjacent to the PML; the magnitudes σ_x and σ_z are the anisotropic conductivities of the PML. The continuity conditions for the wave equation (2.67) are that $\psi(x,z)$, $\left(1-i\sigma_x/\omega\varepsilon_0 n_p^2\right)^{-1}\frac{\partial\psi}{\partial x}$ and $\left(1-i\sigma_z/\omega\varepsilon_0 n_p^2\right)^{-1}\frac{\partial\psi}{\partial z}$ are continuous. It is worthy to note that the wave equation with the anisotropic conductivities in the PML is not the conventional Helmholtz equation in an anisotropic conductive medium. Therefore, the PML is a medium to which the Helmholtz equation is mapped via an anisotropic complex transformation.

One of the remarkable features of the anisotropic complex mapping of Eq. (2.67) in that the resultant equation produces a wave that is perfectly matched with the outgoing waves in the conventional media. In order to verify this point, let us consider a plane wave incident at an infinite interface at $x = 0$ between two PML media. The waves in medium (1) and medium (2) can be written, respectively, as:

$$\psi_1 = e^{-ik_{x1}x - ik_{z1}z} + re^{ik_{x1}x - ik_{z1}z};$$ (2.68a)

$$\psi_2 = te^{-ik_{x2}x - ik_{z2}z},$$ (2.68b)

where r and t are the reflection and the transmission coefficients, respectively, and:

$$k_{xj} = \left(1 - i\sigma_{xj}/\omega\varepsilon_0 n_p^2\right) n_p k_0 \cos\theta_i;$$ (2.69a)

$$k_{zj} = \left(1 - i\sigma_{zj}/\omega\varepsilon_0 n_p^2\right) n_p k_0 \sin\theta_i.$$ (2.69b)

Here, θ_i ($i = 1, 2$) are the angles of reflection and transmission, respectively, k_0 being the wave-number of the plane wave in free space. By matching the boundary conditions between the two PML media, it follows:

$$\left(1 - \sigma_{z1}/\omega\varepsilon_0 n_p^2\right)\sin\theta_1 = \left(1 - i\sigma_{z2}/\omega\varepsilon_0 n_p^2\right)\sin\theta_2,$$ (2.70)

and:

$$r = \frac{\cos\theta_1 - \cos\theta_2}{\cos\theta_1 + \cos\theta_2}.$$ (2.71)

If we choose the conductivities of the two media such that:

$$\sigma_{z1} = \sigma_{z2},$$ (2.72)

then the boundary condition (2.70) gives $\theta_1 = \theta_2$ and following the formula (2.71) it gives $r = 0$. Thus, a plane wave incident on an interface perpendicular to x between two PML experiences no reflection if the anisotropic conductivities in the plane parallel to the interface are equal and in particular for $\sigma_{z1} = \sigma_{z2} = 0$. This conclusion is valid for plane waves of arbitrary incidence angles and frequencies and independent of the conductivities of the media along x. This is certainly true for an interface between a lossless medium with zero conductivity and a PML medium with a finite conductivity along x.

Under the perfectly matching condition, the transmitted field in the PML medium is given by:

$$\psi_2(x,z) = e^{-in_p k_0(x\cos\theta + z\sin\theta)} e^{-\sigma_{x2}n_p^{-1}\cos\theta\sqrt{\mu_0/\varepsilon_0}x}.$$ (2.73)

Therefore, the non-physical wave in the fictitious PML medium matches perfectly with the physical wave in the real free-space adjacent to it along z and decays exponentially along x with an effective attenuation coefficient α_{eff} given by:

$$\alpha_{eff} = \sigma_{x2} n_p^{-1} \sqrt{\mu_0/\varepsilon_0}. \tag{2.74}$$

The directionally dependent conductive layer (Bérenger layer) can be modelled by making a complex transformation of the normal coordinate to the layer surface [15]:

$$x(\rho) = \int_0^\rho \left(1 - i\frac{\sigma(\rho')}{\omega\varepsilon_0 n_p^2}\right) d\rho' \quad (\rho \geq 0), \tag{2.75}$$

where $\sigma(\rho)$ is the conductivity profile of the Bérenger layer. To derive the reflection coefficient of a Bérenger layer, let us start with the plane wave solution of the paraxial wave equation. A plane wave on the form of:

$$\phi(x) = e^{-i\alpha x} = e^{-ik_0 \sin\theta x}, \tag{2.76}$$

transmitting into this layer does not exhibit any reflection, as has been shown before and it propagates as:

$$\phi(x) = e^{-i\alpha x(\rho)}. \tag{2.77}$$

When the wave arrives at the end of the layer, of width d, it is reflected by a perfectly reflecting surface as:

$$\phi(x) = e^{i\alpha x(d)} e^{i\alpha[x(\rho) - x(d)]}. \tag{2.78}$$

The plane wave again crosses the Bérenger layer and returns to the adjacent medium with the amplitude $\left|e^{-2i\alpha x(d)}\right| = r$. The outgoing wave emerging from the Bérenger layer is $re^{-i\alpha x}$, where the reflection coefficient can be written as:

$$r(\theta) = \exp\left\{-2\frac{k_0 \sin\theta}{\omega\varepsilon_0}\int_0^d \sigma(\rho)d\rho\right\}, \tag{2.79}$$

making explicit the dependence between the reflection coefficient of the Bérenger layer and the propagation angle of the plane wave. Let us note that r can be arbitrarily low if $\sigma(\rho)$ is large enough, except for grazing incidence ($\theta = 0$) where $r(0) = 1$.

For the formulation of PML boundary conditions for the scalar FD-BPM [25, 36], let us start with the wave equation (2.67), where we have assumed that $\sigma_z = 0$:

$$\frac{1}{1 - i\sigma_x/\omega\varepsilon_0 n_p^2}\frac{\partial}{\partial x}\left(\frac{1}{1 - i\sigma_x/\omega\varepsilon_0 n_p^2}\frac{\partial\psi}{\partial x}\right) + \frac{\partial^2\psi}{\partial z^2} + n_p^2 k_0^2\psi = 0. \tag{2.80}$$

Assuming the optical field is of the form of the conventional slowly varying field $u(x,z)$:

$$\psi(x,z) = u(x,z)e^{ik_0 n_0 z}, \tag{2.81}$$

and using the SVEA, we obtain:

$$2ik_0n_0\frac{\partial u}{\partial z} = q\frac{\partial}{\partial x}\left(q\frac{\partial u}{\partial x}\right) + k_0^2\left(n_p^2 - n_0^2\right)u, \tag{2.82}$$

where, for the sake of brevity, we have defined the variable q as:

$$q \equiv \frac{1}{1 - i\sigma_x/\omega\varepsilon_0 n_p^2}. \tag{2.83}$$

The expression $q\dfrac{\partial}{\partial x}\left[q\dfrac{\partial u}{\partial x}\right]$ in Eq. (2.82) can be approximated using finite differences (see Appendix A), resulting in:

$$q\frac{\partial}{\partial x}\left[q\frac{\partial}{\partial x}\right] = q\frac{\partial}{\partial x}\left[q_j\frac{u_{j+1/2} - u_{j-1/2}}{\Delta x}\right]$$

$$= \frac{q_j}{\Delta x}\left[\frac{q_{j+1} - q_j}{2}\frac{u_{j+1} - u_j}{\Delta x} - \frac{q_j - q_{j-1}}{2}\frac{u_j - u_{j-1}}{\Delta x}\right] \tag{2.84}$$

$$= \frac{q_j}{2(\Delta x)^2}\left[(q_j + q_{j+1})u_{j-1} - (q_{j-1} + 2q_j + q_{j+1})u_j + (q_j + q_{j-1})u_{j-1}\right]$$

Using this result, the finite difference wave equation (2.82), following the Crank–Nicolson scheme, can now be written accordingly as:

$$\frac{2ik_0n_0}{\Delta z}\left[u_j^{m+1} - u_j^m\right] = (1-\alpha)\frac{T_{j-1}^m u_{j-1}^m - 2R_j^m u_j^m + T_{j+1}^m u_{j+1}^m}{(\Delta x)^2} + (1-\alpha)k_0^2\left[\left(n_{p,j}^m\right)^2 - n_0^2\right]u_j^m$$

$$+ \alpha\frac{T_{j-1}^{m+1} u_{j-1}^{m+1} - 2R_j^{m+1} u_j^{m+1} + T_{j+1}^{m+1} u_{j+1}^{m+1}}{(\Delta x)^2} + \alpha k_0^2\left[\left(n_{p,j}^{m+1}\right)^2 - n_0^2\right]u_j^{m+1} \tag{2.85}$$

where u_j^m denotes the (SVE amplitude) of the optical field at the transverse grid point j and at the longitudinal step m and we have defined the $T_{j\pm1}$ and R_j coefficients as:

$$T_{j\pm1} = \frac{q_j(q_j + q_{j\pm1})}{2}; \tag{2.86a}$$

$$R_j = \frac{q_j(q_{j-1} + 2q_j + q_{j+1})}{4}. \tag{2.86b}$$

Rearranging terms, the coefficients a_j, b_j, c_j and r_j of the tridiagonal system (2.85) are given by:

$$a_j = -\frac{\alpha}{(\Delta x)^2} T_{j-1}^{m+1};$$
(2.87a)

$$b_j = \frac{2ik_0n_0}{\Delta z} + \frac{2\alpha}{(\Delta x)^2} R_j^{m+1} - \alpha k_0^2 \left[\left(n_{p,j}^{m+1} \right)^2 - n_0^2 \right];$$
(2.87b)

$$c_j = -\frac{\alpha}{(\Delta x)^2} T_{j+1}^{m+1};$$
(2.87c)

$$r_j = \frac{(1-\alpha)}{(\Delta x)^2} \left[T_{j-1}^m u_{j-1}^m + T_{j+1}^m u_{j+1}^m \right] + \left\{ \frac{2ik_0n_0}{\Delta z} - \frac{2(1-\alpha)R_j^m}{(\Delta x)^2} + (1-\alpha)k_0^2 \left[\left(n_{p,j}^m \right)^2 - n_0^2 \right] \right\} u_j^m$$
(2.87d)

This system provides an unconditionally stable algorithm, being second-order accurate in both the transversal and longitudinal steps, which can be efficiently solved by the Thomas method.

The results of the tilted Gaussian beam propagating in a homogeneous medium following the FD-BPM algorithm with PML implemented are shown in Figure 2.21. In that example, two PML regions of thickness $\delta = 1$ µm have been added to the computational domain, with a parabolic conductivity profile given by:

$$\sigma(x) = \sigma_{max} \left(\frac{x - L/2 + \delta}{\delta} \right), \qquad x \in \left[\frac{L}{2}, \frac{L}{2} - \delta \right];$$
(2.88a)

$$\sigma(x) = \sigma_{max} \left(\frac{L/2 + \delta - x}{\delta} \right)^2, \qquad x \in \left[-\frac{L}{2} + \delta, -\frac{L}{2} \right].$$
(2.88b)

The maximum conductivity is set to $\sigma_{max} = 5\, \varepsilon_0\omega$. It is observed that this thin PML region is enough to prevent significant reflection from the upper boundary.

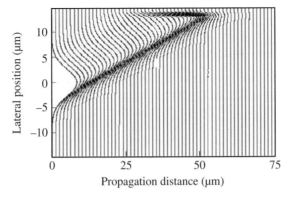

Figure 2.21 Tilted Gaussian beam propagation by means of FD-BPM, using the same parameters as in previous figures when PML boundary conditions are implemented

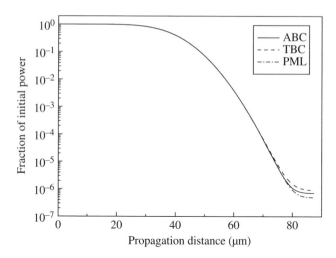

Figure 2.22   Total power in the computational window as a function of the propagation distance, when a tilted Gaussian beam propagates in a homogeneous medium. The three curves represent calculation performed by the implementation of absorbing regions (continuous line), transparent boundary conditions (dashed line) and perfectly matched layers (dotted-dashed line)

Figure 2.22 shows a comparison of the three methods here presented to avoid reflection of radiation from the computational edges, by calculating the fraction of the original energy remaining in the computational window as a function of the propagation distance of the $20°$ tilted Gaussian beam. To provide a consistent comparison between the three methods, the energy in the extra region of absorbers for the ABC and PML methods are excluded in the total. As it can be seen in the figure, the remaining fraction of the initial energy after 80 μm is less than 10^{-6} for all three methods, indicating that all of them can be satisfactorily used for avoiding reflection at the boundaries. However, the use of absorbers is cumbersome and the absorber must be re-tailored for each new problem. Finally, at variance with the TBC algorithm, the ABC and PML methods impose some penalty of computer runtime and storage owing to the necessity of calculating and storing the field at a number of extra absorption zones, although this is less burdensome in the PML method.

2.7   Obtaining the Eigenmodes Using BPM

BPM was developed to simulate the propagation of light in 2D and 3D structures of arbitrary geometry and refractive index, and to furnish accurate description of spatial (near-field) and angular (far-field) properties of the electromagnetic field. Soon after, it was realized that BPM could also be used to obtain the eigenmodes of longitudinal-invariant structures (straight waveguides). One method, proposed by Feit and Fleck [2], consists of calculating the overlap integral of the just-obtained field at every BPM step $u(x,y,z)$ with the initial field $u(x,y,0)$. The peaks of the Fourier transform of the z-dependent overlap integral lie at the propagation constants of the transverse modes. If one propagates again and beats with a given propagation constant, one can extract the field distribution of the eigenmode corresponding to that propagation constant. Another method of calculating the eigenmodes using BPM, initially used by Feit and Fleck [2], was later established by Yevick *et al.* and Hawkins [28, 29], and it is based on the BPM propagation along the imaginary z-axis using the paraxial wave equation. The fundamental mode of a z invariant optical waveguide has the largest propagation constant, thus seeing the most rapid oscillations in phase when travelling down the real axis. In contrast, propagation along the imaginary axis changes these sinusoidal variations in phase into exponential growths in amplitude and thus a sufficiently long imaginary propagation gives the mode with the highest growth; that is, the fundamental mode.

To describe the basis of the BPM for obtaining the modal properties of straight waveguides, let's start with the paraxial full-vector wave equation for the electric field expressed in its compact form:

$$\frac{\partial \boldsymbol{\Psi}}{\partial z} = -i\hat{\mathbf{H}}\boldsymbol{\Psi}, \tag{2.89}$$

where $\boldsymbol{\Psi}(x,y,z)$ is the two transversal component SVE-field vector and we have dropped the subscript 't' for clarity. Here, the complete transversal electric field $\boldsymbol{E}_t(x,y,z)$ has been expressed by the product of the SVE-field $\boldsymbol{\Psi}(x,y,z)$ and a carrier wave moving in the positive z direction, e^{-iKz}:

$$\boldsymbol{E}_t(x,y,z) = \boldsymbol{\Psi}(x,y,z)e^{-iKz}, \tag{2.90}$$

where K is the reference wavenumber ($K = n_0k_0$). On the other hand, $\hat{\mathbf{H}}$ is the associated 2×2 matrix operator defined in Eq. (2.15) for the electric field.

The formal solution of Eq. (2.89) can be expressed as:

$$\boldsymbol{\Psi}(x,y,z) = e^{iz\hat{\mathbf{H}}}\boldsymbol{\Psi}(x,y,0). \tag{2.91}$$

For straight waveguides (uniform along the z direction) in which the refractive index only has dependence on the x and y coordinates, $n = n(x,y)$, any arbitrary input field at $z = 0$ can be represented by the summation of the eigenmodes $f_\nu(x,y)$ of the structure and weighted by their associated power through the modal weights a_ν:

$$\boldsymbol{\Psi}(x,y,0) = \sum_{V=0}^{\infty} a_\nu f_\nu(x,y), \tag{2.92}$$

where the summation includes both the guided and the radiation modes.

The eigenvalue ξ_ν (relative propagation constant) and eigenvector $f_\nu(x,y)$ (transversal field distribution) of the operator $\hat{\mathbf{H}}$ satisfy:

$$\hat{\mathbf{H}}f_\nu(x,y) = \xi_\nu f_\nu(x,y). \tag{2.93}$$

On the other hand, the eigenvalues can be arranged in the order:

$$\xi_0 > \xi_1 .... > \xi_\nu > ..... . \tag{2.94}$$

As $\boldsymbol{\Psi}$ denotes the transversal SVE-field, the relation between the eigenvalue ξ_ν and the propagation constant of νth order β_ν for the paraxial wave equation is:

$$\beta_\nu = n_0 k_0 + \xi_\nu. \tag{2.95}$$

Now, by using Eqs. (2.92) and (2.93), Eq. (2.91) can be rewritten as:

$$\boldsymbol{\Psi}(x,y,z) = \sum_{\nu=0}^{\infty} a_\nu e^{-i\xi_\nu z} f_\nu(x,y). \tag{2.96}$$

2.7.1 The Correlation Function Method

The mode propagation constants and the power associated to each mode can be determined by using BPM, following the method proposed by Feit and Fleck [2, 30], from a Fourier analysis of the complex field amplitude correlation function defined by $P(z) \equiv \langle \boldsymbol{\Psi}*(0)\cdot\boldsymbol{\Psi}(z)\rangle$ where the bracket signifies integration over the waveguide cross section and z represents the axial distance. The function $P(\xi)$, which is the Fourier transform of $P(z)$ with respect to z, should display a set of resonant peaks that identify the guided modes that have been excited by the input source at $z = 0$. The peaks occur at values of ξ that correspond to the mode propagation constants $\beta = K + \xi$ and the heights of the peaks are proportional to the corresponding modes' powers.

The procedure presented here for determining the mode weights and propagation constants (modal spectrum) of a z invariant waveguide is called the correlation-function method. For this purpose, let us form the product $\boldsymbol{\Psi}^*(x,y,0)\cdot\boldsymbol{\Psi}(x,y,z)$ and integrate it over the cross section A of the waveguide. In this way we build the correlation function $P(z)$ given by:

$$P(z) = \iint_A \boldsymbol{\Psi}^*(x,y,0) \cdot \boldsymbol{\Psi}(x,y,z)\,dxdy. \tag{2.97}$$

Using the mode orthogonality, and assuming that the transversal distributions of the modes are normalized, we have:

$$\iint_A f_\mu^*(x,y) \cdot f_\nu(x,y)\,dxdy = \delta_{\mu\nu}. \tag{2.98}$$

Considering the modal expansion (2.96) and the relation (2.98), the correlation function $P(z)$ can be alternatively expressed as a function of the modal weights a_ν and the relative propagation constants ξ_ν in the form:

$$P(z) = \sum_\nu |a_\nu|^2 e^{-i\xi_\nu z}. \tag{2.99}$$

Now, taking the Fourier transform of Eq. (2.99), it yields:

$$P(\xi) = \sum_\nu |a_\nu|^2 \delta(\xi - \xi_\nu), \tag{2.100}$$

where δ is the delta Kronecker function. This expression indicates that the calculated spectrum of $P(z)$ (that is, its Fourier transform) will display a series of resonances with maxima at $\xi = \xi_\nu$, and peak values proportional to the mode weight coefficients a_ν. Moreover, the coefficient a_ν is related to the relative power W_ν carried by the ν-th mode by:

$$W_\nu = |a_\nu|^2, \tag{2.101}$$

and the total power carried in the waveguide is simply given by:

$$W = \sum_\nu W_\nu. \tag{2.102}$$

In practice, as only a finite record of $P(z)$ is available, the resulting resonances in the spectrum $P(\xi)$ will exhibit a finite width and shape that are characteristic of the record propagation length Z. Since in general the resonance peaks do not coincide exactly with the sampled values of ξ, errors will results in the W_ν and ξ_ν values inferred from the maxima in sampled data set for $P(\xi)$. The maximum uncertainty in ξ_ν can be expressed in terms of the sampling interval $\Delta \xi$ along the ξ-axis and the propagation distance Z over which the solution $\Psi(x,y,z)$ is available as:

$$\Delta \xi_\nu = \Delta \xi / 2 = \pi / Z. \tag{2.103}$$

In order to reduce the uncertainty in the determination of the propagation constant it is necessary to increase the propagation length Z but this is at the expense of an increase in computational time.

On the other hand, the magnitudes of the relative propagation constants for the guided modes will be bounded according to:

$$0 < |\xi_\nu| < K\Delta n_{max}, \tag{2.104}$$

where $\Delta n_{max} = n_{max} - n_0$. Therefore, to insure that the modal spectrum is accurately represented, it is necessary for the axial sampling distance Δz to satisfy:

$$\Delta z < \pi / K\Delta n_{max}, \tag{2.105}$$

or expressed as a function of the wavelength:

$$\Delta z < \lambda / 2n_0 \Delta n_{max}. \tag{2.106}$$

In practice, this condition should be satisfied by a factor of ~ 5.

Let us now see in practice how to proceed in obtaining the propagation constants of the modes as well as their modal weights. For that purpose, let us consider a two-dimensional multimode waveguide that is invariant in the propagation direction (straight waveguide). If the excitation light at the input coincides with the distribution field corresponding to a guided mode, we expect this light distribution will propagate without losing its shape and intensity, because the energy transfer between modes is forbidden by the modal orthogonality relationship. If, by contrast, the light injection at $z = 0$ provokes the excitation of various modes, the transversal distribution of the optical field will change as the beam propagates along the waveguide, due to the fact that each mode has a different propagation constant and therefore the relative phase between modes will change as a function of the propagation distance.

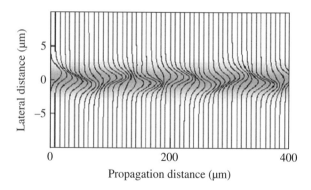

Figure 2.23 Light propagation, using FD-BPM with TBC, in a multimode straight waveguide. Parameters of the simulation: 400 points in the transversal direction; grid spacing $\Delta x = 0.05\,\mu\text{m}$; propagation step $\Delta z = 2.5\,\mu\text{m}$ and wavelength $\lambda = 0.85\,\mu\text{m}$. The field intensities are plotted at every 8 μm

Figure 2.23 shows the evolution of intensity profile of a TE-polarized beam in a two-dimensional straight waveguide. The planar waveguide has a symmetric Gaussian refractive index profile given by:

$$n(x) = n_0 + \Delta n \ e^{-(x/\sigma)^2}, \tag{2.107}$$

with a width of $\sigma = 2\,\mu\text{m}$, substrate index of $n_0 = 1.45$ and a maximum index change of $\Delta n = 0.02$. The wavelength used is $\lambda = 0.85\,\mu\text{m}$ and the simulation has been performed by FD-BPM where TBC has been implemented. At this wavelength, the waveguide structure supports three guided modes. The input light launched at $z = 0$ is a Gaussian beam of 2 μm width, shifted 1 μm with respect to the waveguide centre to provide excitation of both even and odd modes. As can be seen in the figure, the field intensity profile changes as the beam propagates due to the fact that the light distribution at the input does not correspond with any particular guided mode, but several of them have been simultaneously excited having different propagation constants.

Using Eq. (2.97), the correlation function $P(z)$ corresponding to the propagation described in Figure 2.23 is constructed and is presented in Figure 2.24. After a first inspection of this figure, it seems difficult to extract any relevant information about the behaviour of the modal propagation; perhaps the only clear thing is that the evolution of $P(z)$ does not follow a pure sinusoidal function.

Now, a Fourier transform of the correlation function $P(z)$ for a total propagation distance of $Z = 20\,480\,\mu\text{m}$ gives rise to a series of well-defined peaks with different heights (Figure 2.25). The resonance peaks appear at values of the relative propagation constant of $\xi = 0.11\,106$, $0.04\,724$ and $0.00\,705\,\mu\text{m}^{-1}$, with different intensities increasing with ξ. These values correspond to guided modes of the structure, because $\xi > 0$. Table 2.1 shows these values of ξ_ν beside the 'exact' propagation constants obtained using a multilayer analysis [31].

The error found in the propagation constants, referring to their exact values, is close to that expected from the formula (2.103), which predicts an uncertainty of $\Delta \xi_\nu = 0.00015\,\mu\text{m}^{-1}$. In addition, the use of a propagation step of $\Delta z = 2.5\,\mu\text{m}$ assures us that the whole modal spectrum is represented because the condition $\Delta z < \lambda/2n_0\Delta n_{\text{max}} = 14\,\mu\text{m}$ is fulfilled.

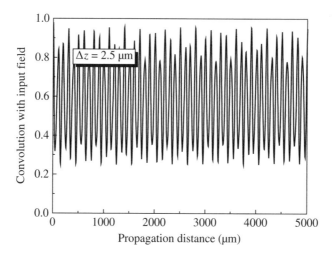

Figure 2.24 Correlation function $P(z)$ corresponding to the propagation described in Figure 2.23

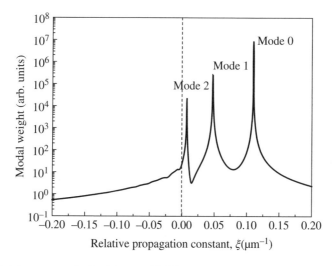

Figure 2.25 Axial spectrum obtained from a 20 480 µm propagation length in a graded index planar waveguide, after launching a Gaussian beam shifted 1 µm from the waveguide centre

Table 2.1 Relative propagation constants determined by the correlation-function method (1) and imaginary distance-BPM (2) beside their 'exact' values calculated by the multilayer method

Mode	$\xi_\nu(1)(\mu m^{-1})$	$\xi_\nu(2)(\mu m^{-1})$	$\xi_\nu(\text{exact})(\mu m^{-1})$
0	0.11106	0.11093	0.11089
1	0.04724	0.04701	0.04700
2	0.00705	0.00673	0.00676

Once the propagation constants of the waveguide modes have been calculated, with a further and simple analysis it is possible in addition to determine the transversal field distribution $f(x,y)$ corresponding to each propagation mode. For this purpose, it is enough to allow the input source to propagate along the straight waveguide, and multiply the optical field at each propagation step by the factor $\exp(i\xi_\mu z)$, where ξ_μ indicates the relative propagation constant (previously calculated) of the μ-th mode from which we want to obtain its transverse field distribution $f_\mu(x,y)$. In this way, if we add these contributions as the wave propagates all the modes, except the μ-th mode, will have a phase that changes with the propagation distance and will cancel for a long enough propagation length. In contrast, the μ-th mode will not have any dependence on z because it has been eliminated after multiplying the optical field by the factor $\exp(i\xi_\mu z)$. Therefore, the transversal field distribution corresponding to the μ-th mode will build up as the perturbation advances along the straight waveguide.

To see how this method works, lets us restrict the problem to a two-dimensional waveguide where the modal field profiles only have dependence on the x coordinate. If we assume that the optical field in the waveguide can only be expressed by the set of m confined modes, thus ignoring the energy carried out by the continuum radiation modes, the optical field can be written as:

$$\Psi(x,y,z) = \sum_{\nu=0}^{m-1} a_\nu f_\nu(x,y) e^{-i\xi_\nu z}. \tag{2.108}$$

For obtaining the field distribution corresponding to the μ-th mode, we build the function $g_\mu(x,y)$ defined as:

$$g_\mu(x,y) = \int_0^Z \Psi(x,y,z) e^{i\xi_\mu z} dz = \int_0^Z \left(\sum_{\nu=0}^{m-1} a_\nu f_\nu(x,y) e^{-i\Delta\xi_{\mu\nu} z} \right) dz, \tag{2.109}$$

where Z is the total propagation length, and $\Delta\xi_{\mu\nu}$ is defined as $(\xi_\nu - \xi_\mu)$. Taking into account the relation (2.108), this function can be split in two terms as:

$$g_\mu(x,y) = a_\mu f_\mu(x,y) Z + \sum_{\substack{\nu=0 \\ \nu\neq\mu}}^{m-1} \left(a_\nu f_\nu(x,y) \int_0^Z e^{-i\Delta\xi_{\mu\nu} z} dz \right). \tag{2.110}$$

The integrals in Eq. (2.110) are periodic functions of z and thus they are limited to a given value depending on the propagation constant of the modes. In particular, their absolute values are limited to $|2/\Delta\xi_{\mu\nu}|$ and therefore the second term in Eq. (2.110) is a bounded value, which depends on the power transported for each mode through their modal weights a_ν. On the contrary, the first term in Eq. (2.110) is a function which grows linearly with the propagation length Z. Therefore, for large propagation length values, the second term can be neglected and the first term will then reproduce the transverse field distribution of the μ-th mode; or in other words, the function $g_\mu(x,y)$ becomes proportional to the field distribution $f_\mu(x,y)$. For obtaining an accurate field distribution, and assuming equally spaced modes and similar modal weights, the propagation length should fulfil the condition:

$$Z \gg m(m-1)/\Delta\xi, \tag{2.111}$$

m being the number of confined modes that support the waveguide and $\Delta\xi$ the difference between the propagation constants of two consecutive modes. Nevertheless, although Z is chosen to be long, it does not guarantee the total elimination of the contribution of a specific mode ν and to accurately build the field distribution of a mode it is necessary to excite it with a high energy fraction.

In order to overcome these problems, alternatively, we can choose a slightly different method to obtain the mode field transversal profiles [31]. First, a function $h_0(x,y)$ is built according to:

$$h_0(x,y) = \int_0^{2\pi/\Delta\xi_{\mu 0}} \Psi(x,y,z)e^{i\xi_{\mu}z}dz, \tag{2.112}$$

where $\Psi(x,y,z)$ is the field calculated by BPM using an arbitrary input field. This function can be expressed in terms of the modal eigenfunctions (Eq. 2.108) and can be split as follows:

$$h_0(x,y) = \int_0^{2\pi/\Delta\beta_{\mu 0}} \left(\sum_{\nu=0}^{m-1} a_\nu f_\nu(x,y)e^{-i\Delta\beta_{\mu\nu}z}\right)dz$$

$$= \int_0^{2\pi/\Delta\beta_{\mu 0}} a_0 f_0(x,y)e^{-i\Delta\beta_{\mu 0}z}dz + \int_0^{2\pi/\Delta\beta_{\mu 0}} \left(\sum_{\nu=1}^{m-1} a_\nu f_\nu(x,y)e^{-i\Delta\beta_{\mu\nu}z}dz\right). \tag{2.113}$$

The first term on the right hand of Equation (2.113) is cancelled exactly because of the proper choice of integral limits. Thus, we are assured that the function $h_0(x,y)$ has no more contribution to the 0-*th* mode field distribution given by $f_0(x,y)$.

Secondly, the function $h_0(x,y)$ previously calculated is chosen as a new input field $\Psi(x,y,z=0)$ and a new propagation sequence is performed. The new input field cannot, therefore, excite the 0-th mode. We repeat the two-step procedure for each mode, except for the μ-th mode, obtaining the sequence of functions $h_1(x,y)$, $h_2(x,y)$ and so on. In this way, the contribution of each propagating mode to the optical field is subtracted and the final result is the procurement of the field distribution of the μ-th mode.

This method is exact because all the modes, except the selected one, are eliminated. Also, this procedure is accurate providing that the propagation constants of the propagating modes are well determined and assuming that the contribution to the radiation modes can be neglected. The main advantage of this method is that the propagation length for the elimination of every mode can be calculated exactly and consequently the algorithm is straightforward, thus saving computational time, which is especially important for modelling three-dimensional structures.

As an example of application, let us consider the two-dimensional straight waveguide presented previously. Figure 2.26 shows the field profiles corresponding to the three TE modes that support the waveguide at $\lambda = 0.85\ \mu m$, calculated by a multilayer approach, and those obtained by the FD-BPM-based method. It can be observed that the field profiles coincide accurately with only very small differences appreciated at the evanescent tails of the higher mode.

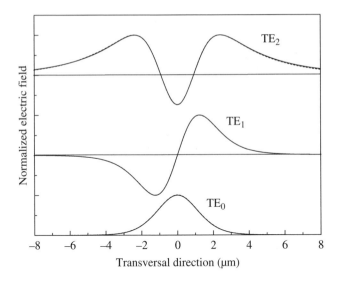

Figure 2.26 Field amplitude corresponding to the three guided modes in a symmetric graded index planar waveguide. Solid line: mode solver (multilayer); dashed line: BPM-convolution method. The field profiles obtained by the Im-Dis-BPM are indistinguishable from those calculated by a multilayer approach

2.7.2 The Imaginary Distance Beam Propagation Method

Although the convolution method can provide the propagation constants and field profiles of the guided modes of any arbitrary straight waveguide, to achieve accurate results long propagation distances are required in the calculation. One alternative and elegant method, also based on BPM, suitable for obtaining the propagation constants and the field profiles of the confined modes of waveguides is the Imaginary Distance Beam Propagation Method (Im-Dis-BPM) [32, 33]. Using this method, highly accurate results are obtained with surprisingly short distances of propagation that are applicable to 2D and 3D structures, which can be implemented for scalar, semi-vectorial and full vectorial approaches.

Let us examine the consequences derived if the field is propagated along the imaginary axis. If we put iz instead of z in Eq. (2.96), it yields:

$$\Psi(x,y,iz) = \sum_{\nu=0}^{\infty} a_\nu e^{\xi_\nu z} f_\nu(x,y). \qquad (2.114)$$

If we chose the reference refractive index as $n_0 = \beta_0/k_0$ (close to β_0/k_0 in practice), then $\xi_0 = 0$ (eigenvalue corresponding with the fundamental mode) and all the other eigenvalues are negative, according to Eq. (2.94). Thus, while the fundamental mode remains unchanged with propagation along the imaginary axis, all the other modes will experiment exponential decay as the field propagates. Then, after propagating a certain distance, the fundamental mode will become dominant and all the higher order modes relatively die down. Mathematically, this result is expressed as:

$$\lim_{z \to \infty} \Psi(x,y,iz) = a_0 e^{\xi_0 z} f_0(x,y),$$ (2.115)

indicating that the transversal distribution of the computed field corresponds to the fundamental mode for sufficiently long propagating distances.

In practice, it is difficult to choose $n_0 = \beta_0/k_0$ simply because β_0 is unknown. Under the circumstances of $n_0 \neq \beta_0/k_0$, the fundamental mode also decays (or grows), but it decays slower (or grows faster) than the higher order modes. As a result, after a certain distance the fundamental mode still will be the dominant mode and Eq. (2.115) still hold.

Once the first mode $f_0(x,y)$ has been obtained, we proceed to eliminate it from the excitation input $\Psi(x,y,0)$ and a new propagation along the imaginary axis is performed to obtain the next mode. The new input $\Psi_{new}(x,y,0)$ is thus given by:

$$\Psi_{new}(x,y,0) = \Psi(x,y,0) - a_0 f_0(x,y),$$ (2.116)

where the a_0 coefficient is calculated by:

$$a_0 = \frac{\iint \Psi(x,y,0) \cdot f_0^*(x,y) dxdy}{\iint |f_0(x,y)|^2 dxdy}.$$ (2.117)

Higher-order modes are successively calculated by recursive elimination of the previously obtained lower-order modes [34].

Based on the growth rate during the imaginary distance propagation, the propagation constant of the most weighted mode, following Eq. (2.112), can be calculated by [32]:

$$\beta_0 = n_0 k_0 + \lim_{z \to \infty} \frac{\ln \left[\iint u(x,y,i(z+\Delta z)) dxdy \right] - \ln \left[\iint u(x,y,iz) dxdy \right]}{\Delta z},$$ (2.118)

u being any of the two transversal components of the SVE vector Ψ.

Figure 2.27 shows the application of the imaginary distance BPM to the waveguide example previously presented. We start the TE propagation with the launch of a 2 µm width Gaussian beam shifted 1 µm (upper figure), performed with a propagation step of $\Delta z = 10$ µm. The lines represent the normalized intensity profiles at each propagation step. After a propagation distance of 80 µm the field reaches the steady state and the zero-order mode is obtained, which represents only eight steps of propagation. After this first run, the contribution of the zero-order mode is subtracted to the original input and it is used for the new input, where the propagation is represented in the middle frame of Figure 2.27. Now, the first-order mode is accurately obtained after a propagation of 120 µm (only 12 steps). The procedure is repeated and the result is plotted at the bottom of the figure. An imaginary distance of 300 µm is enough for an accurate building of the field profile corresponding to the second-order mode, which is performed with 30 propagation steps. The calculated field profiles of the three modes are represented in Figure 2.26, and are indistinguishable from the 'exact' profiles computed by the multilayer approximation. Therefore, only

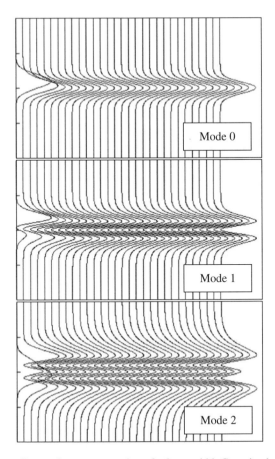

Figure 2.27 Imaginary distance beam propagation of a 2 μm width Gaussian beam shifted 1 μm along a graded index planar waveguide. The intensity profiles are drawn at every propagation step ($\Delta z = 10$ μm). After each new run, the contribution of the previously calculated mode is subtracted for the new input

50 propagation steps are enough to calculate accurately the three propagation modes of the planar waveguide, which can be favourable compared with the 8192 propagation steps needed when using the correlation method. This superior computation efficiency makes the imaginary distance BPM the favourite method for calculating the propagation modes of waveguides, especially when dealing with three-dimensional structures or in vectorial analysis of waveguides.

In addition, the values of the propagation constants of the modes provided by Im-Dis-BPM using the formula in Equation (2.118) are far more accurate than the values obtained by the correlation method (see Table 2.1). The results for the effective indices of the guided TE-modes calculated by the multilayer approach, BPM correlation method and imaginary-distance BPM are summarized in Table 2.2.

Alternatively, once a particular eigenmode is obtained by Im-Dis-BPM, the associated propagation constant can be calculated from the field transversal distribution following a variation type expression, called the Rayleigh quotient [14].

Table 2.2 Effective indices of the confined TE-modes, calculated by the multilayer approach ('exact'), BPM-correlation-function method and imaginary distance BPM

	Mode 0	Mode 1	Mode 2
Multilayer	1.465001	1.456358	1.450914
BPM	1.465024	1.456391	1.450954
Im-Dis-BPM	1.465007	1.456360	1.450910

In the scalar approach, the waveguide modes fulfil the following equation:

$$\frac{\partial^2 u}{\partial x^2} + \frac{\partial^2 u}{\partial y^2} + k_0^2 \left(n^2 - N_{eff}^2 \right) u = 0, \tag{2.119}$$

which can be readily obtained from the scalar wave equation (2.24) by putting $\partial u/\partial z = 0$ and $n_0 = N_{eff}$. Multiplying Eq. (2.119) by the field conjugate u^* and operating double integration over the waveguide transversal section, it yields:

$$\iint u^*(x,y) \left\{ \left[k_0^2 n^2(x,y) - \beta^2 \right] u(x,y) + \frac{\partial^2 u(x,y)}{\partial x^2} + \frac{\partial^2 u(x,y)}{\partial y^2} \right\} dxdy = 0, \tag{2.120}$$

where the effective index of the mode N_{eff} is related with its propagation constant by $\beta = k_0/N_{eff}$. From Eq. (2.120) it follows:

$$\beta^2 = \frac{\iint u^*(x,y) \left\{ k_0^2 n^2(x,y) u(x,y) + \frac{\partial^2 u(x,y)}{\partial x^2} + \frac{\partial^2 u(x,y)}{\partial y^2} \right\} dxdy}{\iint |u(x,y)|^2 dxdy}. \tag{2.121}$$

Alternatively, one can use the following variational formula [35]:

$$\beta^2 = \frac{\iint \left\{ k_0^2 n^2(x,y) |u(x,y)|^2 - \left| \frac{\partial u(x,y)}{\partial x} \right|^2 - \left| \frac{\partial u(x,y)}{\partial y} \right|^2 \right\} dxdy}{\iint |u(x,y)|^2 dxdy}, \tag{2.122}$$

which holds also only for scalar propagation.

For semi-vectorial propagation, the formula has to be modified accordingly to take into account the discontinuities of the fields across the interfaces. The calculation of the propagation constant for quasi-TM modes, using the SVE of the electric field component E_x considering Eq. (2.21a), is given by:

$$\beta^2 = \frac{\iint \Psi_x^* \left\{ k_0^2 n^2 \Psi_x + \frac{\partial}{\partial x} \left(\frac{1}{n^2} \frac{\partial}{\partial x} \left(n^2 \Psi_x \right) \right) + \frac{\partial^2 \Psi}{\partial y^2} \right\} dxdy}{\iint |\Psi_x|^2 dxdy}. \tag{2.123}$$

Similarly, for quasi-TE modes using semi-vectorial approach the formula derived from Eq. (2.21b) to obtain the propagation constant of the eigenmode, based on the SVE of the magnetic field E_y, yields:

$$\beta^2 = \frac{\iint \Psi_y^* \left\{ k_0^2 n^2 \Psi_y + \frac{\partial^2 \Psi_y}{\partial x^2} + \frac{\partial}{\partial y} \left(\frac{1}{n^2} \frac{\partial}{\partial y} \left(n^2 \Psi_y \right) \right) \right\} dxdy}{\iint |\Psi_y|^2 dxdy}. \tag{2.124}$$

In a similar manner, the magnetic field components can be used to calculate the propagation constant of the eigenmodes. For SV-quasi-TM modes, the formula derived from Eq. (2.22b) using the y-component H_y gives:

$$\beta^2 = \frac{\iint \Phi_y^* \left\{ k_0^2 n^2 \Phi_y + n^2 \frac{\partial}{\partial x} \left(\frac{1}{n^2} \frac{\partial \Phi_y}{\partial x} \right) + \frac{\partial^2 \Phi}{\partial y^2} \right\} dxdy}{\iint |\Phi_y|^2 dxdy}, \tag{2.125}$$

and the propagation constant of quasi-TE modes (using Eq. 2.22a) is calculated by:

$$\beta^2 = \frac{\iint \Phi_x^* \left\{ k_0^2 n^2 \Phi_x + \frac{\partial^2 \Phi}{\partial x^2} + n^2 \frac{\partial}{\partial y} \left(\frac{1}{n^2} \frac{\partial \Phi_x}{\partial y} \right) + \right\} dxdy}{\iint |\Phi_x|^2 dxdy}. \tag{2.126}$$

The corresponding formulae for vectorial 2D-BPM (planar waveguides) are readily obtained just by using Eqs. (2.121)–(2.126) by putting $\partial/\partial y = 0$ and also where the integrals are performed only in the x-direction.

References

[1] Taflove, A. and Umashankar, K.R. (1989) Review of FD-TD numerical modeling of electromagnetic wave scattering and radar cross section. *Proceedings of the IEEE* **77**, 682–697.
[2] Feit, M.D. and Fleck, J.A. (1979) Calculation of dispersion in graded-index multimode fibres by a propagating-beam method, *Applied Optics* **18**, 2843–2851.

[3] Huang, W., Xu, C., Chu, S. and Chaudhuri, S.K. (1992) The finite-difference vector beam propagation method: analysis and assessment. *Journal of Lightwave Technology* **10**, 295–305.

[4] Montanari, E., Selleri, S., Vincetti, L. and Zoboli, M. (1998) Finite-element full-vectorial propagation analysis for three-dimensional z-varying optical waveguides. *Journal of Lightwave Technology* **16**, 703–714.

[5] Xu, C.L., Huang, W.P., Chrostowski, J. and Chaudhuri, S.K. (1994) A full-vectorial beam propagation method for anisotropic waveguides. *Journal of Lightwave Technology* **12**, 1926–1931.

[6] Maes, B., Bienstman, P., Baets, R., Hu, B., Sewell, P. and Benson, T. (2008) Modeling comparison of second harmonic generation in high-index-contrast devices. *Optical and Quantum Electronics* **40**, 13–22.

[7] Saitoh, K. and Koshiba, M. (2001) Full-vectorial finite element beam propagation method with perfectly matched layers for anisotropic optical waveguides. *Journal of Lightwave Technology* **19**, 405–413.

[8] Wijeratne, I.N.M., Kejalakshmy, N., Rahman, B.M.A. and Grattan, K.T.V. (2013) Rigorous full-vectorial beam propagation analysis of second-harmonic generation in zinc oxide waveguides. *IEEE Photonics Journal* **5**, doi: 10.1109/JPHOT.2013.2256115.

[9] Hadley, G.R. (1992) Wide-angle beam propagation using Padé approximant operators. *Optics Letters* **17**, 1426–1428.

[10] Tsuji, Y. Koshiba, M. and Tanabe, T. (1997) A wide-angle beam propagation method using a finite element scheme. *Electronics and Communications in Japan* **80**, 18–26.

[11] Rao, H. Scarmozzino, R. and Osgood Jr., R.M. (1999) A bidirectional beam propagation method for multiple dielectric interfaces. *IEEE Photonics Technology Letters* **11**, 830–832.

[12] Yoneta, S., Koshiba, M. and Tsuji, Y. (1999) Combined beam propagation and finite element method for bidirectional optical beam propagation analysis. *Electronics and Communications in Japan*, **82** (10, Pt. 2).

[13] Huang, W.P., Xu, C.L. and Chaudhuri, S.K. (1991) A vector beam propagation method based on H fields. *IEEE Transactions Photonics Technology Letters* **3**, 1117–1120.

[14] Scarmozzino, R., Gopinath, A., Pregla, R. and Helfert, S. (2000) Numerical techniques for modeling guided-wave photonic devices. *IEEE Journal of Selected Topics in Quantum Electronics* **6**, 150–162.

[15] Huang, W.P., Xu, C.L. and Chaudhuri, S.K. (1992) A finite-difference vector beam propagation method for three-dimensional waveguide structures. *IEEE Photonics Technology Letters* **4**, 148–151.

[16] Huang, W.P. and Xu, C.L. (1993) Simulation of three-dimensional optical waveguides by a full-vector beam propagation method. *Journal of Quantum Electronics* **29**, 2639–2649.

[17] Press, W.H., Teukolsky, S.A., Vetterling, W.T. and Flannery, B.P. (1996) *Numerical Recipes in Fortran 77: The Art of Scientific Computing*, Chapter 19. Cambridge University Press, New York.

[18] Press, W.H., Teukolsky, S.A., Vetterling, W.T. and Flannery, B.P. (1996) *Numerical Recipes in Fortran 77: The Art of Scientific Computing*, Chapter 2. Cambridge University Press, New York.

[19] Liu, P.L. and Li, B.J. (1991) Study of form birefringence in waveguide devices using the semivectorial beam propagation method. *IEEE Photonics Technology Letters* **3**, 913–915.

[20] Haus, H.A., Huang, W.P. and Snyder, W. (1989) Coupled-mode formulations. *Optics Letters* **14**, 1222–1224.

[21] Yevick, D., Yu, J. and Yayon, Y. (1995) Optimal absorbing boundary conditions. *Journal of the Optical Society of America A* **12**, 107–110.

[22] Hadley, G.R. (1991) Transparent boundary condition for beam propagation. *Optics Letters* **16**, 624–626.

[23] Hadley, G.R. (1992) Transparent boundary condition for the beam propagation method. *IEEE Journal of Quantum Electronics* **28**, 363–370.

[24] Bérenger, J.P. (1994) A perfectly matched layer for the absorption of electromagnetic waves. *Journal of Computational Physics* **114**, 185–200.

[25] Jiménez, D., Ramírez, C., Pérez-Murano, F. and Guzmán, A. (1999) Implementation of Bérenger layers as boundary conditions for the beam propagation method: applications to integrated waveguides. *Optics Communications* **159**, 43–48.

[26] Scarmozzino, R. and Osgood, R.M. (1991) Comparison of finite-difference and Fourier-transform solutions of the parabolic wave equation with emphasis on integrated-optics applications. *Journal of the Optical Society of America A* **8**, 724–731.

[27] Huang, W.P., Xu, C.L., Lui, W. and Yokoyama, K. (1996) The perfectly matched layer (PML) boundary condition for the beam propagation method. *IEEE Photonics Technology Letters* **8**, 649–651.

[28] Hermansson, B. and Yevick, D. (1987) Numerical analyses of the modal eigenfunctions of chirped and unchirped stripe geometry laser arrays. *Journal of the Optical Society of America A* **4**, 379–390.

[29] Hawkins, R.J. (1987) Propagation properties of single-mode dielectric waveguide structures: a path integral approach. *Applied Optics* **26**, 1183–1188.

[30] Feit, M.D. and Fleck, J.A. (1980) Computation of mode properties in optical fiber waveguides by a propagating beam method, *Applied Optics* **19**, 1154–1164.

[31] Lifante, G. (2003) *Integrated Photonics: Fundamentals*. John Wiley & Sons, Ltd, Chichester.

[32] Xu, C.L., Huang, W.P. and Chaudhuri, S.K. (1993) Efficient and accurate vector mode calculations by beam propagation method. *Journal of Lightwave Technology* **11**, 1209–1215.

[33] Jüngling, S. and Chen, J.C. (1994) A study and optimization of eigenmode calculations using the imaginary-distance beam-propagation method. *IEEE Journal of Quantum Electronics* **30**, 2098–2105.

[34] Chen, J.C. and Jüngling, S. (1994) Computation of high-order waveguide modes by imaginary-distance beam propagation method. *Optical and Quantum Electronics* **26**, S199–S205.

[35] Huang, W.P. and Haus, H.A. (1991) A simple variational approach to optical rib waveguides. *Journal of Lightwave Technology* **9**, 56–61.

[36] Vassallo, C. and Collino, F. (1996) Highly efficient absorbing boundary conditions for the beam propagation method. *Journal of Lightwave Technology* **14**, 1570–1577.

3

Vectorial and Three-Dimensional Beam Propagation Techniques

Introduction

In the previous chapter, it was shown that the vectorial nature of the electromagnetic waves propagating in an inhomogeneous medium dictates that the transverse electric field (TE) components are polarization-dependent and, in general, coupled with each other. The vectorial properties of the beams become important when the variation of the refractive index over the transverse cross section of the waveguide structure is abrupt and/or the index differences at the discontinuities are large. The vectorial character of the light beams breaks the degeneracy between the x- and y-polarized waves because of $P_{xx} \neq P_{yy}$ (or $Q_{xx} \neq Q_{yy}$) and induces coupling between the two polarizations through P_{xy} and P_{yx} (or through Q_{xy} and Q_{yx}).

For two-dimensional optical structures (planar waveguides) the polarization coupling terms associated with crossing operators P_{xy} and P_{yx} (or Q_{xy} and Q_{yx}) vanish, indicating that there is no coupling between the two polarized waves. In this case the vector waves may be decomposed into the TE (y-polarized transverse electric field) and TM (y-polarized transverse magnetic field) waves, which can be treated and solved separately [1]. On the other hand, for three-dimensional structures (channel waveguides) the coupling between the two polarizations occurs, so that the fields are, in general, hybrid [2].

It is important to note that the equations for the TE and TM fields (E_t and H_t fields) are decoupled so they can be solved independently. Any of the two formulations (based either on the electric or magnetic fields) can be chosen to simulate the optical propagation of vectorial waves in inhomogeneous media. A comparison of the results using both formalisms indicates excellent agreement between them [3]. In this chapter we will present the BPM (beam propagation method) based on finite differences for two-dimensional structures and then FD-BPM (finite difference-beam propagation method) algorithms will be applied to simulate light propagation in three-dimensional optical waveguides. In both cases, formalisms based either on the electric or magnetic field transversal components will be developed.

Beam Propagation Method for Design of Optical Waveguide Devices, First Edition. Ginés Lifante Pedrola.
© 2016 John Wiley & Sons, Ltd. Published 2016 by John Wiley & Sons, Ltd.

3.1   Two-Dimensional Vectorial Beam Propagation Method

A picture of a two-dimensional waveguide is presented in Figure 3.1, where the optical structure is invariant along the y-direction and the light propagates mainly along the z direction. At
variance with 3D structures, light in two-dimensional waveguide structures can propagate as
pure TE or TM modes. Light propagation can be simulated using electric field formalism or a
description based on the magnetic field. We will see that for two-dimensional structures the
starting difference equation results in a pure tridiagonal system of equation [4], which can
be solved quite efficiently by the Thomas algorithm, as we have shown in the previous chapter.

3.1.1   Formulation Based on the Electric Field

For the two-dimensional case, as the problem is y invariant, the derivatives with respect to the
y coordinate in the paraxial wave equations (2.6) can be dropped. Doing that, the first salient
feature is that the operators in Eq. (2.11) containing cross terms vanish and the vectorial wave
equations (2.10) reduce to:

$$2in_0k_0\frac{\partial \Psi_x}{\partial z} = P_{xx}\Psi_x; \tag{3.1a}$$

$$2in_0k_0\frac{\partial \Psi_y}{\partial z} = P_{yy}\Psi_y, \tag{3.1b}$$

and the differential operators P_{ij} simplify to:

$$P_{xx}\Psi_x = \frac{\partial}{\partial x}\left(\frac{1}{n^2}\frac{\partial}{\partial x}\left(n^2\Psi_x\right)\right) + k_0^2\left(n^2 - n_0^2\right)\Psi_x; \tag{3.2a}$$

$$P_{yy}\Psi_y = \frac{\partial^2 \Psi_y}{\partial x^2} + k_0^2\left(n^2 - n_0^2\right)\Psi_y. \tag{3.2b}$$

It is worth noting that these two differential equations are completely decoupled so they can be
solved independently, which physically means that the fields Ψ_x and Ψ_y propagate independently (as TM and TE polarized modes, respectively).

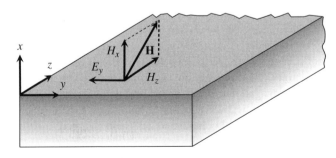

Figure 3.1   TE-propagation through 2D structures, where the electric field has only a non-null
component (y component)

3.1.1.1 TE-Propagation

If the light propagates with TE polarization, only the y component of the electric field is present ($E = [0, E_y, 0]$), while the magnetic field has x and z components ($H = [H_x, 0, H_z]$) (Figure 3.1). Therefore, the BPM description for TE propagation in two-dimensions based on the SVE (slowly varying envelope) electric field results on solving the following partial derivative equation:

$$2in_0k_0\frac{\partial\Psi_y}{\partial z} = \frac{\partial^2\Psi_y}{\partial x^2} + k_0^2\left(n^2 - n_0^2\right)\Psi_y. \tag{3.3}$$

Let us now derive the finite-difference approximation of this differential equation.

Since the tangential electric field component E_y is continuous across the index interface, the finite-difference approximation of the first derivative with respect to the x coordinate is simply given by (see Appendix A):

$$\frac{\partial\Psi_y}{\partial x} \approx \frac{\Psi_y(i+1/2,m) - \Psi_y(i-1/2,m)}{\Delta x}, \tag{3.4}$$

where Ψ_y is the SVE of the electromagnetic field component E_y:

$$E_y(x,z) = \Psi_y(x,z)e^{-iKz}, \tag{3.5}$$

and the i and m indices stand for the transversal and longitudinal positions as:

$$\Psi_y(i,m) \equiv \Psi_y(x = i\Delta x, z = m\Delta z). \tag{3.6}$$

From the third Maxwell's equation, the following relations can be obtained:

$$\frac{\partial E_y}{\partial x} = -i\omega\mu_0 H_z \quad \text{or} \quad \frac{\partial\Psi_y}{\partial x} = -i\omega\mu_0\Phi_z; \tag{3.7a}$$

$$\frac{\partial E_y}{\partial z} = i\omega\mu_0 H_x, \tag{3.7b}$$

being Φ_z the SVE of the electromagnetic field component H_z. Now, as the tangential component of the magnetic field H_z is continuous across the index interface, the relation (3.7a) implies that the first derivative with respect to the electric field component is also continuous across the x direction and thus the finite difference approximation of the second derivative can be expressed as (see Appendix A):

$$\frac{\partial^2\Psi_y}{\partial x^2} \approx \frac{\Psi_y(i-1,m) - 2\Psi_y(i,m) + \Psi_y(i+1,m)}{(\Delta x)^2}. \tag{3.8}$$

Also, the E_y component is continuous across the possible index discontinuities along the z direction. Therefore, the following approximation holds (see Appendix A):

$$\frac{\partial \Psi_y}{\partial z} \approx \frac{\Psi_y(i, m+1) - \Psi_y(i, m)}{\Delta z}. \tag{3.9}$$

With these finite-difference approaches, the usual Crank–Nicolson method implemented to Eq. (3.3) is as follows:

$$
\begin{aligned}
\frac{2ik_0 n_0}{\Delta z} \left[u_j(z + \Delta z) - u_j(z) \right] &= (1-\alpha) \frac{u_{j-1}(z) - 2u_j(z) + u_{j+1}(z)}{(\Delta x)^2} \\
&+ (1-\alpha)k_0^2 \left[n_j^2(z) - n_0^2 \right] u_j(z) + \alpha \frac{u_{j-1}(z + \Delta z) - 2u_j(z + \Delta z) + u_{j+1}(z + \Delta z)}{(\Delta x)^2} \\
&+ \alpha k_0^2 \left[n_j^2(z + \Delta z) - n_0^2 \right] u_j(z + \Delta z)
\end{aligned} \tag{3.10}
$$

where $u_j(z)$ represents the slowly varying amplitude Ψ_y of the electric field at the nodal position $x = j\Delta x$ and at the longitudinal position z. Rearranging terms in this equation, one obtains:

$$a_j u_{j-1}(z + \Delta z) + b_j u_j(z + \Delta z) + c_j u_{j+1}(z + \Delta z) = r_j(z), \tag{3.11}$$

where the a_j, b_j, c_j and r_j coefficients are defined by:

$$a_j = -\frac{\alpha}{(\Delta x)^2}; \tag{3.12a}$$

$$b_j = \frac{2ik_0 n_0}{\Delta z} + \frac{2\alpha}{(\Delta x)^2} - \alpha k_0^2 \left[n_j^2(z + \Delta z) - n_0^2 \right]; \tag{3.12b}$$

$$c_j = -\frac{\alpha}{(\Delta x)^2}; \tag{3.12c}$$

$$r_j = \frac{(1-\alpha)}{(\Delta x)^2} \left[u_{j-1}(z) + u_{j+1}(z) \right] + \left\{ \frac{2ik_0 n_0}{\Delta z} - \frac{2(1-\alpha)}{(\Delta x)^2} + (1-\alpha)k_0^2 \left[n_j^2(z) - n_0^2 \right] \right\} u_j(z). \tag{3.12d}$$

Equation (3.11), besides the definitions in Eq. (3.12a–d), forms a tridiagonal system identical to that obtained for the scalar BPM in two dimensions. Therefore, the solution provided by the scalar approach for planar structures is in fact the exact solution for the E_y component for TE-polarized beams. In consequence, the algorithm developed in Section 2.4 can be directly used to model TE-propagation in 2D structures.

As an example, Figure 3.2 shows the TE-propagation based on the E_y field along an asymmetric step-index planar waveguide, which is monomode at the working wavelength of 1.55 μm. The cover, film and substrate indices are $n_c = 1$, $n_f = 1.95$, $n_s = 1.45$, respectively, the film thickness being $d = 0.6$ μm. To start the propagation, at the input a 0.5 μm width Gaussian TE-polarized beam is launched. The simulation is performed by using a grid spacing of $\Delta x = 0.005$ μm, with 1000 mesh points in the transversal direction and a longitudinal step of $\Delta z = 0.2$ μm. The scheme parameter is set to $\alpha = 0.50001$ and the reference index is chosen

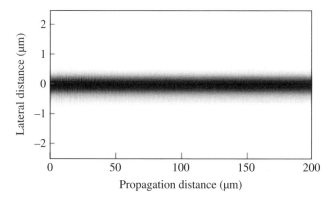

Figure 3.2 Propagation of a launched TE-polarized Gaussian beam, based on the E_y-BPM-formulation, through an asymmetric step-index planar waveguide

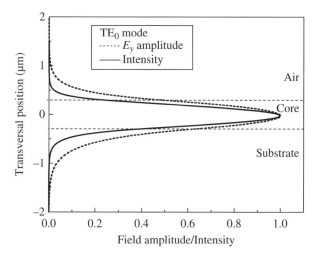

Figure 3.3 Dashed line: field distribution for the E_y component of the fundamental TE mode in an asymmetric step-index waveguide at 1.55 μm, obtained by imaginary-distance-BPM based on the y component of the electric field. Continuous line: intensity distribution of the mode

to be equal to the substrate refractive index, $n_0 = 1.45$. It can be observed from the figure that the beam accommodates to the guided mode of the structure along the first ~50 μm and then it remains unchanged as expected for a steady-state mode modelled by a non-dissipative propagation scheme.

Based on this TE-formulation, and using propagation along the imaginary z-axis, the fundamental mode of the 2D-waveguide is calculated. The results of this propagation are shown in Figure 3.3, where the field amplitude (u_j) and field intensity ($|u_j|^2$) is plotted. Let us note the continuity of the field and its derivative across the index discontinuities. The effective index of the guided-mode N_{eff} can be obtained from the field distribution by means of Eq. (2.124), which gives a value of $N_{eff} = 1.774255$ and compares favourably with the exact value computed by the multilayer approach ($N_{eff} = 1.774241$). The difference between these two values is due to the discretization inherent to the FD method (FD, finite difference). This gives confidence on the used propagation scheme of the BPM.

3.1.1.2   TM-Propagation (Derivation Using the Electric Field Component E_x)

In TM-propagation, the magnetic field has only y-component, and the electric field has no component along the y-axis ($\boldsymbol{E} = [E_x, 0, E_z]$, $\boldsymbol{H} = [0, H_y, 0]$), as shown in Figure 3.4.

The starting Eq. (2.10a) for the non-null transversal component of the electric field Ψ_x, after dropping the terms involving derivatives with respect to the y coordinate, is:

$$2in_0k_0\frac{\partial \Psi_x}{\partial z} = \frac{\partial}{\partial x}\left(\frac{1}{n^2}\frac{\partial}{\partial x}\left(n^2\Psi_x\right)\right) + k_0^2\left(n^2 - n_0^2\right)\Psi_x. \tag{3.13}$$

Let us see now that the expression in Eq. (3.13) is arranged in such a form that the finite difference scheme can be applied directly. First, let us note that, although the electric field component E_x is not continuous across the index interface, the continuity can be imposed in the displacement vector component D_x (continuity of the normal component of $\boldsymbol{D}$); that is, continuity of n^2E_x. Therefore, the finite-difference approximation for the first derivative in Eq. (3.13) can be straightforwardly obtained as follows:

$$\frac{\partial}{\partial x}\left(n^2\Psi_x\right) \approx \frac{n_{j+1/2}^2 u_{j+1/2} - n_{j-1/2}^2 u_{j-1/2}}{\Delta x}, \tag{3.14}$$

where u_j represents the discretized version of the SVE-field Ψ_x at the transversal position $x = j\Delta x$. To obtain the finite difference approximation for the derivative on the right-hand part of Eq. (3.13), first we obtain the next relations from the Maxwell's equations:

$$\frac{\partial H_y}{\partial x} = i\omega\varepsilon_0 n^2 E_z; \tag{3.15a}$$

$$\frac{\partial H_y}{\partial z} = -i\omega\varepsilon_0 n^2 E_x. \tag{3.15b}$$

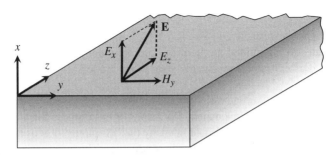

Figure 3.4 Electromagnetic field components in TM-propagation through 2D structures. Only the y component of the magnetic field is present

Now, we should use the fact that the magnetic field component H_x is continuous across the index interface. Let us develop the left hand side of Eq. (3.15b) in terms of the SVE field Φ_y:

$$\frac{\partial H_y}{\partial z} = \frac{\partial\left(\Phi_y e^{-iKz}\right)}{\partial z} = \frac{\partial\Phi_y}{\partial z}e^{-iKz} - iK\Phi_y e^{-iKz} = e^{-iKz}\left(\frac{\partial\Phi_y}{\partial z} - iK\Phi_y\right). \tag{3.16}$$

Now applying the SVE approximation: $\left|\dfrac{\partial\Phi_y}{\partial z}\right| \ll K|\Phi_y|$, we get:

$$\frac{\partial H_y}{\partial z} \approx -iKe^{-iKz}\Phi_y. \tag{3.17}$$

Using this result and taking the partial derivative with respect to the x coordinate of expression (3.15b) we have:

$$-iKe^{-iKz}\frac{\partial\Phi_y}{\partial x} \approx -i\omega\varepsilon_0 e^{-iKz}\frac{\partial\left(n^2\Psi_x\right)}{\partial x}, \tag{3.18}$$

which implies:

$$K\frac{\partial\Phi_y}{\partial x} \approx \omega\varepsilon_0\frac{\partial\left(n^2\Psi_x\right)}{\partial x}. \tag{3.19}$$

On the other hand, the relation (3.15a) can be written as:

$$\frac{1}{n^2}\frac{\partial H_y}{\partial x} = i\omega\varepsilon_0 E_z. \tag{3.20}$$

This relation implies that, as the E_z component is continuous across the x interface, the quantity $\dfrac{1}{n^2}\dfrac{\partial H_y}{\partial x}$ is also continuous across that index discontinuity. Looking at Eq. (3.19), we now see that the quantity $\dfrac{1}{n^2}\dfrac{\partial\left(n^2\Psi_x\right)}{\partial x}$ is continuous along the x direction. Thus, the first term on the right hand in Eq. (3.13) can be approximated as (see Appendix A):

$$\frac{\partial}{\partial x}\left[\frac{1}{n^2}\frac{\partial}{\partial x}\left(n^2\Psi_x\right)\right] \approx \frac{1}{\Delta x}\left[\frac{1}{n_{j+1/2}^2}\left(\frac{\partial(n^2 u)}{\partial x}\right)_{j+1/2} - \frac{1}{n_{j-1/2}^2}\left(\frac{\partial(n^2 u)}{\partial x}\right)_{j-1/2}\right] \approx$$

$$\frac{1}{\Delta x}\left[\frac{2}{n_{j+1}^2+n_j^2}\left(\frac{n_{j+1}^2 u_{j+1}-n_j^2 u_j}{\Delta x}\right) - \frac{2}{n_j^2+n_{j-1}^2}\left(\frac{n_j^2 u_j-n_{j-1}^2 u_{j-1}}{\Delta x}\right)\right] \tag{3.21}$$

$$= \frac{2}{(\Delta x)^2}\left[\frac{n_{j-1}^2}{n_j^2+n_{j-1}^2}u_{j-1} - \left(\frac{n_j^2}{n_{j+1}^2+n_j^2}+\frac{n_j^2}{n_j^2+n_{j-1}^2}\right)u_j + \frac{n_{j+1}^2}{n_{j+1}^2+n_j^2}u_{j+1}\right].$$

The finite difference wave equation, following the Crank–Nicolson scheme, can now be written accordingly:

$$\frac{2ik_0n_0}{\Delta z}\left[u_j^{m+1}-u_j^m\right]=(1-\alpha)\frac{T_{j-1}^m u_{j-1}^m-2R_j^m u_j^m+T_{j+1}^m u_{j+1}^m}{(\Delta x)^2}+(1-\alpha)k_0^2\left[\left(n_j^m\right)^2-n_0^2\right]u_j^m$$

$$+\alpha\frac{T_{j-1}^{m+1} u_{j-1}^{m+1}-2R_j^{m+1} u_j^{m+1}+T_{j+1}^{m+1} u_{j+1}^{m+1}}{(\Delta x)^2}+\alpha k_0^2\left[\left(n_j^{m+1}\right)^2-n_0^2\right]u_j^{m+1}$$

$$(3.22)$$

where u_j^m represents the SVE field at the transversal position $x=j\Delta x$ and at the longitudinal position $z=m\Delta z$. For the sake of compactness, in the last equation we have defined the $T_{j\pm1}$ and R_j coefficients as:

$$T_{j\pm1}=\frac{2n_{j\pm1}^2}{n_j^2+n_{j\pm1}^2};\qquad(3.23a)$$

$$R_j=\frac{n_j^2}{n_{j+1}^2+n_j^2}+\frac{n_j^2}{n_j^2+n_{j-1}^2}=2-\frac{T_{j+1}+T_{j-1}}{2}.\qquad(3.23b)$$

With these definitions, Eq. (3.22) forms again a tridiagonal system of linear equations (similar to that in Eq. 3.10), with the following coefficients:

$$a_j=-\frac{\alpha}{(\Delta x)^2}T_{j-1}^{m+1};\qquad(3.24a)$$

$$b_j=\frac{2ik_0n_0}{\Delta z}+\frac{2\alpha}{(\Delta x)^2}R_j^{m+1}-\alpha k_0^2\left[\left(n_j^{m+1}\right)^2-n_0^2\right];\qquad(3.24b)$$

$$c_j=-\frac{\alpha}{(\Delta x)^2}T_{j+1}^{m+1};\qquad(3.24c)$$

$$r_j=\frac{(1-\alpha)}{(\Delta x)^2}\left[T_{j-1}^m\ u_{j-1}^m+T_{j+1}^m\ u_{j+1}^m\right]+\left\{\frac{2ik_0n_0}{\Delta z}-\frac{2(1-\alpha)R_j^m}{(\Delta x)^2}+(1-\alpha)k_0^2\left[\left(n_j^m\right)^2-n_0^2\right]\right\}u_j^m.$$

$$(3.24d)$$

This system equation, combined with the Thomas method, provides an unconditionally stable algorithm, being second-order accurate in both the transversal and longitudinal steps.

Figure 3.5 shows the propagation of a TM-polarized Gaussian beam through a planar step-index waveguide at $\lambda=1.55\ \mu m$. The parameters of the structure, as well as the simulation parameters, are those indicated for Figure 3.2. After a distance of $\sim150\ \mu m$, the input Gaussian beam accommodates to the field profile corresponding for the fundamental TM mode of the structure. Following the described formalism, the propagation is stable, even for long propagation steps. Also, when using a scheme parameter close to $\alpha=0.5$, the algorithm is energy conservative.

Using Im-Dis-BPM based of the E_x formulation, the field transversal distribution (u_j) and intensity ($n_j^2|u_j|^2$) are obtained and the results are presented in Figure 3.6. Let us note the field

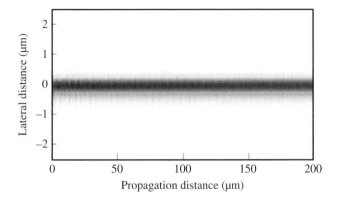

Figure 3.5   Propagation of a TM-polarized Gaussian beam based on the E_x-BPM-formulation

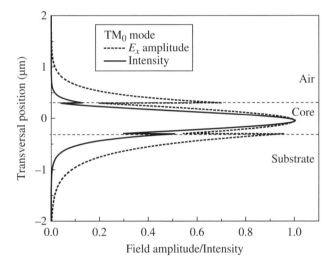

Figure 3.6   Dashed line: field distribution for the E_x component of the fundamental TM mode obtained by imaginary distance BPM based on the x component of the electric field. Continuous line: intensity distribution of the mode

discontinuity of E_x at the index interfaces, where the jump is a direct consequence of the continuity imposed to $n^2 E_x$. From the field profile, an effective index of $N_{eff} = 1.684620$ is calculated, which is very close to the exact value of 1.684600 obtained by a multilayer approach. The difference is due to the discretization of the computational window along the transversal direction, as will be shown next.

In general, to achieve accurate results in the numerical simulation, the grid spacing Δx of the computational window chosen to perform the FD-BPM simulation should be small enough to reproduce the refractive index profile of the waveguide. For a given structure, the accuracy of the simulation is increased as the grid spacing reduces and can be evaluated by computing the modal effective index of a z invariant waveguide as the transversal cell size reduces. Figure 3.7 shows the errors of the calculated effective indexes for the fundamental TE and TM modes of

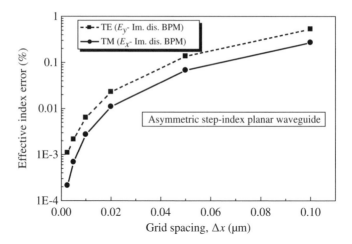

Figure 3.7   Error ($100^*\Delta N_{eff}/N_{eff}$) of the mode effective index calculated by imaginary-distance BPM for TE- and TM-polarized modes on an asymmetric step-index planar waveguide as a function of the transversal grid spacing, using FD-BPM based on the electric field components.

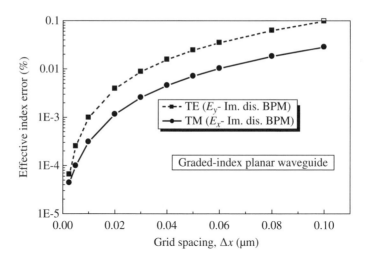

Figure 3.8   Error ($100^*\Delta N_{eff}/N_{eff}$) of the mode effective index calculated by imaginary distance BPM for TE- and TM-polarized beams on a graded-index planar waveguide as a function of the transversal grid size, using FD-BPM based on the electric field components

the previously examined waveguide using E-field based-BPM. The errors for TE and TM modes are similar and drastically reduce as the grid spacing decreases, reaching a value lower than 0.01% for a cell size of $\Delta x = 0.01$ μm. In general, graded index profiles provide more accurate results than waveguides with strong index discontinuities for a given grid spacing. As an example, Figure 3.8 shows the errors in the calculated effective indices for TE and TM fundamental modes in a straight waveguide with a Gaussian index profile. For a grid spacing of $\Delta x = 0.01$ μm, the errors found in both polarizations are lower than 0.001%.

3.1.2 Formulation Based on the Magnetic Field

The description of the light propagation in planar waveguides based on the magnetic field transversal components starts by considering the Eqs. (2.12a) and (2.12b), where the crossing terms vanish due to the non-dependence of the problem with respect to the y coordinate:

$$2in_0k_0\frac{\partial\Phi_x}{\partial z} = \frac{\partial^2\Phi_x}{\partial x^2} + k_0^2\left(n^2-n_0^2\right)\Phi_x; \tag{3.25a}$$

$$2in_0k_0\frac{\partial\Phi_y}{\partial z} = n^2\frac{\partial}{\partial x}\left(\frac{1}{n^2}\frac{\partial\Phi_y}{\partial x}\right) + k_0^2\left(n^2-n_0^2\right)\Phi_y. \tag{3.25b}$$

As in the case of the electric field equations, these two equations are decoupled in two-dimensional structures, allowing the classification into pure TE or TM modes.

3.1.2.1 TE-Propagation. $(0, E_y, 0)$, $(H_x, 0, H_z)$

The magnetic field vector for TE propagation has both x and z components for planar structures. The equation needed to track the propagation is in this case the equation for the transversal x component of the magnetic field:

$$2in_0k_0\frac{\partial\Phi_x}{\partial z} = \frac{\partial^2\Phi_x}{\partial x^2} + k_0^2\left(n^2-n_0^2\right)\Phi_x. \tag{3.26}$$

Developing the derivative with respect to the z coordinate of the electric field component E_y in terms of the SVE field:

$$\frac{\partial E_y}{\partial z} = \frac{\partial\left(\Psi_y e^{-iKz}\right)}{\partial z} = e^{-iKz}\left(\frac{\partial\Psi_y}{\partial z} - iK\Psi_y\right) \approx -iKe^{-iKz}\Psi_y. \tag{3.27}$$

Using now the relation (3.7a) it follows that:

$$\Phi_x \approx -\frac{K}{\omega\mu_0}\Psi_y. \tag{3.28}$$

Thus, for TE propagation in 2D structures within the SVE approximation the two formulations based on the electric or magnetic field transversal components are identical.

3.1.2.2 TM-Propagation. $(E_x, 0, E_z)$, $(0, H_y, 0)$

As we have seen in Figure 3.4, for TM-propagation in planar structures the magnetic field vector only has a y component. The starting equation for the non-null transversal component of the magnetic field is:

$$2in_0k_0\frac{\partial\Phi_y}{\partial z} = n^2\frac{\partial}{\partial x}\left(\frac{1}{n^2}\frac{\partial\Phi_y}{\partial x}\right) + k_0^2\left(n^2-n_0^2\right)\Phi_y. \tag{3.29}$$

The continuity of H_y implies that the operation $\dfrac{\partial \Phi_y}{\partial x}$ can be performed in a straightforward way using the finite difference approximation.

On the other hand, using Eq. (3.15a) and keeping in mind the continuity of the electric field component E_z, it is deduced that the quantity $\dfrac{1}{n^2}\dfrac{\partial H_y}{\partial x}$ is also continuous across the index interface, thus allowing us to express the derivative $\dfrac{\partial}{\partial x}\left[\dfrac{1}{n^2}\dfrac{\partial \Phi_y}{\partial x}\right]$ in a conventional way as follows (see Appendix A):

$$
n^2 \frac{\partial}{\partial x}\left[\frac{1}{n^2}\frac{\partial \Phi_y}{\partial x}\right] \approx n_j^2 \frac{1}{\Delta x}\left[\frac{1}{n_{j+1/2}^2}\left(\frac{\partial u}{\partial x}\right)_{j+1/2} - \frac{1}{n_{j-1/2}^2}\left(\frac{\partial u}{\partial x}\right)_{j-1/2}\right] \approx
$$

$$
\frac{n_j^2}{\Delta x}\left[\frac{2}{n_{j+1}^2+n_j^2}\frac{u_{j+1}-u_j}{\Delta x} - \frac{2}{n_j^2+n_{j-1}^2}\frac{u_j-u_{j-1}}{\Delta x}\right] \tag{3.30}
$$

$$
= \frac{2}{(\Delta x)^2}\left[\frac{n_j^2}{n_j^2+n_{j-1}^2}u_{j-1} - \left(\frac{n_j^2}{n_{j+1}^2+n_j^2}+\frac{n_j^2}{n_j^2+n_{j-1}^2}\right)u_j + \frac{n_j^2}{n_{j+1}^2+n_j^2}u_{j+1}\right].
$$

The finite difference wave equation, following the Crank–Nicolson scheme, can now be written accordingly to:

$$
\frac{2ik_0 n_0}{\Delta z}\left[u_j^{m+1}-u_j^m\right] = (1-\alpha)\frac{T_{j-1}^m u_{j-1}^m - 2R_j^m u_j^m + T_{j+1}^m u_{j+1}^m}{(\Delta x)^2} + (1-\alpha)k_0^2\left[\left(n_j^m\right)^2 - n_0^2\right]u_j^m
$$

$$
+ \alpha\frac{T_{j-1}^{m+1} u_{j-1}^{m+1} - 2R_j^{m+1} u_j^{m+1} + T_{j+1}^{m+1} u_{j+1}^{m+1}}{(\Delta x)^2} + \alpha k_0^2\left[\left(n_j^{m+1}\right)^2 - n_0^2\right]u_j^{m+1}
$$

$$
\tag{3.31}
$$

where we have defined the $T_{j\pm1}$ and R_j coefficients as:

$$
T_{j\pm1} = \frac{2n_j^2}{n_j^2+n_{j\pm1}^2}; \tag{3.32a}
$$

$$
R_j = \frac{n_j^2}{n_{j+1}^2+n_j^2}+\frac{n_j^2}{n_j^2+n_{j-1}^2} = \frac{T_{j+1}+T_{j-1}}{2}. \tag{3.32b}
$$

Here, u_j^m denotes the (SVE amplitude) y-component of the magnetic field H_y at the transverse grid point j and at the longitudinal step m. Rearranging terms, the coefficients a_j, b_j, c_j and r_j of the tridiagonal system are given by:

$$
a_j = -\frac{\alpha}{(\Delta x)^2}T_{j-1}^{m+1}; \tag{3.33a}
$$

$$b_j = \frac{2ik_0n_0}{\Delta z} + \frac{2\alpha}{(\Delta x)^2}R_j^{m+1} - \alpha k_0^2 \left[\left(n_j^{m+1} \right)^2 - n_0^2 \right];$$

(3.33b)

$$c_j = -\frac{\alpha}{(\Delta x)^2}T_{j+1}^{m+1};$$

(3.33c)

$$r_j = \frac{(1-\alpha)}{(\Delta x)^2}\left[T_{j-1}^m\, u_{j-1}^m + T_{j+1}^m\, u_{j+1}^m \right] + \left\{ \frac{2ik_0n_0}{\Delta z} - \frac{2(1-\alpha)R_j^m}{(\Delta x)^2} + (1-\alpha)k_0^2 \left[\left(n_j^m \right)^2 - n_0^2 \right] \right\}u_j^m.$$

(3.33d)

and this system, combined with the Thomas method, provides an unconditionally stable algorithm, being second order accurate in both transversal and longitudinal steps.

Figure 3.9 shows the propagation of a TM-polarized beam through the step-index planar waveguide at $\lambda = 1.55$ μm, following the formalism based on the magnetic field component. As expected, the simulation of the beam propagation is stable, even for long propagation steps. At a distance of ~100 μm the input beam accommodates to the guided fundamental mode and after this longitudinal position the energy in the computational region remains unchanged, indicating that when using a scheme parameter of $\alpha = 0.5$, the algorithm is energy conservative. The field transversal distribution (u_j) and intensity ($|u_j|^2/n_j^2$) of the guided mode, obtained by Im-Dis-BPM, are plotted in Figure 3.10. Let us note that while the optical field (H_y component) is continuous across the index interfaces (at $x = \pm 0.3$ μm), its derivative is discontinuous at these positions. Also it is noted that the intensity distribution presents jumps at the index boundaries, and that this intensity profile is identical to that obtained using the modal description based on the electric field transversal component (see Figure 3.6), indicating the validity of the optical propagation description based on either the electric or magnetic fields. As expected, the

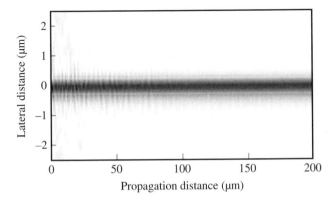

Figure 3.9 Propagation of a TM-polarized Gaussian beam based on the H_y BPM formulation, through an asymmetric step-index planar waveguide. Structure and simulations parameters are identical to those described in Figure 3.2

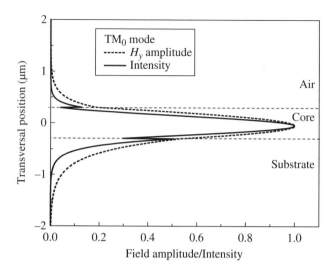

Figure 3.10   Dashed line: field distribution for the H_y component of the fundamental TM mode obtained by imaginary-distance-BPM based on the H_y component of the magnetic field. Continuous line: modal intensity distribution

effective index of the TM-mode calculated using the H_y component ($N_{eff} = 1.684615$) is coincident to that obtained using electric field BPM description.

3.2   Three-Dimensional BPM Based on the Electric Field

We have seen that the one-way propagation of the vectorial electromagnetic fields is governed by two sets of coupled equations for the envelope of the TE and TM fields. By solving these two sets of coupled equations independently using finite-difference schemes, the propagation of vectorial guided waves can be simulated as an initial value problem. Here, the alternating direction implicit (ADI) method [5] is adopted to simulate the full-vectorial 3D beam propagation. The numerical simulation of a 3D problem requires solving a large matrix equation; by using the ADI-method, the 3D problem is reduced to solving a 2D problem twice. In the ADI-BPM the calculation for one step ($\psi(x,y,z) \to \psi(x,y,z + \Delta z)$) is split into two sub-steps ($\psi(x,y,z) \to \psi(x,y,z + \Delta z/2)$) and ($\psi(x,y,z + \Delta z/2) \to \psi(x,y,z + \Delta z)$). These two sub-steps are treated successively in the x and y directions, which involves solving tridiagonal matrix equations twice. Thus, high numerical efficiency is attained. In this formalism, the vectorial wave equation is discretized in the longitudinal direction via the Crank–Nicolson scheme and it is evaluated step by step with the aid of the ADI method. The algorithm used and the finite difference expressions of the differential operators provide second order accuracy along the propagation direction and also assure truncation errors of second order in the transversal direction [6, 7].

We recall the full vectorial wave equations for the transversal SVE electric field components:

$$2in_0k_0\frac{\partial \Psi_x}{\partial z} = P_{xx}\Psi_x + P_{xy}\Psi_y; \tag{3.34a}$$

$$2in_0k_0\frac{\partial \Psi_y}{\partial z}=P_{yy}\Psi_y+P_{yx}\Psi_x, \tag{3.34b}$$

where the differential operators P_{ij} were defined as:

$$P_{xx}\Psi_x \equiv \frac{\partial}{\partial x}\left(\frac{1}{n^2}\frac{\partial}{\partial x}\left(n^2\Psi_x\right)\right)+\frac{\partial^2\Psi_x}{\partial y^2}+k_0^2\left(n^2-n_0^2\right)\Psi_x; \tag{3.35a}$$

$$P_{yy}\Psi_y \equiv \frac{\partial^2\Psi_y}{\partial x^2}+\frac{\partial}{\partial y}\left(\frac{1}{n^2}\frac{\partial}{\partial y}\left(n^2\Psi_y\right)\right)+k_0^2\left(n^2-n_0^2\right)\Psi_y; \tag{3.35b}$$

$$P_{xy}\Psi_y \equiv \frac{\partial}{\partial x}\left(\frac{1}{n^2}\frac{\partial}{\partial y}\left(n^2\Psi_y\right)\right)-\frac{\partial^2\Psi_y}{\partial x\partial y}; \tag{3.35c}$$

$$P_{yx}\Psi_x \equiv \frac{\partial}{\partial y}\left(\frac{1}{n^2}\frac{\partial}{\partial x}\left(n^2\Psi_x\right)\right)-\frac{\partial^2\Psi_x}{\partial y\partial x}. \tag{3.35d}$$

Equation (3.34a,b) can be written in matrix form as:

$$2in_0k_0\frac{\partial}{\partial z}\begin{bmatrix}\Psi_x \\ \Psi_y\end{bmatrix}=\begin{bmatrix}P_{xx} & P_{xy} \\ P_{yx} & P_{yy}\end{bmatrix}\begin{bmatrix}\Psi_x \\ \Psi_y\end{bmatrix}. \tag{3.36}$$

To achieve an algorithm of second-order accuracy in the longitudinal direction using the ADI method, the matrix operator in Eq. (3.36) is decomposed according to:

$$\begin{bmatrix}P_{xx} & P_{xy} \\ P_{yx} & P_{yy}\end{bmatrix}=\begin{bmatrix}A_x+A_y & C \\ D & B_x+B_y\end{bmatrix}. \tag{3.37}$$

Here, A_x and A_y denote the x- and y-dependent parts of the operator P_{xx}, respectively, with:

$$A_x\Psi_x \equiv \frac{\partial}{\partial x}\left(\frac{1}{n^2}\frac{\partial}{\partial x}\left(n^2\Psi_x\right)\right)+\frac{1}{2}k_0^2\left(n^2-n_0^2\right)\Psi_x; \tag{3.38a}$$

$$A_y\Psi_x \equiv \frac{\partial^2\Psi_x}{\partial y^2}+\frac{1}{2}k_0^2\left(n^2-n_0^2\right)\Psi_x. \tag{3.38b}$$

Similarly, the operators B_x and B_y denote the x- and y-dependent parts of P_{yy}, respectively, defined by:

$$B_x\Psi_y \equiv \frac{\partial^2\Psi_y}{\partial x^2}+\frac{1}{2}k_0^2\left(n^2-n_0^2\right)\Psi_y; \tag{3.39a}$$

$$B_y\Psi_y \equiv \frac{\partial}{\partial y}\left(\frac{1}{n^2}\frac{\partial}{\partial y}\left(n^2\Psi_y\right)\right)+\frac{1}{2}k_0^2\left(n^2-n_0^2\right)\Psi_y. \tag{3.39b}$$

The differential operators C and D denote the cross-coupling terms, and are just the P_{xy} and P_{yx} operators, respectively, given by:

$$C\Psi_y \equiv \frac{\partial}{\partial x}\left(\frac{1}{n^2}\frac{\partial}{\partial y}\left(n^2\Psi_y\right)\right) - \frac{\partial^2\Psi_y}{\partial x\partial y}; \tag{3.40a}$$

$$D\Psi_x \equiv \frac{\partial}{\partial y}\left(\frac{1}{n^2}\frac{\partial}{\partial x}\left(n^2\Psi_x\right)\right) - \frac{\partial^2\Psi_x}{\partial y\partial x}. \tag{3.40b}$$

For the sake of readability, we will denote u and v to be the x and y components of the slowly varying electric field (Ψ_x and Ψ_y), respectively. The finite difference scheme of the defined operator $A_x u$ is implemented as [7] (see Appendix A):

$$A_x u = \frac{T_{i-1,j}u_{i-1,j} - 2R_{i,j}u_{i,j} + T_{i+1,j}u_{i+1,j}}{(\Delta x)^2} + \frac{1}{2}k_0^2\left[n_{i,j}^2 - n_0^2\right]u_{i,j}, \tag{3.41}$$

where the coefficients $T_{i\pm1,j}$ and $R_{i,j}$ are defined by:

$$T_{i\pm1,j} \equiv \frac{2n_{i\pm1,j}^2}{n_{i,j}^2 + n_{i\pm1,j}^2}; \tag{3.42a}$$

$$R_{i,j} \equiv \frac{n_{i,j}^2}{n_{i+1,j}^2 + n_{i,j}^2} + \frac{n_{i,j}^2}{n_{i,j}^2 + n_{i-1,j}^2} = 2 - \frac{T_{i+1,j} + T_{i-1,j}}{2}; \tag{3.42b}$$

Additionally, the finite difference expressions of the operators $A_y u$, $B_x v$ and $B_y v$ are given by [7] (see Appendix A):

$$A_y u = \frac{u_{i,j-1} - 2u_{i,j} + u_{i,j+1}}{(\Delta y)^2} + \frac{1}{2}k_0^2\left[n_{i,j}^2 - n_0^2\right]u_{i,j}; \tag{3.43a}$$

$$B_x v = \frac{v_{i-1,j} - 2v_{i,j} + v_{i+1,j}}{(\Delta x)^2} + \frac{1}{2}k_0^2\left[n_{i,j}^2 - n_0^2\right]v_{i,j}; \tag{3.43b}$$

$$B_y v = \frac{T_{i,j-1}v_{i,j-1} - 2R_{i,j}v_{i,j} + T_{i,j+1}v_{i,j+1}}{(\Delta y)^2} + \frac{1}{2}k_0^2\left[n_{i,j}^2 - n_0^2\right]v_{i,j}, \tag{3.43c}$$

where in this case the coefficients $T_{i,j\pm1}$ and $R_{i,j}$ are defined by:

$$T_{i,j\pm1} \equiv \frac{2n_{i,j\pm1}^2}{n_{i,j}^2 + n_{i,j\pm1}^2}; \tag{3.44a}$$

$$R_{i,j} \equiv \frac{n_{i,j}^2}{n_{i,j+1}^2 + n_{i,j}^2} + \frac{n_{i,j}^2}{n_{i,j}^2 + n_{i,j-1}^2} = 2 - \frac{T_{i,j+1} + T_{i,j-1}}{2}. \tag{3.44b}$$

Finally, the finite difference formulations of the operators Cv and Du are implemented as [7] (see Appendix A):

$$Cv = \frac{1}{4\Delta x\Delta y}\left(S1_{i,j}v_{i+1,j+1} - S2_{i,j}v_{i+1,j-1} - S3_{i,j}v_{i-1,j+1} + S4_{i,j}v_{i-1,j-1}\right); \tag{3.45a}$$

$$Du = \frac{1}{4\Delta x\Delta y}\left(Z1_{i,j}u_{i+1,j+1} - Z2_{i,j}u_{i-1,j+1} - Z3_{i,j}u_{i+1,j-1} + Z4_{i,j}u_{i-1,j-1}\right), \tag{3.45b}$$

where the $S1_{i,j}$, $S2_{i,j}$, $S3_{i,j}$, $S4_{i,j}$ and $Z1_{i,j}$, $Z2_{i,j}$, $Z3_{i,j}$, $Z4_{i,j}$ coefficients are defined as:

$$S1_{i,j} \equiv \frac{n_{i+1,j+1}^2}{n_{i+1,j}^2} - 1; \tag{3.46a}$$

$$S2_{i,j} \equiv \frac{n_{i+1,j-1}^2}{n_{i+1,j}^2} - 1; \tag{3.46b}$$

$$S3_{i,j} \equiv \frac{n_{i-1,j+1}^2}{n_{i-1,j}^2} - 1; \tag{3.46c}$$

$$S4_{i,j} \equiv \frac{n_{i-1,j-1}^2}{n_{i-1,j}^2} - 1; \tag{3.46d}$$

$$Z1_{i,j} \equiv \frac{n_{i+1,j+1}^2}{n_{i,j+1}^2} - 1; \tag{3.46e}$$

$$Z2_{i,j} \equiv \frac{n_{i-1,j+1}^2}{n_{i,j+1}^2} - 1; \tag{3.46f}$$

$$Z3_{i,j} \equiv \frac{n_{i+1,j-1}^2}{n_{i,j-1}^2} - 1; \tag{3.46g}$$

$$Z4_{i,j} \equiv \frac{n_{i-1,j-1}^2}{n_{i,j-1}^2} - 1; \tag{3.46h}$$

The application of the ADI method to the wave equation (3.36) using the operator decomposition (3.37) will be applied first to the semi-vectorial formulation where the cross-coupling terms can be neglected. Then the scalar approach will be presented as a special case where the polarization dependence of the 3D-problem can be ignored. Finally, the ADI method is applied to the full vectorial wave equation, which includes polarization dependence and polarization coupling. Table 3.1 shows a summary of the different approaches that can be applied to 3D-BPM, with different levels of approach to the polarization character of the simulated beam light.

Table 3.1 Levels of approach to the polarization character of the simulated light in 3D-BPM numerical formalisms

	Scalar	Semi-vectorial	Full vectorial
Polarization dependence	—	×	×
Polarization coupling	—	—	×
Field components	1	1	2
Modes	Unpolarized	x polarized or y polarized	Hybrid

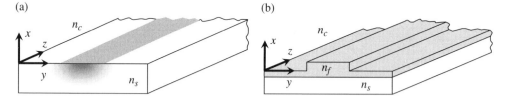

Figure 3.11 (a) Diffused channel waveguides in glasses show in general smooth index changes along the x and y directions, which allows neglecting coupling between orthogonal polarizations. (b) Rib-type waveguides in semiconductors show abrupt and strong index contrast at the boundaries, and therefore coupling between polarizations is present

3.2.1 Semi-Vectorial Formulation

The coupling between the x and y field components due to the waveguide structure is usually very weak, and the coupling terms can be ignored with good accuracy in most cases. Even in the presence of large index discontinuities such as semiconductor-air interfaces, this coupling is small (see Figure 3.11). The numerical scheme that neglects the coupling terms between orthogonal polarizations but keeps the other polarization dependent terms is called a semi-vectorial-beam propagation method (SV-BPM).

Regarding Eq. (3.34a, 3.34b), the cross-coupling terms in equation (C and D operators) are assumed negligible in the semi-vectorial formulation and the wave equations for the x and y components of the envelope fields are decoupled, resulting in:

$$2in_0k_0\frac{\partial \Psi_x}{\partial z} = \left(A_x + A_y\right)\Psi_x; \tag{3.47a}$$

$$2in_0k_0\frac{\partial \Psi_y}{\partial z} = \left(B_x + B_y\right)\Psi_y. \tag{3.47b}$$

These two uncoupled equations can be individually evaluated via the standard ADI method. When the major component of the electric field is in the y direction (following the geometry shown in Figure 3.11), the modes are called y-polarized modes or quasi-TE modes (to be consistent with the nomenclature used for 2D waveguides). On the contrary, if the major component of the electric field is E_x, they are called x-polarized modes or quasi-TM modes.

3.2.1.1 Quasi-TM Propagation

For Eq. (3.47a), the discretization form using the standard Crank–Nicolson scheme becomes:

$$2in_0k_0\frac{u^{m+1}-u^m}{\Delta z}=\left(A_x+A_y\right)\frac{u^{m+1}+u^m}{2}, \tag{3.48}$$

where u denotes the field component Ψ_x. Rearranging terms yields:

$$\left(\frac{2in_0k_0}{\Delta z}-\frac{A_x+A_y}{2}\right)u^{m+1}=\left(\frac{2in_0k_0}{\Delta z}+\frac{A_x+A_y}{2}\right)u^m, \tag{3.49}$$

which can be put in a more convenient form as:

$$\left(1+\frac{i\Delta z}{4n_0k_0}A_x+\frac{i\Delta z}{4n_0k_0}A_y\right)u^{m+1}=\left(1-\frac{i\Delta z}{4n_0k_0}A_x-\frac{i\Delta z}{4n_0k_0}A_y\right)u^m. \tag{3.50}$$

In order to implement the ADI method, we add second-order error terms in Δz to Eq. (3.50), yielding:

$$\left(1+\frac{i\Delta z}{4n_0k_0}A_x+\frac{i\Delta z}{4n_0k_0}A_y+\left(\frac{i\Delta z}{4n_0k_0}\right)^2A_xA_y\right)u^{m+1}\approx$$
$$\left(1-\frac{i\Delta z}{4n_0k_0}A_x-\frac{i\Delta z}{4n_0k_0}A_y+\left(\frac{i\Delta z}{4n_0k_0}\right)^2A_xA_y\right)u^m, \tag{3.51}$$

which allows us to factorize the operators in the form:

$$\left(1+\frac{i\Delta z}{4n_0k_0}A_x\right)\left(1+\frac{i\Delta z}{4n_0k_0}A_y\right)u^{m+1}=\left(1-\frac{i\Delta z}{4n_0k_0}A_x\right)\left(1-\frac{i\Delta z}{4n_0k_0}A_y\right)u^m. \tag{3.52}$$

Finally, we rewrite this equation as:

$$\left(\frac{4in_0k_0}{\Delta z}-A_x\right)\left(\frac{4in_0k_0}{\Delta z}-A_y\right)u^{m+1}=\left(\frac{4in_0k_0}{\Delta z}+A_x\right)\left(\frac{4in_0k_0}{\Delta z}+A_y\right)u^m. \tag{3.53}$$

To perform the propagation following the ADI method, this equation is solved in two sub-steps. First, the optical field at the transverse plane m is used to calculate the intermediate field at the intermediate position $m+1/2$, following the scheme:

$$\left(\frac{4in_0k_0}{\Delta z}-A_y\right)u^{m+1/2}=\left(\frac{4in_0k_0}{\Delta z}+A_y\right)u^m. \tag{3.54}$$

Then, to complete the propagation, a second step is performed, using the field previously calculated $u^{m+1/2}$ to finally obtain the field u^{m+1}:

$$\left(\frac{4in_0k_0}{\Delta z}-A_x\right)u^{m+1}=\left(\frac{4in_0k_0}{\Delta z}+A_x\right)u^{m+1/2}. \tag{3.55}$$

The great advantage of the ADI method is that the two sub-steps involve solving tridiagonal systems and thus the method has good, efficient performance, requiring moderate computational effort.

Now, we will describe the numerical schemes applied to the ADI method for the implementation of 3D-semi-vectorial SV-BPM. Regarding the first sub-step, the finite difference expressions of Eq. (3.54) can be written as:

$$
\frac{4in_0k_0}{\Delta z}u_{i,j}^{m+1/2} - \frac{u_{i,j-1}^{m+1/2}-2u_{i,j}^{m+1/2}+u_{i,j+1}^{m+1/2}}{(\Delta y)^2} - \frac{1}{2}k_0^2\left[\left(n_{i,j}^{m+1/2}\right)^2-n_0^2\right]u_{i,j}^{m+1/2}
$$
$$
= \frac{4in_0k_0}{\Delta z}u_{i,j}^{m} + \frac{u_{i,j-1}^{m}-2u_{i,j}^{m}+u_{i,j+1}^{m}}{(\Delta y)^2} + \frac{1}{2}k_0^2\left[\left(n_{i,j}^{m}\right)^2-n_0^2\right]u_{i,j}^{m}.
$$

(3.56)

The superscript m denotes the step along the longitudinal direction. The coefficients of this tridiagonal system are:

$$
a_j = -\frac{1}{(\Delta y)^2};
$$

(3.57a)

$$
b_j = \frac{4in_0k_0}{\Delta z} + \frac{2}{(\Delta y)^2} - \frac{1}{2}k_0^2\left[\left(n_{i,j}^{m+1/2}\right)^2-n_0^2\right];
$$

(3.57b)

$$
c_j = -\frac{1}{(\Delta y)^2};
$$

(3.57c)

$$
r_j = \frac{1}{(\Delta y)^2}\left[u_{i,j-1}^{m}+u_{i,j+1}^{m}\right] + \left\{\frac{4in_0k_0}{\Delta z} - \frac{2}{(\Delta y)^2} + \frac{1}{2}k_0^2\left[\left(n_{i,j}^{m}\right)^2-n_0^2\right]\right\}u_{i,j}^{m}.
$$

(3.57d)

In a very similar way, the *second sub-step* (Eq. (3.55)) derived from the application of the ADI method yields:

$$
\frac{4in_0k_0}{\Delta z}u_{i,j}^{m+1} - \frac{T_{i-1,j}^{m+1}u_{i-1,j}^{m+1}-2R_{i,j}^{m+1}u_{i,j}^{m+1}+T_{i+1,j}^{m+1}u_{i+1,j}^{m+1}}{(\Delta x)^2} - \frac{1}{2}k_0^2\left[\left(n_{i,j}^{m+1}\right)^2-n_0^2\right]u_{i,j}^{m+1}
$$
$$
= \frac{4in_0k_0}{\Delta z}u_{i,j}^{m+1/2} + \frac{T_{i-1,j}^{m+1/2}u_{i-1,j}^{m+1/2}-2R_{i,j}^{m+1/2}u_{i,j}^{m+1/2}+T_{i+1,j}^{m+1/2}u_{i+1,j}^{m+1/2}}{(\Delta x)^2}
$$
$$
+ \frac{1}{2}k_0^2\left[\left(n_{i,j}^{m+1/2}\right)^2-n_0^2\right]u_{i,j}^{m+1/2}
$$

(3.58)

The coefficients T and R are those defined in Eq. (3.42a,b). The coefficients for the tridiagonal system are now given by:

$$
a_i = -\frac{1}{(\Delta x)^2}T_{i-1,j}^{m+1};
$$

(3.59a)

$$b_i = \frac{4in_0k_0}{\Delta z} + \frac{2}{(\Delta x)^2}R_{i,j}^{m+1} - \frac{1}{2}k_0^2\left[\left(n_{i,j}^{m+1}\right)^2 - n_0^2\right]; \qquad (3.59b)$$

$$c_i = -\frac{1}{(\Delta x)^2}T_{i+1,j}^{m+1}; \qquad (3.59c)$$

$$r_i = \frac{1}{(\Delta x)^2}\left[T_{i-1,j}^{m+1/2}\,u_{i-1,j}^{m+1/2} + T_{i+1,j}^{m+1/2}\,u_{i+1,j}^{m+1/2}\right]$$
$$+\left\{\frac{4in_0k_0}{\Delta z} - \frac{2R_{i,j}^{m+1/2}}{(\Delta x)^2} + \frac{1}{2}k_0^2\left[\left(n_{i,j}^{m+1/2}\right)^2 - n_0^2\right]\right\}u_{i,j}^{m+1/2} \qquad (3.59d)$$

Following the semi-vectorial approach using the ADI-method here, the propagation constant and the field profile of the fundamental quasi-TM mode of a rib semiconductor waveguide is calculated. This example will serve to compare with the results obtained using a full vectorial-beam propagation method (FV-BPM) and to estimate the error of neglecting coupling between orthogonal polarizations. Also, this optical waveguide is a good example of polarization dependent structure, where the polarization direction of the light is of relevance.

The geometry of the rib waveguide is shown in Figure 3.12 (note the directions of the x and y axis, chosen to be consistent with the 2D waveguide analysis). The refractive indices of the cover, core and substrate are $n_1 = 1$, $n_2 = 3.44$, $n_3 = 3.36$ and the wavelength used is $\lambda = 1.55\,\mu m$. The height and the width of the rib are $h = 0.2\,\mu m$ and $w = 3\,\mu m$. The thickness of side slab is $d = 0.8\,\mu m$. The simulation is performed with transversal step sizes of $\Delta x = 0.02\,\mu m$ and $\Delta y = 0.1\,\mu m$, the window dimension being $L_x \times L_y = 4 \times 15\,\mu m^2$ (200×150 grid points in the computational region).

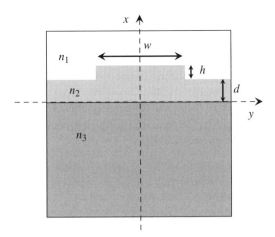

Figure 3.12 Semiconductor rib waveguide, showing the geometry and the relevant parameters

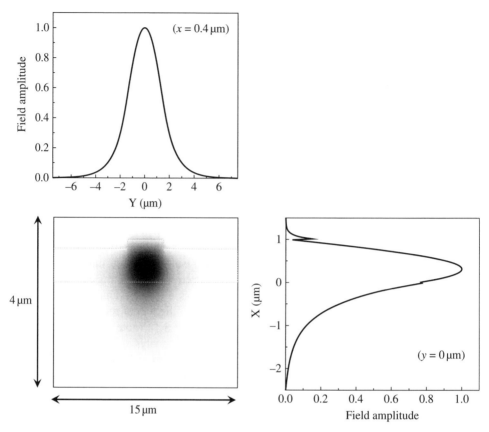

Figure 3.13   Field amplitude of the fundamental quasi-TM mode (x-polarized mode) obtained by semi-vectorial BPM based on the transverse electric field component E_x

Figure 3.13 shows the transversal field profile of the fundamental quasi-TM mode obtained by SV-BPM based on the E_x formulation. The upper graph is a plot of the E_x amplitude along the y-direction at $x = 0.4\,\mu m$ (centre of the slab region). The right graph shows the field profile along the x direction at $y = 0$, where it the discontinuity of the electric field is noted at the index discontinuities, being more pronounced at the rib–air interface ($x = 1\,\mu m$). The effective index calculated by the field distribution (using the Rayleigh quotient formula (2.123)) obtained after the imaginary distance BPM is $N_{eff}(TM) = 3.38916$, which is coincident to the value derived by the growth rate following the formula (2.118).

3.2.1.2   Quasi-TE Propagation

For quasi-TE propagation, the major transversal component of the electric field is Ψ_y. Therefore, the wave equation to be solved is Eq. (3.47b), which its discretization form using the Crank–Nicolson scheme is:

$$2in_0k_0\frac{v^{m+1}-v^m}{\Delta z}=(B_x+B_y)\frac{v^{m+1}+v^m}{2},\tag{3.60}$$

where, for the sake of clarity, v denotes the SVE-field component Ψ_y. Proceeding in a similar way as it was done before, and after adding second-order error terms, it yields:

$$\left(\frac{4in_0k_0}{\Delta z}-B_x\right)\left(\frac{4in_0k_0}{\Delta z}-B_y\right)v^{m+1}=\left(\frac{4in_0k_0}{\Delta z}+B_x\right)\left(\frac{4in_0k_0}{\Delta z}+B_y\right)v^m.\tag{3.61}$$

To perform the propagation the operator is decomposed in to two and the solution of this equation is split in two sub-steps. In the first step, the optical field at the transverse plane m (longitudinal position z) is used to calculate the intermediate field at the intermediate position $m+1/2$ by solving the following equation:

$$\left(\frac{4in_0k_0}{\Delta z}-B_y\right)v^{m+1/2}=\left(\frac{4in_0k_0}{\Delta z}+B_y\right)v^m.\tag{3.62}$$

The propagation for a distance Δz is completed by performing a second step using the previously obtained field $v^{m+1/2}$ to calculate the field v^{m+1} at the longitudinal position $z+\Delta z$:

$$\left(\frac{4in_0k_0}{\Delta z}-B_x\right)v^{m+1}=\left(\frac{4in_0k_0}{\Delta z}+B_x\right)v^{m+1/2}.\tag{3.63}$$

Regarding the first sub-step (Eq. (3.62)), the finite difference wave equation is given by:

$$\frac{4in_0k_0}{\Delta z}v_{i,j}^{m+1/2}-\frac{T_{i,j-1}^{m+1/2}v_{i,j-1}^{m+1/2}-2R_{i,j}^{m+1/2}v_{i,j}^{m+1/2}+T_{i,j+1}^{m+1/2}v_{i,j+1}^{m+1/2}}{(\Delta y)^2}-\frac{1}{2}k_0^2\left[\left(n_{i,j}^{m+1/2}\right)^2-n_0^2\right]v_{i,j}^{m+1/2}$$

$$=\frac{4in_0k_0}{\Delta z}v_{i,j}^m+\frac{T_{i,j-1}^m v_{i,j-1}^m-2R_{i,j}^m v_{i,j}^m+T_{i,j+1}^m v_{i,j+1}^m}{(\Delta y)^2}+\frac{1}{2}k_0^2\left[\left(n_{i,j}^m\right)^2-n_0^2\right]v_{i,j}^m$$

$$\tag{3.64}$$

The coefficients T and R are those defined in Eq. (3.44a,b). The coefficients of this set of linear equations are:

$$a_j=-\frac{1}{(\Delta y)^2}T_{i,j-1}^{m+1/2};\tag{3.65a}$$

$$b_j=\frac{4in_0k_0}{\Delta z}+\frac{2}{(\Delta y)^2}R_{i,j}^{m+1/2}-\frac{1}{2}k_0^2\left[\left(n_{i,j}^{m+1/2}\right)^2-n_0^2\right];\tag{3.65b}$$

$$c_j=-\frac{1}{(\Delta y)^2}T_{i,j+1}^{m+1/2};\tag{3.65c}$$

$$r_j = \frac{1}{(\Delta y)^2} \left[T_{i,j-1}^m \, v_{i,j-1}^m + T_{i,j+1}^m \, v_{i,j+1}^m \right] + \left\{ \frac{4in_0k_0}{\Delta z} - \frac{2R_{i,j}^m}{(\Delta y)^2} + \frac{1}{2}k_0^2 \left[\left(n_{i,j}^m \right)^2 - n_0^2 \right] \right\} v_{i,j}^m. \quad (3.65d)$$

The second sub-step (Eq. (3.63)) derived from the application of the ADI method is expressed in its finite difference form as:

$$\frac{4in_0k_0}{\Delta z} v_{i,j}^{m+1} - \frac{v_{i-1,j}^{m+1} - 2v_{i,j}^{m+1} + v_{i+1,j}^{m+1}}{(\Delta x)^2} - \frac{1}{2}k_0^2 \left[\left(n_{i,j}^{m+1} \right)^2 - n_0^2 \right] v_{i,j}^{m+1}$$

$$= \frac{4in_0k_0}{\Delta z} v_{i,j}^{m+1/2} + \frac{v_{i-1,j}^{m+1/2} - 2v_{i,j}^{m+1/2} + v_{i+1,j}^{m+1/2}}{(\Delta x)^2} + \frac{1}{2}k_0^2 \left[\left(n_{i,j}^{m+1/2} \right)^2 - n_0^2 \right] v_{i,j}^{m+1/2} \quad (3.66)$$

This tridiagonal matrix gives the following coefficients for the Thomas algorithm:

$$a_i = -\frac{1}{(\Delta x)^2}; \quad (3.67a)$$

$$b_i = \frac{4in_0k_0}{\Delta z} + \frac{2}{(\Delta x)^2} - \frac{1}{2}k_0^2 \left[\left(n_{i,j}^{m+1} \right)^2 - n_0^2 \right]; \quad (3.67b)$$

$$c_i = -\frac{1}{(\Delta x)^2}; \quad (3.67c)$$

$$r_i = \frac{1}{(\Delta x)^2} \left[v_{i-1,j}^{m+1/2} + v_{i+1,j}^{m+1/2} \right] + \left\{ \frac{4in_0k_0}{\Delta z} - \frac{2}{(\Delta x)^2} + \frac{1}{2}k_0^2 \left[\left(n_{i,j}^{m+1/2} \right)^2 - n_0^2 \right] \right\} v_{i,j}^{m+1/2} \quad (3.67d)$$

The semi-vectorial SV-BPM is applied to the calculation of the quasi-TE fundamental mode of the rib waveguide previously described. Figure 3.14 shows the field amplitude of the E_y transversal component obtained by imaginary distance BPM at 1.55 μm, using grid spacing of 0.02 and 0.1 μm along the vertical and horizontal directions (x- and y-axis), respectively. Now, at variance with the quasi-TM mode, the field profile along $y = 0$ is continuous at the index interfaces for TE polarized waves. Also, a value of N_{eff}(TE) = 3.39378 is obtained, which is different to that found for the fundamental quasi-TM mode and that clearly manifests the form-induced birefringent of this high-index contrast guiding structure. Thus, the SV-BPM takes into account the vectorial nature of the electromagnetic waves in 3D guiding structures, where the polarization of the light is considered.

In the next example, the coupling length of two-coupled rib waveguides presented in Figure 3.15 is analysed. The single rib structure is that shown in previous examples and the working wavelength is also set to 1.55 μm. SV-BPM is used to determine the coupling length of the double-rib waveguide by calculating the effective modal indices of the even (symmetric) and odd (anti-symmetric) eigenmodes of the structure as functions of the separation between the guides, and for the two orthogonal polarized modes (TM or quasi x-polarized mode and TE or quasi-y-polarized mode).

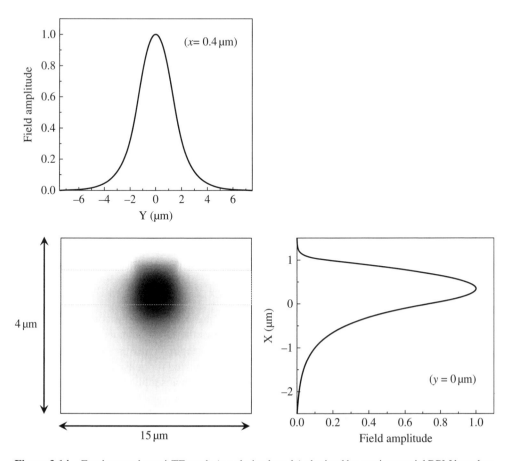

Figure 3.14 Fundamental quasi-TE mode (y-polarized mode) obtained by semi-vectorial BPM based on the transverse electric field component E_y

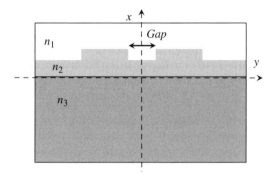

Figure 3.15 Coupled rib waveguides

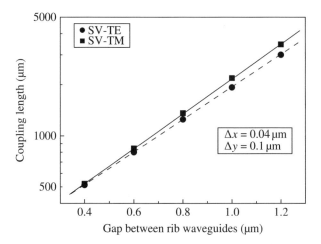

Figure 3.16 Coupling length of two-identical rib-waveguides as a function of their separation, for quasi-TE and quasi-TM polarization

Once the effective indices of the double-structure are calculated, the coupling length is obtained by using the relation (2.52). Figure 3.16 shows the results obtained by SV-TE and SV-TM 3D-BPM analysis. As expected, the coupling length increases exponentially with the waveguides' separation. Also, the coupling length calculated for quasi-TM modes is longer than the obtained for quasi-TE modes. This difference is because the evanescent field of the modes depends on their polarization, which is particularly relevant for strong index contrast structure. Thus, it is clear that a correct analysis of directional coupler in semiconductors must be carried out using numerical approaches that take into account the polarization character of the light, as is the case of the SV-BPM presented here.

3.2.2 Scalar Approach

The scalar approximation consists of taking into consideration that the polarization dependence of the problem is neglected, which implies that all derivatives of the field are continuous across all dielectric interfaces. Thus, in the scalar approach there is no distinction between quasi-TE or quasi-TM polarization and one can consider a single scalar field for the description of the optical propagation.

To obtain the wave equation for performing scalar 3D-BPM it is enough to consider one of the Eqs. (3.47a) or (3.47b), and put $T_{i,j} = R_{i,j} = 1$ in the differential operators. Nevertheless, we will present the algorithm for scalar-BPM using the scheme parameter α for controlling the propagation, as it was done in 2D-BPM. In this case the wave equation that governs the SVE scalar field $u(x,y,z)$ is given by:

$$2in_0k_0\frac{\partial u}{\partial z} = Gu,\tag{3.68}$$

where the differential operator G is given by:

$$Gu \equiv \frac{\partial^2 u}{\partial x^2} + \frac{\partial^2 u}{\partial y^2} + k_0^2\left(n^2 - n_0^2\right)u. \tag{3.69}$$

In order to solve this differential equation using the ADI method, the differential operator G is decomposed in to two terms:

$$G = G_x + G_y, \tag{3.70}$$

defined as follows:

$$G_x u \equiv \frac{\partial^2 u}{\partial x^2} + \frac{1}{2}k_0^2\left(n^2 - n_0^2\right)u; \tag{3.71a}$$

$$G_y u \equiv \frac{\partial^2 u}{\partial y^2} + \frac{1}{2}k_0^2\left(n^2 - n_0^2\right)u. \tag{3.71b}$$

The discretization form of Eq. (3.68) after applying the Crank–Nicolson scheme yields:

$$2in_0 k_0 \frac{u^{m+1} - u^m}{\Delta z} = \alpha\left(G_x + G_y\right)u^{m+1} + (1-\alpha)\left(G_x + G_y\right)u^m, \tag{3.72}$$

and rearranging terms:

$$\left(\frac{2in_0 k_0}{\Delta z} - \alpha\left(G_x + G_y\right)\right)u^{m+1} = \left(\frac{2in_0 k_0}{\Delta z} + (1-\alpha)\left(G_x + G_y\right)\right)u^m, \tag{3.73}$$

or:

$$\left(1 + \frac{i\alpha\Delta z}{2n_0 k_0}G_x + \frac{i\alpha\Delta z}{2n_0 k_0}G_y\right)u^{m+1} = \left(1 - \frac{i(1-\alpha)\Delta z}{2n_0 k_0}G_x - \frac{i(1-\alpha)\Delta z}{2n_0 k_0}G_y\right)u^m. \tag{3.74}$$

In order to implement the ADI method, we add second-order error terms to Eq. (3.74), resulting in:

$$\left(1 + \frac{i\alpha\Delta z}{2n_0 k_0}G_x + \frac{i\alpha\Delta z}{2n_0 k_0}G_y + \left(\frac{i\alpha\Delta z}{2n_0 k_0}\right)^2 G_x G_y\right)u^{m+1} \approx$$
$$\left(1 - \frac{i\alpha\Delta z}{2n_0 k_0}G_x - \frac{i\alpha\Delta z}{2n_0 k_0}G_y + \left(\frac{i\alpha\Delta z}{2n_0 k_0}\right)^2 G_x G_y\right)u^m \tag{3.75}$$

The introduction of the extra term permits factorizing the operators, which allows us to rewrite this equation as:

$$\left(1 + \frac{i\alpha\Delta z}{2n_0 k_0}G_x\right)\left(1 + \frac{i\alpha\Delta z}{2n_0 k_0}G_y\right)u^{m+1} = \left(1 - \frac{i(1-\alpha)\Delta z}{2n_0 k_0}G_x\right)\left(1 - \frac{i(1-\alpha)\Delta z}{2n_0 k_0}G_y\right)u^m. \tag{3.76}$$

Finally, this equation is put in a more convenient form as:

$$\left(\frac{2in_0k_0}{\Delta z}-\alpha G_x\right)\left(\frac{2in_0k_0}{\Delta z}-\alpha G_y\right)u^{m+1}=\left(\frac{2in_0k_0}{\Delta z}+(1-\alpha)G_x\right)\left(\frac{2in_0k_0}{\Delta z}+(1-\alpha)G_y\right)u^m.$$

(3.77)

As was done previously, this equation is solved in two sub-steps by splitting the operators. First, the optical field at the transverse plane m is used to calculate the intermediate field at the intermediate position $m+1/2$, following the scheme:

$$\left(\frac{2in_0k_0}{\Delta z}-\alpha G_y\right)u^{m+1/2}=\left(\frac{2in_0k_0}{\Delta z}+(1-\alpha)G_y\right)u^m.$$

(3.78)

Then, to complete the propagation, the second step is performed, using the field previously calculated $u^{m+1/2}$ to finally obtain the field u^{m+1}:

$$\left(\frac{2in_0k_0}{\Delta z}-\alpha G_x\right)u^{m+1}=\left(\frac{2in_0k_0}{\Delta z}+(1-\alpha)G_x\right)u^{m+1/2}.$$

(3.79)

Once more, the ADI method in each of the two sub-steps involves solving tridiagonal systems, being thus quite numerically efficient.

The finite difference approximations of the wave equation (3.78) (first sub-step) following the Crank–Nicolson scheme is given by:

$$\frac{2in_0k_0}{\Delta z}u_{i,j}^{m+1/2}-\alpha\frac{u_{i,j-1}^{m+1/2}-2u_{i,j}^{m+1/2}+u_{i,j+1}^{m+1/2}}{(\Delta y)^2}-\frac{\alpha}{2}k_0^2\left[\left(n_{i,j}^{m+1/2}\right)^2-n_0^2\right]u_{i,j}^{m+1/2}$$

$$=\frac{2in_0k_0}{\Delta z}u_{i,j}^m+(1-\alpha)\frac{u_{i,j-1}^m-2u_{i,j}^m+u_{i,j+1}^m}{(\Delta y)^2}+\frac{(1-\alpha)}{2}k_0^2\left[\left(n_{i,j}^m\right)^2-n_0^2\right]u_{i,j}^m$$

(3.80)

where the superscript m denotes the step along the longitudinal direction. The coefficients of this tridiagonal system are:

$$a_j=-\frac{\alpha}{(\Delta y)^2};$$

(3.81a)

$$b_j=\frac{2in_0k_0}{\Delta z}+\frac{2\alpha}{(\Delta y)^2}-\frac{\alpha}{2}k_0^2\left[\left(n_{i,j}^{m+1/2}\right)^2-n_0^2\right];$$

(3.81b)

$$c_j=-\frac{\alpha}{(\Delta y)^2};$$

(3.81c)

$$r_j=\frac{(1-\alpha)}{(\Delta y)^2}\left[u_{i,j-1}^m+u_{i,j+1}^m\right]+\left\{\frac{2in_0k_0}{\Delta z}-\frac{2(1-\alpha)}{(\Delta y)^2}+\frac{(1-\alpha)}{2}k_0^2\left[\left(n_{i,j}^m\right)^2-n_0^2\right]\right\}u_{i,j}^m.$$

(3.81d)

The finite difference corresponding to the second sub-step (Eq. (3.79)) derived from the application of the ADI method yields:

$$
\frac{2in_0k_0}{\Delta z}u_{i,j}^{m+1} - \alpha\frac{u_{i-1,j}^{m+1} - 2u_{i,j}^{m+1} + u_{i+1,j}^{m+1}}{(\Delta x)^2} - \frac{\alpha}{2}k_0^2\left[\left(n_{i,j}^{m+1}\right)^2 - n_0^2\right]u_{i,j}^{m+1}
$$

$$
= \frac{2in_0k_0}{\Delta z}u_{i,j}^{m+1/2} + (1-\alpha)\frac{u_{i-1,j}^{m+1/2} - 2u_{i,j}^{m+1/2} + u_{i+1,j}^{m+1/2}}{(\Delta x)^2} + \frac{(1-\alpha)}{2}k_0^2\left[\left(n_{i,j}^{m+1/2}\right)^2 - n_0^2\right]u_{i,j}^{m+1/2}
$$

$$(3.82)$$

From where the Thomas algorithm can be applied with the following tridiagonal system coefficients:

$$
a_i = -\frac{\alpha}{(\Delta x)^2};
\tag{3.83a}
$$

$$
b_i = \frac{2in_0k_0}{\Delta z} + \frac{2\alpha}{(\Delta x)^2} - \frac{\alpha}{2}k_0^2\left[\left(n_{i,j}^{m+1}\right)^2 - n_0^2\right];
\tag{3.83b}
$$

$$
c_i = -\frac{\alpha}{(\Delta x)^2};
\tag{3.83c}
$$

$$
r_i = \frac{(1-\alpha)}{(\Delta x)^2}\left[u_{i-1,j}^m + u_{i+1,j}^m\right] + \left\{\frac{2in_0k_0}{\Delta z} - \frac{2(1-\alpha)}{(\Delta x)^2} + \frac{(1-\alpha)}{2}k_0^2\left[\left(n_{i,j}^{m+1/2}\right)^2 - n_0^2\right]\right\}u_{i,j}^m.
\tag{3.83d}
$$

The scalar approach to simulate light propagation in waveguides is a good approach to simulate waveguide optical circuits in glasses. Usually, waveguides in multicomponent glasses are fabricated by diffusion processes, which give rise to graded refractive index profiles. Thus low index contrast structures are obtained, where the scalar approximation is a good approach to the problem. Here, the light propagation in a graded-index channel waveguide forming an S-bend is simulated by using the scalar 3D-BPM formulation. The curved region is formed by two opposite circle arcs, which provides a lateral shift with respect to the input straight waveguide (Figure 3.17).

The refractive index of the waveguide beneath the surface ($x < 0$) is given by:

$$
n(x,y) = \begin{cases} n_s + \Delta n \cdot \exp\left[-(x^2/d^2)\right]\exp\left[-(y+W/2)^2/d^2\right] & y \leq -W/2 \\[2mm] n_s + \Delta n \cdot \exp\left[-(x^2/d^2)\right] & -W/2 < y < W/2 \\[2mm] n_s + \Delta n \cdot \exp\left[-(x^2/d^2)\right]\exp\left[-(y-W/2)^2/d^2\right] & y \geq W/2 \end{cases}
\tag{3.84}
$$

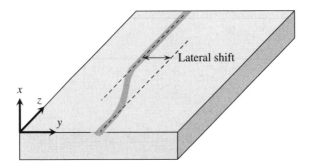

Figure 3.17   Geometry of the diffused channel waveguide forming an S-bend

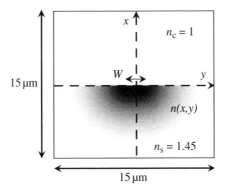

Figure 3.18   Refractive index map of the diffused channel waveguide

where $n_s = 1.45$ is the refractive index of the substrate, $\Delta n = 0.025$ the index increase due to the diffusion fabrication process, $d = 3\ \mu m$ the diffusion depth and $W = 1\ \mu m$ is the width of the mask used to perform the diffusion. The cover is assumed to be air, and thus $n(x,y) = 1$ for $x \geq 0$ (Figure 3.18).

The light propagation along the channel waveguide is performed using a working wavelength of $\lambda = 1.55\ \mu m$. The fundamental mode, obtained by imaginary distance 3D-BPM, is injected at the input and the fraction of guided power at the end of the curved region is evaluated by the overlapping integral between the calculated field at that position and the previously calculated waveguide eigenmode (see Appendices C and D). Figure 3.19 presents the results provided by scalar 3D-BPM of the resultant guided power as a function of the propagation step, for a transition length region of 800 μm and a lateral shift of 12.5 μm. The numerical simulation uses grid spacing of $\Delta x = \Delta y = 0.4\ \mu m$ and a computational window of 40×150 points along the x and y directions, respectively. The computational region is wide enough to include not only the structure, but also the evanescent field of the waveguide mode. At the edges of the domain, TBC (transparent boundary condition) boundary conditions are implemented, that allow the radiation mode to escape from the computational window.

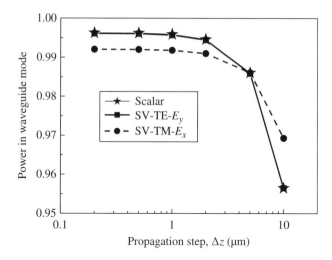

Figure 3.19   Calculated waveguide mode power that remains in the channel waveguide after the S-bend region, as a function of the propagation step used in the BPM simulations

The results in Figure 3.19 indicate that excellent accuracy is obtained by using propagation steps shorter than $\Delta z \approx 1\,\mu m$. Even with propagation a propagation step as large as $10\,\mu m$ the error in the guided power is only $\sim 4\%$ with respect to the asymptotic value. In order to see the validity of the scalar approach in this example, in the same graph the results of semi-vectorial quasi-x- and quasi-y-polarized BPM approaches have been plotted. The values provided by the scalar formalism are almost coincident with those calculated by the SV-quasi-TE (SV, semi-vectorial) formalism, and differs a $\sim 0.4\%$ with respect to the results obtained using SV-quasi-TM propagation. Thus, the scalar 3D-BPM is a suitable approach to simulate beam propagation in low-index contrast structures, such as optical waveguide circuits in dielectric materials fabricated by diffusion processes. Using these results, the scalar 3D-BPM is used to evaluate the losses induced by the curved waveguide as a function of the S-length extension (Figure 3.20) for a lateral shift of $12.5\,\mu m$. From that figure, it can be deduced that to maintain low losses in the S-bend, the transition must be longer than $400\,\mu m$.

Based on the S-bend waveguides, power splitters can be designed. Figure 3.21 shows a top view of the refractive index map of a Y-branch power splitter, having a transition length of $L = 800\,\mu m$ (from $z = 200$–$1000\,\mu m$) and a separation between branches of $D = 25\,\mu m$. The transversal index profile of the single waveguide (input and branches waveguides) has the functional form of expression (3.84) and is drawn in Figure 3.22.

The guided power at the parallel branches, after the Y-branch transition region, is calculated as a function of the Y-branch transition length L for a fixed separation between arms of $D = 25\,\mu m$. The simulation window uses grid spacing of $\Delta x = \Delta y = 0.4\,\mu m$, with 40×150 points along the x and y directions, respectively and the propagation step is $\Delta z = 1\,\mu m$. TBCs are applied to the computational window edges. Figure 3.23 indicates that a guided power greater than 95% is reached by designing a Y-branch transition length L longer than $700\,\mu m$. Shorter Y-branches lead to wide angles between arms, which provoke high losses at the transition.

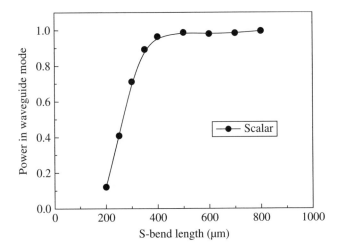

Figure 3.20   Calculated waveguide mode power after the S-bend region as a function of the transition S-bend length

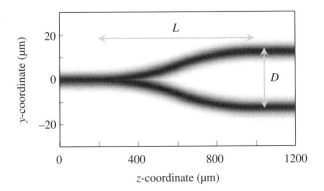

Figure 3.21   Refractive index map (below the surface at $x=0$), showing the geometry of the Y-branch power splitter. The transition starts at $z=200\,\mu m$ and ends at $z=1000\,\mu m$. The branches' separation is $D=25\,\mu m$

3.2.3   Full Vectorial BPM

In order to implement the ADI method for FV-BPM [6], we split the matrix operator in Eq. (3.37) as the sum of two matrix operators as follows:

$$2in_0k_0\frac{\partial}{\partial z}\begin{bmatrix}\Psi_x\\\Psi_y\end{bmatrix}=\begin{bmatrix}A_x+A_y & C\\D & B_x+B_y\end{bmatrix}\begin{bmatrix}\Psi_x\\\Psi_y\end{bmatrix}=\left(\begin{bmatrix}A_x & C\\0 & B_x\end{bmatrix}+\begin{bmatrix}A_y & 0\\D & B_y\end{bmatrix}\right)\begin{bmatrix}\Psi_x\\\Psi_y\end{bmatrix}. \qquad (3.85)$$

After discretizing this differential equation using the standard the Crank–Nicolson scheme, we obtain:

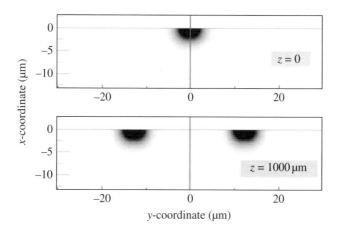

Figure 3.22 Transversal refractive index maps at $z = 0$ (upper figure, input single waveguide) and at $z = 1000\,\mu$m (bottom figure, parallel waveguides)

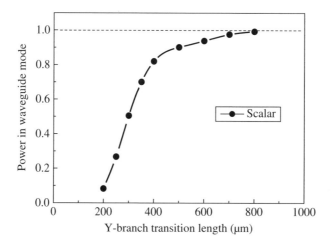

Figure 3.23 Power in waveguide mode at the output of the Y-branch as a function of the transition length L, for fixed arms' separation of $D = 25\,\mu$m

$$\frac{2in_0k_0}{\Delta z}\left(\begin{bmatrix} u \\ v \end{bmatrix}^{m+1} - \begin{bmatrix} u \\ v \end{bmatrix}^{m}\right) = \frac{1}{2}\left(\begin{bmatrix} A_x & C \\ 0 & B_x \end{bmatrix} + \begin{bmatrix} A_y & 0 \\ D & B_y \end{bmatrix}\right)\left(\begin{bmatrix} u \\ v \end{bmatrix}^{m+1} + \begin{bmatrix} u \\ v \end{bmatrix}^{m}\right), \qquad (3.86)$$

where we have denoted u and v as the x and y components of the electric field envelope, respectively. Adopting the ADI method, introducing second order error terms in Δz, the last equation can be written as:

$$\begin{bmatrix} u \\ v \end{bmatrix}^{m+1} = \frac{\left(1 - \dfrac{i\Delta z}{4n_0k_0}\begin{bmatrix} A_x & C \\ 0 & B_x \end{bmatrix}\right)}{\left(1 + \dfrac{i\Delta z}{4n_0k_0}\begin{bmatrix} A_x & C \\ 0 & B_x \end{bmatrix}\right)} \cdot \frac{\left(1 - \dfrac{i\Delta z}{4n_0k_0}\begin{bmatrix} A_y & 0 \\ D & B_y \end{bmatrix}\right)}{\left(1 + \dfrac{i\Delta z}{4n_0k_0}\begin{bmatrix} A_y & 0 \\ D & B_y \end{bmatrix}\right)} \cdot \begin{bmatrix} u \\ v \end{bmatrix}^{m}. \qquad (3.87)$$

In this form, this equation can be split in two artificial sub-steps to perform a single propagation step of length Δz, according to the sequence:

$$\begin{bmatrix} u \\ v \end{bmatrix}^{m+1/2} = \frac{\left(1 - \dfrac{i\Delta z}{4n_0 k_0} \begin{bmatrix} A_y & 0 \\ D & B_y \end{bmatrix}\right)}{\left(1 + \dfrac{i\Delta z}{4n_0 k_0} \begin{bmatrix} A_y & 0 \\ D & B_y \end{bmatrix}\right)} \cdot \begin{bmatrix} u \\ v \end{bmatrix}^{m}; \tag{3.88a}$$

$$\begin{bmatrix} u \\ v \end{bmatrix}^{m+1} = \frac{\left(1 - \dfrac{i\Delta z}{4n_0 k_0} \begin{bmatrix} A_x & C \\ 0 & B_x \end{bmatrix}\right)}{\left(1 + \dfrac{i\Delta z}{4n_0 k_0} \begin{bmatrix} A_x & C \\ 0 & B_x \end{bmatrix}\right)} \cdot \begin{bmatrix} u \\ v \end{bmatrix}^{m+1/2}. \tag{3.88b}$$

Equation (3.88a) is indeed the formal expression of the following equation:

$$\left(1 + \frac{i\Delta z}{4n_0 k_0} \begin{bmatrix} A_y & 0 \\ D & B_y \end{bmatrix}\right) \cdot \begin{bmatrix} u \\ v \end{bmatrix}^{m+1/2} = \left(1 - \frac{i\Delta z}{4n_0 k_0} \begin{bmatrix} A_y & 0 \\ D & B_y \end{bmatrix}\right) \cdot \begin{bmatrix} u \\ v \end{bmatrix}^{m}. \tag{3.89}$$

Now, this vectorial equation is separated into its two components. The first component is expressed as:

$$\left(1 + \frac{i\Delta z}{4n_0 k_0} A_y\right) u^{m+1/2} = \left(1 - \frac{i\Delta z}{4n_0 k_0} A_y\right) u^{m}. \tag{3.90a}$$

The second component is:

$$\left(1 + \frac{i\Delta z}{4n_0 k_0} B_y\right) v^{m+1/2} + \frac{i\Delta z}{4n_0 k_0} D u^{m+1/2} = \left(1 - \frac{i\Delta z}{4n_0 k_0} B_y\right) v^{m} - \frac{i\Delta z}{4n_0 k_0} D u^{m}. \tag{3.90b}$$

Equation (3.90a) allows us to obtain the x component of the intermediate field $u^{m+1/2}$ from the known field u^{m}. Then, Eq. (3.90b) is solved to find the y component of the intermediate field $v^{m+1/2}$ from the known fields u^{m}, $u^{m+1/2}$ and v^{m}. This concatenated sequence allows us to perform the first sub-step of the propagation from m to $(m + 1/2)$.

To complete the propagation along a distance Δz, it is necessary to carry out a second step. For that purpose, let us consider Eq. (3.88b), which can be expressed in the form:

$$\left(1 + \frac{i\Delta z}{4n_0 k_0} \begin{bmatrix} A_x & C \\ 0 & B_x \end{bmatrix}\right) \cdot \begin{bmatrix} u \\ v \end{bmatrix}^{m+1} = \left(1 - \frac{i\Delta z}{4n_0 k_0} \begin{bmatrix} A_x & C \\ 0 & B_x \end{bmatrix}\right) \cdot \begin{bmatrix} u \\ v \end{bmatrix}^{m+1/2}. \tag{3.91}$$

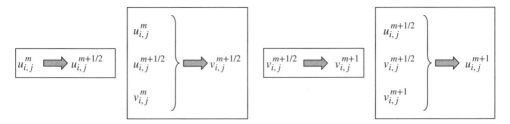

Figure 3.24 Sequence of the four steps to perform a single propagation step Δz using full-vectorial BPM, following the ADI method

Once again, this matrix equation is decomposed in its two components, resulting in:

$$\left(1 + \frac{i\Delta z}{4n_0k_0}A_x\right)u^{m+1} + \frac{i\Delta z}{4n_0k_0}Cv^{m+1} = \left(1 - \frac{i\Delta z}{4n_0k_0}A_x\right)u^{m+1/2} - \frac{i\Delta z}{4n_0k_0}Cv^{m+1/2}; \quad (3.92a)$$

$$\left(1 + \frac{i\Delta z}{4n_0k_0}B_x\right)v^{m+1} = \left(1 - \frac{i\Delta z}{4n_0k_0}B_x\right)v^{m+1/2}. \quad (3.92b)$$

Using Eq. (3.92b) allows us to calculate the field v^{m+1} from the known field $v^{m+1/2}$. Once this field is obtained, Eq. (3.92a) is used to find the field u^{m+1} from the known fields $u^{m+1/2}$, $v^{m+1/2}$ and v^{m+1}, completing thus the whole step from m to $m+1$. Figure 3.24 summarizes the sequence for performing a single longitudinal step.

Adopting this sequential ADI scheme the stability is assured and the four equations in finite differences derived from them are all tridiagonal, thus being quite an efficient way of solving the vectorial propagation of light in three-dimensional structures [6]. Let us see now how to proceed with each of these differential equations in terms of their finite difference versions.

3.2.3.1 First Component of the First Step

Regarding the first sub-step, the finite difference wave equation (3.90a), following the standard Crank–Nicolson scheme, can be rewritten in a more convenient form as:

$$\left(\frac{4in_0k_0}{\Delta z} - A_y\right)u^{m+1/2} = \left(\frac{4in_0k_0}{\Delta z} + A_y\right)u^m. \quad (3.93)$$

The finite difference form of the operator $A_y u$ has already been shown in Eq. (3.43a). Using that finite difference expression, Eq. (3.93) is finally given by:

$$\frac{4in_0k_0}{\Delta z}u_{i,j}^{m+1/2} - \frac{u_{i,j-1}^{m+1/2} - 2u_{i,j}^{m+1/2} + u_{i,j+1}^{m+1/2}}{(\Delta y)^2} - \frac{1}{2}k_0^2\left[\left(n_{i,j}^{m+1/2}\right)^2 - n_0^2\right]u_{i,j}^{m+1/2} = \frac{4in_0k_0}{\Delta z}u_{i,j}^m$$
$$+ \frac{u_{i,j-1}^m - 2u_{i,j}^m + u_{i,j+1}^m}{(\Delta y)^2} + \frac{1}{2}k_0^2\left[\left(n_{i,j}^m\right)^2 - n_0^2\right]u_{i,j}^m \quad (3.94)$$

From this equation, the coefficients of the tridiagonal linear system remain as:

$$a_j = -\frac{1}{(\Delta y)^2};$$

(3.95a)

$$b_j = \frac{4in_0k_0}{\Delta z} + \frac{2}{(\Delta y)^2} - \frac{1}{2}k_0^2\left[\left(n_{i,j}^{m+1/2}\right)^2 - n_0^2\right];$$

(3.95b)

$$c_j = -\frac{1}{(\Delta y)^2};$$

(3.95c)

$$r_j = \frac{1}{(\Delta y)^2}\left[u_{i,j-1}^m + u_{i,j+1}^m\right] + \left\{\frac{4in_0k_0}{\Delta z} - \frac{2}{(\Delta y)^2} + \frac{1}{2}k_0^2\left[\left(n_{i,j}^m\right)^2 - n_0^2\right]\right\}u_{i,j}^m.$$

(3.95d)

3.2.3.2 Second Component of the First Step

To establish the finite difference approximation of Eq. (3.90b), we put it in the form:

$$\left(\frac{4in_0k_0}{\Delta z} - B_y\right)v^{m+1/2} - Du^{m+1/2} = \left(\frac{4in_0k_0}{\Delta z} + B_y\right)v^m + Du^m,$$

(3.96)

and we use the finite difference forms of the operators $B_y v$ and Du as they were previously defined in Eqs. (3.43c) and (3.43b). Using these expressions, the finite difference equation corresponding to Eq. (3.96) is given by:

$$\frac{4in_0k_0}{\Delta z}v_{i,j}^{m+1/2} - \frac{T_{i,j-1}^{m+1/2}v_{i,j-1}^{m+1/2} - 2R_{i,j}^{m+1/2}v_{i,j}^{m+1/2} + T_{i,j+1}^{m+1/2}v_{i,j+1}^{m+1/2}}{(\Delta y)^2}$$

$$-\frac{1}{2}k_0^2\left[\left(n_{i,j}^{m+1/2}\right)^2 - n_0^2\right]v_{i,j}^{m+1/2}$$

$$= \frac{4in_0k_0}{\Delta z}v_{i,j}^m + \frac{T_{i,j-1}^m v_{i,j-1}^m - 2R_{i,j}^m v_{i,j}^m + T_{i,j+1}^m v_{i,j+1}^m}{(\Delta y)^2} + \frac{1}{2}k_0^2\left[\left(n_{i,j}^m\right)^2 - n_0^2\right]v_{i,j}^m$$

(3.97)

$$+ \frac{1}{4\Delta x\Delta y}\left(Z1_{i,j}^m u_{i+1,j+1}^m - Z2_{i,j}^m u_{i-1,j+1}^m - Z3_{i,j}^m u_{i+1,j-1}^m + Z4_{i,j}^m u_{i-1,j-1}^m\right)$$

$$+ \frac{1}{4\Delta x\Delta y}\left(Z1_{i,j}^{m+1/2} u_{i+1,j+1}^{m+1/2} - Z2_{i,j}^{m+1/2} u_{i-1,j+1}^{m+1/2} - Z3_{i,j}^{m+1/2} u_{i+1,j-1}^{m+1/2} + Z4_{i,j}^{m+1/2} u_{i-1,j-1}^{m+1/2}\right)$$

the coefficients being:

$$a_j = -\frac{1}{(\Delta y)^2}T_{i,j-1}^{m+1/2};$$

(3.98a)

$$b_j = \frac{4in_0k_0}{\Delta z} + \frac{2}{(\Delta y)^2}R_{i,j}^{m+1/2} - \frac{1}{2}k_0^2\left[\left(n_{i,j}^{m+1/2}\right)^2 - n_0^2\right]; \tag{3.98b}$$

$$c_j = -\frac{1}{(\Delta y)^2}T_{i,j+1}^{m+1/2}; \tag{3.98c}$$

$$r_j = \frac{1}{(\Delta y)^2}\left[T_{i,j-1}^m v_{i,j-1}^m + T_{i,j+1}^m v_{i,j+1}^m\right] + \left\{\frac{4in_0k_0}{\Delta z} - \frac{2R_{i,j}^m}{(\Delta y)^2} + \frac{1}{2}k_0^2\left[\left(n_{i,j}^m\right)^2 - n_0^2\right]\right\}v_{i,j}^m$$

$$+ \frac{1}{4\Delta x\Delta y}\left(Z1_{i,j}^m u_{i+1,j+1}^m - Z2_{i,j}^m u_{i-1,j+1}^m - Z3_{i,j}^m u_{i+1,j-1}^m + Z4_{i,j}^m u_{i-1,j-1}^m\right)$$

$$+ \frac{1}{4\Delta x\Delta y}\left(Z1_{i,j}^{m+1/2} u_{i+1,j+1}^{m+1/2} - Z2_{i,j}^{m+1/2} u_{i-1,j+1}^{m+1/2} - Z3_{i,j}^{m+1/2} u_{i+1,j-1}^{m+1/2} + Z4_{i,j}^{m+1/2} u_{i-1,j-1}^{m+1/2}\right)$$

$$\tag{3.98d}$$

The coefficients T, R and $Z1$–$Z4$ are those defined in Eqs. (3.44a,b) and (3.46a,b).

3.2.3.3 Second Component of the Second Step

Using Eq. (3.92b) in the form:

$$\left(\frac{4in_0k_0}{\Delta z} - B_x\right)v^{m+1} = \left(\frac{4in_0k_0}{\Delta z} + B_x\right)v^{m+1/2}, \tag{3.99}$$

its finite-difference approach is given by:

$$\frac{4in_0k_0}{\Delta z}v_{i,j}^{m+1} - \frac{v_{i-1,j}^{m+1} - 2v_{i,j}^{m+1} + v_{i+1,j}^{m+1}}{(\Delta x)^2} - \frac{1}{2}k_0^2\left[\left(n_{i,j}^{m+1}\right)^2 - n_0^2\right]v_{i,j}^{m+1} = \frac{4in_0k_0}{\Delta z}v_{i,j}^{m+1/2}$$

$$+ \frac{v_{i-1,j}^{m+1/2} - 2v_{i,j}^{m+1/2} + v_{i+1,j}^{m+1/2}}{(\Delta x)^2} - \frac{1}{2}k_0^2\left[\left(n_{i,j}^{m+1/2}\right)^2 - n_0^2\right]v_{i,j}^{m+1/2} \tag{3.100}$$

and the tridiagonal system coefficients take the expressions:

$$a_i = -\frac{1}{(\Delta x)^2}; \tag{3.101a}$$

$$b_i = \frac{4in_0k_0}{\Delta z} + \frac{2}{(\Delta x)^2} - \frac{1}{2}k_0^2\left[\left(n_{i,j}^{m+1}\right)^2 - n_0^2\right]; \tag{3.101b}$$

$$c_i = -\frac{1}{(\Delta x)^2}; \tag{3.101c}$$

$$r_i = \frac{1}{(\Delta x)^2}\left[v_{i-1,j}^{m+1/2} + v_{i+1,j}^{m+1/2}\right] + \left\{\frac{4in_0k_0}{\Delta z} - \frac{2}{(\Delta x)^2} + \frac{1}{2}k_0^2\left[\left(n_{i,j}^{m+1/2}\right)^2 - n_0^2\right]\right\}v_{i,j}^{m+1/2}. \tag{3.101d}$$

3.2.3.4 First Component of the Second Step

The last step to complete the propagation is to solve Eq. (3.92a), which is written in the form:

$$\left(\frac{4in_0k_0}{\Delta z} - A_x\right)u^{m+1} - Cv^{m+1} = \left(\frac{4in_0k_0}{\Delta z} + A_x\right)u^{m+1/2} + Cv^{m+1/2}, \tag{3.102}$$

and its finite difference expression is given by:

$$\frac{4in_0k_0}{\Delta z}u_{i,j}^{m+1} - \frac{T_{i-1,j}^{m+1}u_{i-1,j}^{m+1} - 2R_{i,j}^{m+1}u_{i,j}^{m+1} + T_{i+1,j}^{m+1}u_{i+1,j}^{m+1}}{(\Delta x)^2} - \frac{1}{2}k_0^2\left[\left(n_{i,j}^{m+1}\right)^2 - n_0^2\right]u_{i,j}^{m+1}$$

$$= \frac{4in_0k_0}{\Delta z}u_{i,j}^{m+1/2} + \frac{T_{i-1,j}^{m+1/2}u_{i-1,j}^{m+1/2} - 2R_{i,j}^{m+1/2}u_{i,j}^{m+1/2} + T_{i+1,j}^{m+1/2}u_{i+1,j}^{m+1/2}}{(\Delta x)^2}$$

$$+ \frac{1}{2}k_0^2\left[\left(n_{i,j}^{m+1/2}\right)^2 - n_0^2\right]u_{i,j}^{m+1/2} \tag{3.103}$$

$$+ \frac{1}{4\Delta x\Delta y}\left(S1_{i,j}^{m+1/2}v_{i+1,j+1}^{m+1/2} - S2_{i,j}^{m+1/2}v_{i+1,j-1}^{m+1/2} - S3_{i,j}^{m+1/2}v_{i-1,j+1}^{m+1/2} + S4_{i,j}^{m+1/2}v_{i-1,j-1}^{m+1/2}\right)$$

$$+ \frac{1}{4\Delta x\Delta y}\left(S1_{i,j}^{m+1}v_{i+1,j+1}^{m+1} - S2_{i,j}^{m+1}v_{i+1,j-1}^{m+1} - S3_{i,j}^{m+1}v_{i-1,j+1}^{m+1} + S4_{i,j}^{m+1}v_{i-1,j-1}^{m+1}\right)$$

The coefficients T, R and $S1$–$S4$ are those defined in Eqs. (3.42a,b) and (3.46a–h). From this finite difference equation, the coefficients of the tridiagonal system to be solved by the Thomas method are given by:

$$a_i = -\frac{1}{(\Delta x)^2}T_{i-1,j}^{m+1}; \tag{3.104a}$$

$$b_i = \frac{4in_0k_0}{\Delta z} + \frac{2}{(\Delta x)^2}R_{i,j}^{m+1} - \frac{1}{2}k_0^2\left[\left(n_{i,j}^{m+1}\right)^2 - n_0^2\right]; \tag{3.104b}$$

$$c_i = -\frac{1}{(\Delta x)^2}T_{i+1,j}^{m+1}; \tag{3.104c}$$

$$r_i = \frac{1}{(\Delta x)^2}\left[T_{i-1,j}^{m+1/2}u_{i-1,j}^{m+1/2} + T_{i+1,j}^{m+1/2}u_{i+1,j}^{m+1/2}\right]$$

$$+ \left\{\frac{4in_0k_0}{\Delta z} - \frac{2R_{i,j}^{m+1/2}}{(\Delta x)^2} + \frac{1}{2}k_0^2\left[\left(n_{i,j}^{m+1/2}\right)^2 - n_0^2\right]\right\}u_{i,j}^{m+1/2}$$

$$\tag{3.104d}$$

$$+ \frac{1}{4\Delta x\Delta y}\left(S1_{i,j}^{m+1/2}v_{i+1,j+1}^{m+1/2} - S2_{i,j}^{m+1/2}v_{i+1,j-1}^{m+1/2} - S3_{i,j}^{m+1/2}v_{i-1,j+1}^{m+1/2} + S4_{i,j}^{m+1/2}v_{i-1,j-1}^{m+1/2}\right)$$

$$+ \frac{1}{4\Delta x\Delta y}\left(S1_{i,j}^{m+1}v_{i+1,j+1}^{m+1} - S2_{i,j}^{m+1}v_{i+1,j-1}^{m+1} - S3_{i,j}^{m+1}v_{i-1,j+1}^{m+1} + S4_{i,j}^{m+1}v_{i-1,j-1}^{m+1}\right)$$

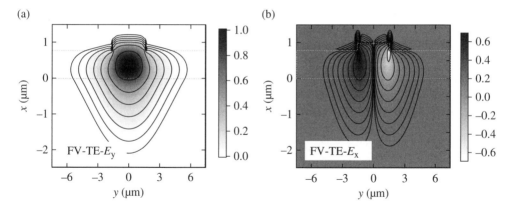

Figure 3.25 Maps of the field amplitude of the electric field y component (a) and x component (b) of the quasi-TE fundamental mode of the rib waveguide shown in Figure 3.12

(In Appendix E the FV-BPM algorithm using the scheme parameter is also developed.)

The rib waveguide structure examined in the previous section is now analysed by means of FV-BPM. To obtain the quasi-y-polarized mode, a y-polarized Gaussian beam is launched into the structure. By using imaginary distance FV-BPM, the y component of the electric field evolves until the fundamental mode is obtained (Figure 3.25a). Also, although initially there is no x-component field, due to the cross coupling terms of the full-wave equation an x-polarized field grows until it reaches the mode pattern corresponding to the fundamental mode (Figure 3.25b). The ratio of the maximum field amplitudes between the y and x components is around $\sim$2.2, the E_x-field being concentrated at the index discontinuities (corners of the rib structure). Although the waveguide structure presents strong index discontinuities, the mode is predominantly y polarized and therefore it is called the quasi-y-polarized mode or quasi-TE mode.

Now launching an x-polarized Gaussian beam, the fundamental quasi-TM mode is obtained by performing the FV-BPM along the imaginary z-axis. The field amplitudes corresponding with the x and y components of the fundamental mode are plotted in Figure 3.26. Now the major component corresponds with the x-electric field component, the y component being the minor component, with a ratio between the maximum amplitudes of $\sim$8.1. This mode is also called a quasi-x-polarized mode

The influence of the propagation step on the correct field distributions obtained by Im-Dis-BPM can be evaluated by calculating the correlation between the calculated field $u(x,y)$ and the field eigenmode $\Psi(x,y)$, defined by Ref. [8] (see Appendix C):

$$\Gamma \equiv \frac{\iint u(x,y)\Psi^*(x,y)dxdy}{\sqrt{\iint |u(x,y)|^2 dxdy}\sqrt{\iint |\Psi(x,y)|^2 dxdy}}.$$ (3.105)

Using this normalized overlap integral, the error in the calculated field can be then established as:

$$Error \equiv 1 - |\Gamma|^2.$$ (3.106)

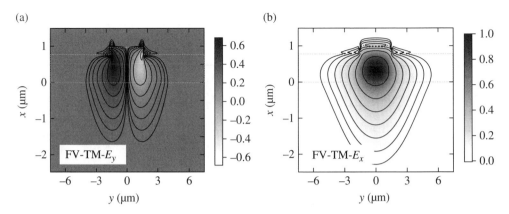

Figure 3.26 Maps of the field amplitude of the electric field y-component (a) and x component (b) of the quasi-TM fundamental mode of the rib waveguide shown in Figure 3.12

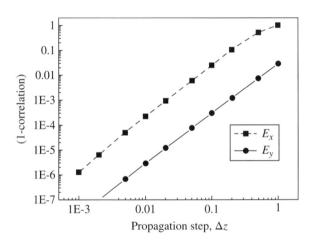

Figure 3.27 Error in the field distribution for the x and y component of the electric field of the fundamental mode in the rib waveguide shown in Figure 3.12, calculated by propagation along the imaginary z-axis using FV-BPM formulation, as a function of the propagation step

Figure 3.27 shows the error found in the evaluation of the major and minor components by imaginary-distance BPM, as a function of the propagation step used to perform the simulation. As the propagation step decreases, the error decreases and for a given propagation step the achieved accuracy for the major component is always superior that for the minor component. Also, decreasing the propagation step by an order of magnitude improves the calculated fields by at least 2 orders of magnitude.

The diffused waveguide presented in Figure 3.18, which was analysed by scalar-3D-BPM, is now examined by FV-BPM. This formalism also allows us to obtain the minor field component

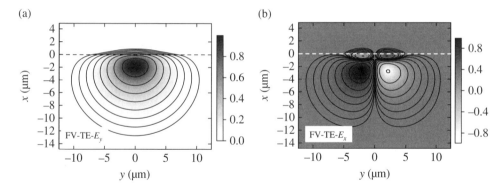

Figure 3.28 Maps of the field amplitude of the electric field y component (a) and x component (b) of the quasi-TE fundamental mode of the diffused waveguide shown in Figure 3.18

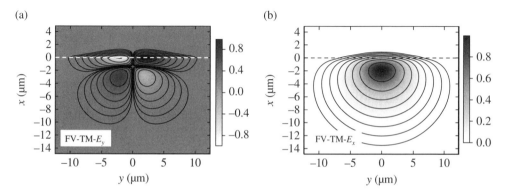

Figure 3.29 Maps of the field amplitude of the electric field y component (a) and x component (b) of the quasi-TM fundamental mode of the diffused waveguide shown in Figure 3.18

of low-index contrast structures. The fundamental modes for quasi-TE and quasi-TM polarizations are presented in Figures 3.28 and 3.29, respectively, where both major and minor components are plotted. Let us note that the field map for the E_y component (major component) of the TE mode is very similar to the E_x component (major component) of the TM-mode, indicating that the polarization dependence of low-contrast-index structures is very weak. This low-influence on the polarization can be also deduced from the ratios between the major and minor field components. These ratios are ∼480 for quasi-TE and ∼160 for quasi-TM fundamental modes, and thus the contributions of the minor components are almost negligible (let us compare with the values of 2.2 and 8.1 found for the rib structure previously examined).

As a second example, Figure 3.30 shows the electric field pattern of the fundamental HE_{11}-mode of a step-index circular fibre, which corresponds to a standard single mode fibre at $\lambda = 1.55\,\mu m$. The refractive indices of the core and cladding are $n_1 = 1.4504$ and $n_2 = 1.4447$, respectively, the radius being $a = 4\,\mu m$. The fundamental mode is obtained by propagating an arbitrary field along the imaginary z-axis using FV-BPM. The width of the squared computational window is $37.5\,\mu m$ and 150 discretization points are chosen in each direction, resulting

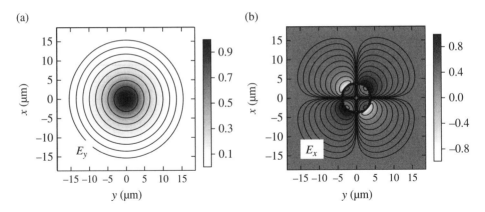

Figure 3.30 Maps of the electric field amplitude of the y component (a) and x component (b) of the HE_{11} fundamental mode of a step-index circular fibre

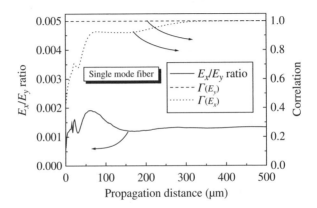

Figure 3.31 Evolution of the ratio between the maximum amplitudes of the minor and the major components of the electric field during the propagation (continuous line). Dashed and dotted lines represent the correlation between the computed field components and the HE_{11} eigenmode of the single-mode step index optical fibre as a function of the propagation distance

in a grid spacing of $\Delta x = \Delta y = 0.25\ \mu m$. The FV-BPM correctly generates both the major and the minor field components, being the ratio between the maximum amplitudes of $E_y/E_x \sim 740$. The effective index calculated by both the field growth rate and the Rayleigh quotient is $N_{eff}(HE_{11}) = 1.447238$.

The propagation along the real z-axis is now examined. For that purpose, a y-polarized beam (corresponding to the exact field distribution of the fundamental mode) is launched into the single-mode step-index optical fibre and the propagation is performed using a longitudinal step of $\Delta z = 0.1\ \mu m$. Figure 3.31 shows the ratio between the maximum values of both components (continuous line). The minor component growths quickly in the first $\sim 20\ \mu m$ and after a propagation length of $\sim 300\ \mu m$ the ratio between the amplitudes of the minor and major components almost reaches its asymptotic value (~ 0.00135).

The steady state can alternatively be monitored by computing the correlation between the actual field distribution and the corresponding eigenmode defined in formula (3.105). In the same Figure 3.31 the modulus of the correlation for the major and minor components of the electric field has been plotted as a function of the propagation distance (dashed and dotted lines, respectively). As expected, the field distribution of the major component remains unchanged during the propagation. On the other hand, as the minor component increases its amplitude with the advance in propagation (continuous line), it accommodates to the field eigenmode and for distances beyond $\sim300\,\mu m$ the correlation reaches a value close to unity (dotted line).

3.3 Three-Dimensional BPM Based on the Magnetic Field

As already mentioned, the description of light propagation in waveguide structures can also be numerically simulated by means of the transverse magnetic field components [3, 8]. Here, we derive the FD-BPM algorithms based on the magnetic field components under the approximations of semi-vectorial (neglecting polarization coupling) and full vectorial approaches, both of them making use of the ADI method.

The starting equations to describe the light propagation using the SVE transversal magnetic field components are given by [3]:

$$2in_0k_0\frac{\partial\Phi_x}{\partial z} = Q_{xx}\Phi_x + Q_{xy}\Phi_y; \tag{3.107a}$$

$$2in_0k_0\frac{\partial\Phi_y}{\partial z} = Q_{yy}\Phi_y + Q_{yx}\Phi_x, \tag{3.107b}$$

where the differential operators Q_{ij} were defined as:

$$Q_{xx}\Phi_x \equiv \frac{\partial^2\Phi_x}{\partial x^2} + n^2\frac{\partial}{\partial y}\left(\frac{1}{n^2}\frac{\partial\Phi_x}{\partial y}\right) + k_0^2\left(n^2 - n_0^2\right)\Phi_x; \tag{3.108a}$$

$$Q_{yy}\Phi_y \equiv n^2\frac{\partial}{\partial x}\left(\frac{1}{n^2}\frac{\partial\Phi_y}{\partial x}\right) + \frac{\partial^2\Phi_y}{\partial y^2} + k_0^2\left(n^2 - n_0^2\right)\Phi_y; \tag{3.108b}$$

$$Q_{xy}\Phi_y \equiv -n^2\frac{\partial}{\partial y}\left(\frac{1}{n^2}\frac{\partial\Phi_y}{\partial x}\right) + \frac{\partial^2\Phi_y}{\partial y\partial x}; \tag{3.108c}$$

$$Q_{yx}\Phi_x \equiv -n^2\frac{\partial}{\partial x}\left(\frac{1}{n^2}\frac{\partial\Phi_x}{\partial y}\right) + \frac{\partial^2\Phi_x}{\partial x\partial y}. \tag{3.108d}$$

Equation (3.124) can be put in a more compact form as:

$$2in_0k_0\frac{\partial}{\partial z}\begin{bmatrix}\Phi_x\\\Phi_y\end{bmatrix} = \begin{bmatrix}Q_{xx} & Q_{xy}\\Q_{yx} & Q_{yy}\end{bmatrix}\begin{bmatrix}\Phi_x\\\Phi_y\end{bmatrix}. \tag{3.109}$$

Now, we will proceed in a similar way as we did in Section 3.2 for the BPM based on the electric field. To proceed with, first the matrix operator in Eq. (3.109) is expressed as:

$$
\begin{bmatrix} Q_{xx} & Q_{xy} \\ Q_{yx} & Q_{yy} \end{bmatrix} = \begin{bmatrix} A_x + A_y & C \\ D & B_x + B_y \end{bmatrix}, \tag{3.110}
$$

where A_x and A_y denote the x- and y-dependent parts of the operator Q_{xx}, respectively, with:

$$
A_x \Phi_x \equiv \frac{\partial^2 \Phi_x}{\partial x^2} + \frac{1}{2} k_0^2 \left(n^2 - n_0^2 \right) \Phi_x; \tag{3.111a}
$$

$$
A_y \Phi_x \equiv n^2 \frac{\partial}{\partial y} \left(\frac{1}{n^2} \frac{\partial \Phi_x}{\partial y} \right) + \frac{1}{2} k_0^2 \left(n^2 - n_0^2 \right) \Phi_x. \tag{3.111b}
$$

Also, the operators B_x and B_y contain the x- and y-dependent parts of Q_{yy}, respectively, with:

$$
B_x \Phi_y \equiv n^2 \frac{\partial}{\partial x} \left(\frac{1}{n^2} \frac{\partial \Phi_y}{\partial x} \right) + \frac{1}{2} k_0^2 \left(n^2 - n_0^2 \right) \Phi_y; \tag{3.112a}
$$

$$
B_y \Phi_y \equiv \frac{\partial^2 \Phi_y}{\partial y^2} + \frac{1}{2} k_0^2 \left(n^2 - n_0^2 \right) \Phi_y. \tag{3.112b}
$$

Additionally, the differential operators C and D denote the cross-coupling terms and they are just the Q_{xy} and Q_{yx} operators, respectively:

$$
C \Phi_y \equiv -n^2 \frac{\partial}{\partial y} \left(\frac{1}{n^2} \frac{\partial \Phi_y}{\partial x} \right) + \frac{\partial^2 \Phi_y}{\partial y \partial x}; \tag{3.113a}
$$

$$
D \Phi_x \equiv -n^2 \frac{\partial}{\partial x} \left(\frac{1}{n^2} \frac{\partial \Phi_x}{\partial y} \right) + \frac{\partial^2 \Phi_x}{\partial x \partial y}. \tag{3.113b}
$$

As we did in the previous section, here we will denote u and v as the x- and y-components of the SV-envelope of the magnetic field. Using this nomenclature, the finite difference expressions of the operators $A_x u$ and $A_y u$ are implemented as:

$$
A_x u = \frac{u_{i-1,j} - 2 u_{i,j} + u_{i+1,j}}{(\Delta x)^2} + \frac{1}{2} k_0^2 \left[n_{i,j}^2 - n_0^2 \right] u_{i,j}; \tag{3.114a}
$$

$$
A_y u = \frac{T_{i,j-1} u_{i,j-1} - 2 R_{i,j} u_{i,j} + T_{i,j+1} u_{i,j+1}}{(\Delta y)^2} + \frac{1}{2} k_0^2 \left[n_{i,j}^2 - n_0^2 \right] u_{i,j}, \tag{3.114b}
$$

where the coefficients T and R are given by:

$$
T_{i,j \pm 1} \equiv \frac{2 n_{i,j}^2}{n_{i,j}^2 + n_{i,j \pm 1}^2}; \tag{3.115a}
$$

$$R_{i,j} \equiv \frac{n_{i,j}^2}{n_{i,j+1}^2 + n_{i,j}^2} + \frac{n_{i,j}^2}{n_{i,j}^2 + n_{i,j-1}^2} = \frac{T_{i,j+1} + T_{i,j-1}}{2}. \tag{3.115b}$$

On the other hand, the finite difference expressions of the operators $B_x v$ and $B_y v$ are implemented as:

$$B_x v = \frac{T_{i-1,j} v_{i-1,j} - 2R_{i,j} v_{i,j} + T_{i+1,j} v_{i+1,j}}{(\Delta x)^2} + \frac{1}{2} k_0^2 \left[n_{i,j}^2 - n_0^2 \right] v_{i,j}; \tag{3.116a}$$

$$B_y v = \frac{v_{i,j-1} - 2v_{i,j} + v_{i,j+1}}{(\Delta y)^2} + \frac{1}{2} k_0^2 \left[n_{i,j}^2 - n_0^2 \right] v_{i,j}, \tag{3.116b}$$

where we have defined the coefficients T and R by:

$$T_{i \pm 1,j} \equiv \frac{2 n_{i,j}^2}{n_{i,j}^2 + n_{i \pm 1,j}^2}; \tag{3.117a}$$

$$R_{i,j} \equiv \frac{n_{i,j}^2}{n_{i+1,j}^2 + n_{i,j}^2} + \frac{n_{i,j}^2}{n_{i,j}^2 + n_{i-1,j}^2} = \frac{T_{i+1,j} + T_{i-1,j}}{2}. \tag{3.117b}$$

Also, the finite difference forms of the C and D operators are expressed as:

$$Cv = \frac{1}{4 \Delta x \Delta y} \left(S1_{i,j} v_{i+1,j+1} - S2_{i,j} v_{i+1,j-1} - S3_{i,j} v_{i-1,j+1} + S4_{i,j} v_{i-1,j-1} \right); \tag{3.118a}$$

$$Du = \frac{1}{4 \Delta x \Delta y} \left(Z1_{i,j} u_{i+1,j+1} - Z2_{i,j} u_{i-1,j+1} - Z3_{i,j} u_{i+1,j-1} + Z4_{i,j} u_{i-1,j-1} \right), \tag{3.118b}$$

and the coefficients $S1$–$S4$ and $Z1$–$Z4$ are defined by:

$$S1_{i,j} \equiv 1 - \frac{n_{i,j}^2}{n_{i,j+1}^2}; \tag{3.119a}$$

$$S2_{i,j} \equiv 1 - \frac{n_{i,j}^2}{n_{i,j-1}^2}; \tag{3.119b}$$

$$S3_{i,j} \equiv 1 - \frac{n_{i,j}^2}{n_{i,j+1}^2}; \tag{3.119c}$$

$$S4_{i,j} \equiv 1 - \frac{n_{i,j}^2}{n_{i,j-1}^2}; \tag{3.119d}$$

$$Z1_{i,j} \equiv 1 - \frac{n_{i,j}^2}{n_{i+1,j}^2}; \tag{3.119e}$$

$$Z2_{i,j} \equiv 1 - \frac{n_{i,j}^2}{n_{i-1,j}^2}; \tag{3.119f}$$

$$Z3_{i,j} \equiv 1 - \frac{n_{i,j}^2}{n_{i+1,j}^2}; \tag{3.119g}$$

$$Z4_{i,j} \equiv 1 - \frac{n_{i,j}^2}{n_{i-1,j}^2}. \tag{3.119h}$$

To obtain a semi-vectorial formulation based on the magnetic field, the ADI method is applied to the wave equation (3.109) using the operator decomposition (3.110) and ignoring the cross-coupling terms. Later, the application of the ADI method to the full vectorial wave equation, now including the cross-coupling terms, will serve to obtain the full-vectorial FD-BPM algorithm based on the magnetic field components.

3.3.1 Semi-Vectorial Formulation

In the semi-vectorial formulation the cross-coupling terms (C and D) are assumed to be negligible and the wave equation (3.109) simplifies to:

$$2in_0k_0\frac{\partial \Phi_x}{\partial z} = \left(A_x + A_y\right)\Phi_x; \tag{3.120a}$$

$$2in_0k_0\frac{\partial \Phi_y}{\partial z} = \left(B_x + B_y\right)\Phi_y, \tag{3.120b}$$

where the $A_{x,y}$ and $B_{x,y}$ differential operators are those defined in Eqs. (3.111a,b) and (3.112a,b).

These two decoupled equations are the basics of the SV-BPM, which can be solved separately as quasi-TE and quasi-TM propagation.

3.3.1.1 Quasi-TE Propagation

Following the geometry of Figure 3.11, for quasi-TE propagation the major component of the magnetic field is the H_x component, where the minor H_y component is neglected. For this case, the wave equation to be solved is Eq. (3.120a). After introducing second order error in the propagation step and applying the ADI method, this results in:

$$\left(\frac{4in_0k_0}{\Delta z} - A_x\right)\left(\frac{4in_0k_0}{\Delta z} - A_y\right)u^{m+1} = \left(\frac{4in_0k_0}{\Delta z} + A_x\right)\left(\frac{4in_0k_0}{\Delta z} + A_y\right)u^m. \tag{3.121}$$

Here we have denoted u as the SVE of the magnetic field component Φ_x. To perform the propagation, this equation is solved in two sub-steps. First, the optical field at the transverse plane m is used to calculate the intermediate field at the fictitious position $m + 1/2$, following the next scheme:

$$\left(\frac{4in_0k_0}{\Delta z} - A_y\right)u^{m+1/2} = \left(\frac{4in_0k_0}{\Delta z} + A_y\right)u^m. \tag{3.122}$$

The propagation is completed by performing a second step, using the previously calculated field $u^{m+1/2}$ to finally obtain the field u^{m+1}:

$$\left(\frac{4in_0k_0}{\Delta z} - A_x\right)u^{m+1} = \left(\frac{4in_0k_0}{\Delta z} + A_x\right)u^{m+1/2}. \tag{3.123}$$

The finite-difference approximation of Eq. (3.122) allows us to perform the first sub-step, which yields:

$$\frac{4in_0k_0}{\Delta z}u_{i,j}^{m+1/2} - \frac{T_{i,j-1}^{m+1/2}u_{i,j-1}^{m+1/2} - 2R_{i,j}^{m+1/2}u_{i,j}^{m+1/2} + T_{i,j+1}^{m+1/2}u_{i,j+1}^{m+1/2}}{(\Delta y)^2} - \frac{k_0^2}{2}\left[\left(n_{i,j}^{m+1/2}\right)^2 - n_0^2\right]u_{i,j}^{m+1/2}$$

$$= \frac{4in_0k_0}{\Delta z}u_{i,j}^m + \frac{T_{i,j-1}^m u_{i,j-1}^m - 2R_{i,j}^m u_{i,j}^m + T_{i,j+1}^m u_{i,j+1}^m}{(\Delta y)^2} + \frac{1}{2}k_0^2\left[\left(n_{i,j}^m\right)^2 - n_0^2\right]u_{i,j}^m$$

$$\tag{3.124}$$

$T_{i,j}$ and $R_{i,j}$ being the coefficients defined in Eq. (3.115a,b). The coefficients of this tridiagonal system are:

$$a_j = -\frac{1}{(\Delta y)^2}T_{i,j-1}^{m+1/2}; \tag{3.125a}$$

$$b_j = \frac{4in_0k_0}{\Delta z} + \frac{2}{(\Delta y)^2}R_{i,j}^{m+1/2} - \frac{1}{2}k_0^2\left[\left(n_{i,j}^{m+1/2}\right)^2 - n_0^2\right]; \tag{3.125b}$$

$$c_j = -\frac{1}{(\Delta y)^2}T_{i,j+1}^{m+1/2}; \tag{3.125c}$$

$$r_j = \frac{T_{i,j-1}^m u_{i,j-1}^m + T_{i,j+1}^m u_{i,j+1}^m}{(\Delta y)^2} + \left\{\frac{4in_0k_0}{\Delta z} - \frac{2R_{i,j}^m}{(\Delta y)^2} + \frac{k_0^2}{2}\left[\left(n_{i,j}^m\right)^2 - n_0^2\right]\right\}u_{i,j}^m. \tag{3.125d}$$

The second sub-step is now performed by the application of the finite-difference approximations to Eq. (3.123), giving:

$$\frac{4in_0k_0}{\Delta z}u_{i,j}^{m+1} - \frac{u_{i-1,j}^{m+1} - 2u_{i,j}^{m+1} + u_{i+1,j}^{m+1}}{(\Delta x)^2} - \frac{1}{2}k_0^2\left[\left(n_{i,j}^{m+1}\right)^2 - n_0^2\right]u_{i,j}^{m+1}$$

$$= \frac{4in_0k_0}{\Delta z}u_{i,j}^{m+1/2} + \frac{u_{i-1,j}^{m+1/2} - 2u_{i,j}^{m+1/2} + u_{i+1,j}^{m+1/2}}{(\Delta x)^2} + \frac{1}{2}k_0^2\left[\left(n_{i,j}^{m+1/2}\right)^2 - n_0^2\right]u_{i,j}^{m+1/2} \tag{3.126}$$

the tridiagonal system coefficients being:

$$a_i = -\frac{1}{(\Delta x)^2}; \tag{3.127a}$$

$$b_i = \frac{4in_0k_0}{\Delta z} + \frac{2}{(\Delta x)^2} - \frac{1}{2}k_0^2\left[\left(n_{i,j}^{m+1}\right)^2 - n_0^2\right]; \tag{3.127b}$$

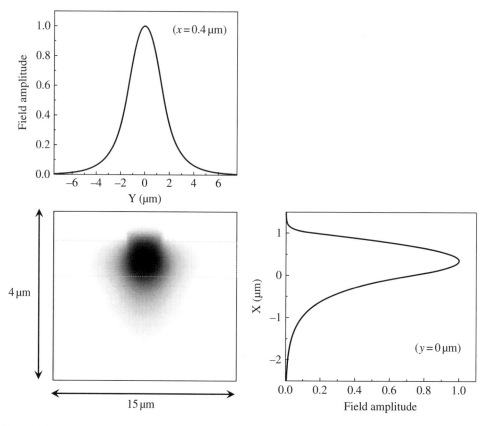

Figure 3.32   Magnetic field amplitude of the fundamental quasi-TE mode corresponding to the structure depicted in Figure 3.12, obtained by semi-vectorial BPM based on the transverse magnetic field component H_x

$$c_i = -\frac{1}{(\Delta x)^2};$$
(3.127c)

$$r_i = \frac{1}{(\Delta x)^2}\left[u_{i-1,j}^{m+1/2} + u_{i+1,j}^{m+1/2}\right] + \left\{\frac{4in_0k_0}{\Delta z} - \frac{2}{(\Delta x)^2} + \frac{k_0^2}{2}\left[\left(n_{i,j}^{m+1/2}\right)^2 - n_0^2\right]\right\}u_{i,j}^{m+1/2}.$$
(3.127d)

Figure 3.32 shows the transversal field profile of the fundamental quasi-TE mode of the high index contrast rib waveguide shown in Figure 3.12, obtained by SV-BPM based on the H_x-formulation at 1.55 μm, using grid spacing of 0.02 and 0.1 μm along the vertical and horizontal directions (x- and y-axis), respectively. The upper graph is a plot of the H_x amplitude along the y direction at $x = 0.4$ μm (centre of the slab region), while right graph shows the field profile along the x direction at $y = 0$, where the continuity of the magnetic field at the index discontinuities is noted. The effective index of the mode, calculated either by its field distribution (using the Rayleigh quotient) or by the growth rate, is coincident to that obtained previously using SV-BPM formalism based on the electric field.

3.3.1.2   Quasi-TM Propagation

For quasi-TM propagation, the major component of the magnetic field is the H_y component, resulting in a negligible H_x component. For this case, the wave equation to be solved is Eq. (3.120b), which after introducing second-order error terms in propagation, becomes:

$$\left(\frac{4in_0k_0}{\Delta z}-B_x\right)\left(\frac{4in_0k_0}{\Delta z}-B_y\right)v^{m+1}=\left(\frac{4in_0k_0}{\Delta z}+B_x\right)\left(\frac{4in_0k_0}{\Delta z}+B_y\right)v^m,\tag{3.128}$$

where v denotes the SVE of the magnetic field component Φ_y. Again, this equation is decomposed in to two sub-steps. The optical field at the transverse plane m is first used to calculate the intermediate field at the intermediate fictitious position $m+1/2$, following the equation:

$$\left(\frac{4in_0k_0}{\Delta z}-B_y\right)v^{m+1/2}=\left(\frac{4in_0k_0}{\Delta z}+B_y\right)v^m.\tag{3.129}$$

Then to complete the propagation, the second step is performed, using the field previously calculated $v^{m+1/2}$ to obtain finally the field v^{m+1}:

$$\left(\frac{4in_0k_0}{\Delta z}-B_x\right)v^{m+1}=\left(\frac{4in_0k_0}{\Delta z}+B_x\right)v^{m+1/2}.\tag{3.130}$$

To perform the first sub-step, Eq. (3.129) is expressed by its finite-difference approximation, yielding:

$$\frac{4in_0k_0}{\Delta z}v_{i,j}^{m+1/2}-\frac{v_{i,j-1}^{m+1/2}-2v_{i,j}^{m+1/2}+v_{i,j+1}^{m+1/2}}{(\Delta y)^2}-\frac{1}{2}k_0^2\left[\left(n_{i,j}^{m+1/2}\right)^2-n_0^2\right]v_{i,j}^{m+1/2}$$

$$=\frac{4in_0k_0}{\Delta z}v_{i,j}^m+\frac{v_{i,j-1}^m-2v_{i,j}^m+v_{i,j+1}^m}{(\Delta y)^2}+\frac{1}{2}k_0^2\left[\left(n_{i,j}^m\right)^2-n_0^2\right]v_{i,j}^m\tag{3.131}$$

The tridiagonal system coefficients in this case are given by:

$$a_j=-\frac{1}{(\Delta y)^2};\tag{3.132a}$$

$$b_j=\frac{4in_0k_0}{\Delta z}+\frac{2}{(\Delta y)^2}-\frac{1}{2}k_0^2\left[\left(n_{i,j}^{m+1/2}\right)^2-n_0^2\right];\tag{3.132b}$$

$$c_j=-\frac{1}{(\Delta y)^2};\tag{3.132c}$$

$$r_j=\frac{1}{(\Delta y)^2}\left[v_{i,j-1}^m+v_{i,j+1}^m\right]+\left\{\frac{4in_0k_0}{\Delta z}-\frac{2}{(\Delta y)^2}+\frac{1}{2}k_0^2\left[\left(n_{i,j}^m\right)^2-n_0^2\right]\right\}v_{i,j}^m.\tag{3.132d}$$

In a very similar way, we proceed with the second sub-step (Eq. (3.130)) derived from the application of the ADI method:

$$
\frac{4in_0k_0}{\Delta z}v_{i,j}^{m+1} - \frac{T_{i-1,j}^{m+1}v_{i-1,j}^{m+1} - 2R_{i,j}^{m+1}v_{i,j}^{m+1} + T_{i+1,j}^{m+1}v_{i+1,j}^{m+1}}{(\Delta x)^2} - \frac{1}{2}k_0^2\left[\left(n_{i,j}^{m+1}\right)^2 - n_0^2\right]v_{i,j}^{m+1}
$$
$$
= \frac{4in_0k_0}{\Delta z}v_{i,j}^{m+1/2} + \frac{T_{i-1,j}^{m+1/2}v_{i-1,j}^{m+1/2} - 2R_{i,j}^{m+1/2}v_{i,j}^{m+1/2} + T_{i+1,j}^{m+1/2}v_{i+1,j}^{m+1/2}}{(\Delta x)^2} + \frac{k_0^2}{2}\left[\left(n_{i,j}^{m+1/2}\right)^2 - n_0^2\right]v_{i,j}^{m+1/2}
$$

$$(3.133)$$

where the T_{ij} and R_{ij} coefficients are defined in Eq. (3.117a,b). The coefficients of the linear system are:

$$
a_i = -\frac{1}{(\Delta x)^2}T_{i-1,j}^{m+1};
$$

$$(3.134a)$$

$$
b_i = \frac{4in_0k_0}{\Delta z} + \frac{2}{(\Delta x)^2}R_{i,j}^{m+1} - \frac{1}{2}k_0^2\left[\left(n_{i,j}^{m+1}\right)^2 - n_0^2\right];
$$

$$(3.134b)$$

$$
c_i = -\frac{1}{(\Delta x)^2}T_{i+1,j}^{m+1};
$$

$$(3.134c)$$

$$
r_i = \frac{T_{i-1,j}^{m+1/2}v_{i-1,j}^{m+1/2} + T_{i+1,j}^{m+1/2}v_{i+1,j}^{m+1/2}}{(\Delta x)^2} + \left\{\frac{4in_0k_0}{\Delta z} - \frac{2R_{i,j}^{m+1/2}}{(\Delta x)^2} + \frac{k_0^2}{2}\left[\left(n_{i,j}^{m+1/2}\right)^2 - n_0^2\right]\right\}v_{i,j}^{m+1/2}.
$$

$$(3.134d)$$

The semi-vectorial FD-BPM developed here is applied now to the calculation of the quasi-TM fundamental mode of the rib waveguide previously described. Figure 3.33 shows the field amplitude of the H_y transversal component obtained by imaginary distance BPM. Again, the field profile along $y = 0$ is continuous at the index interfaces also for TM polarized waves, although a discontinuity in its derivative is clearly seen at $x = 1$ μm, dictated by the boundary conditions imposed to the magnetic field component. A value of $N_{eff}(\text{TM}) = 3.38919$ is obtained either by the field distribution or by the growth rate. This effective index differs only a 0.001% with respect to the value calculated using BPM formulation based on the electric field component, confirming the equivalence between both formulations.

3.3.2 Full Vectorial BPM

In order to implement the ADI method for FV-BPM, we rewrite Eq. (3.109) as:

$$
2in_0k_0\frac{\partial}{\partial z}\begin{bmatrix}\Phi_x\\\Phi_y\end{bmatrix} = \begin{bmatrix}A_x + A_y & C\\D & B_x + B_y\end{bmatrix}\begin{bmatrix}\Phi_x\\\Phi_y\end{bmatrix} = \left(\begin{bmatrix}A_x & C\\0 & B_x\end{bmatrix} + \begin{bmatrix}A_y & 0\\D & B_y\end{bmatrix}\right)\begin{bmatrix}\Phi_x\\\Phi_y\end{bmatrix}.
$$

$$(3.135)$$

After discretizing with the standard Crank–Nicolson scheme, we obtain:

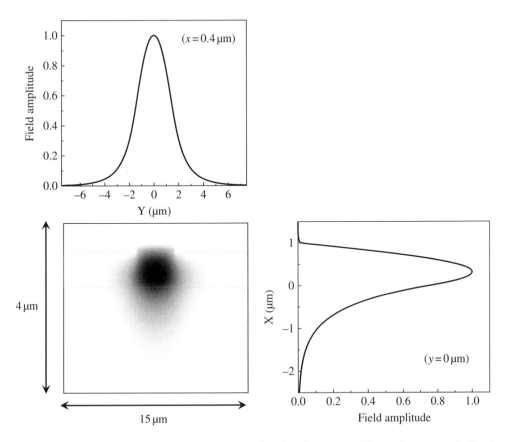

Figure 3.33 Field amplitude of the fundamental quasi-TM mode, corresponding to the structure depicted in Figure 3.12, obtained by semi-vectorial BPM based on the transverse magnetic field component H_y

$$\frac{2in_0k_0}{\Delta z}\left(\begin{bmatrix}u\\v\end{bmatrix}^{m+1}-\begin{bmatrix}u\\v\end{bmatrix}^m\right)=\frac{1}{2}\left(\begin{bmatrix}A_x&C\\0&B_x\end{bmatrix}+\begin{bmatrix}A_y&0\\D&B_y\end{bmatrix}\right)\left(\begin{bmatrix}u\\v\end{bmatrix}^{m+1}+\begin{bmatrix}u\\v\end{bmatrix}^m\right),\qquad(3.136)$$

where we have denoted u and v to be the x and y components of the envelope magnetic field, respectively, and the differential operators $A_{x,y}$, $B_{x,y}$, C and D are given in Eqs. (3.111a,b)–(3.113a,b). After introducing second-order error terms in Δz, and by making use of the ADI method, the last equation becomes:

$$\begin{bmatrix}u\\v\end{bmatrix}^{m+1}=\frac{\left(1-\dfrac{i\Delta z}{4n_0k_0}\begin{bmatrix}A_x&C\\0&B_x\end{bmatrix}\right)}{\left(1+\dfrac{i\Delta z}{4n_0k_0}\begin{bmatrix}A_x&C\\0&B_x\end{bmatrix}\right)}\cdot\frac{\left(1-\dfrac{i\Delta z}{4n_0k_0}\begin{bmatrix}A_y&0\\D&B_y\end{bmatrix}\right)}{\left(1+\dfrac{i\Delta z}{4n_0k_0}\begin{bmatrix}A_y&0\\D&B_y\end{bmatrix}\right)}\cdot\begin{bmatrix}u\\v\end{bmatrix}^m.\qquad(3.137)$$

This equation is now split in two sub-steps to perform a single propagation step of length Δz, according to the following scheme:

$$
\begin{bmatrix} u \\ v \end{bmatrix}^{m+1/2} = \frac{\left(1 - \dfrac{i\Delta z}{4n_0 k_0}\begin{bmatrix} A_y & 0 \\ D & B_y \end{bmatrix}\right)}{\left(1 + \dfrac{i\Delta z}{4n_0 k_0}\begin{bmatrix} A_y & 0 \\ D & B_y \end{bmatrix}\right)} \cdot \begin{bmatrix} u \\ v \end{bmatrix}^m ; \tag{3.138a}
$$

$$
\begin{bmatrix} u \\ v \end{bmatrix}^{m+1} = \frac{\left(1 - \dfrac{i\Delta z}{4n_0 k_0}\begin{bmatrix} A_x & C \\ 0 & B_x \end{bmatrix}\right)}{\left(1 + \dfrac{i\Delta z}{4n_0 k_0}\begin{bmatrix} A_x & C \\ 0 & B_x \end{bmatrix}\right)} \cdot \begin{bmatrix} u \\ v \end{bmatrix}^{m+1/2} . \tag{3.138b}
$$

The sequence to solve Eqs. (3.138a) and (3.138b) is sketched in Figure 3.24, which is equivalent to the procedure used in the FV-BPM electric field formulation.

3.3.2.1   First Step

Equation (3.138a) is the formal expression of the following equation:

$$
\left(1 + \frac{i\Delta z}{4n_0 k_0}\begin{bmatrix} A_y & 0 \\ D & B_y \end{bmatrix}\right) \cdot \begin{bmatrix} u \\ v \end{bmatrix}^{m+1/2} = \left(1 - \frac{i\Delta z}{4n_0 k_0}\begin{bmatrix} A_y & 0 \\ D & B_y \end{bmatrix}\right) \cdot \begin{bmatrix} u \\ v \end{bmatrix}^m . \tag{3.139}
$$

Now, this vectorial equation is separated into its two components. The first component is expressed as:

$$
\left(1 + \frac{i\Delta z}{4n_0 k_0}A_y\right) u^{m+1/2} = \left(1 - \frac{i\Delta z}{4n_0 k_0}A_y\right) u^m . \tag{3.140}
$$

Similarly, the second component of Eq. (3.138a) is:

$$
\left(1 + \frac{i\Delta z}{4n_0 k_0}B_y\right) v^{m+1/2} + \frac{i\Delta z}{4n_0 k_0}D u^{m+1/2} = \left(1 - \frac{i\Delta z}{4n_0 k_0}B_y\right) v^m - \frac{i\Delta z}{4n_0 k_0}D u^m . \tag{3.141}
$$

Equation (3.138a,b) allows us to obtain the x-component of the intermediate field, $u^{m+1/2}$, from the known field u^m at the previous longitudinal step m. Then, Eq. (3.141) is solved to obtain the y-component of the intermediate field $v^{m+1/2}$ from the known fields u^m, $u^{m+1/2}$ and v^m. Thus, these two equations allow us to perform the first sub-step, as shown in Figure 3.24.

To perform the first component of the first step, wave equation (3.140) is expressed in its finite difference form as:

$$
\frac{4 i n_0 k_0}{\Delta z} u_{i,j}^{m+1/2} - \frac{T_{i,j-1}^{m+1/2} u_{i,j-1}^{m+1/2} - 2 R_{i,j}^{m+1/2} u_{i,j}^{m+1/2} + T_{i,j+1}^{m+1/2} u_{i,j+1}^{m+1/2}}{(\Delta y)^2} - \frac{k_0^2}{2}\left[\left(n_{i,j}^{m+1/2}\right)^2 - n_0^2\right] u_{i,j}^{m+1/2}
$$
$$
= \frac{4 i n_0 k_0}{\Delta z} u_{i,j}^m + \frac{T_{i,j-1}^m u_{i,j-1}^m - 2 R_{i,j}^m u_{i,j}^m + T_{i,j+1}^m u_{i,j+1}^m}{(\Delta y)^2} + \frac{1}{2}k_0^2\left[\left(n_{i,j}^m\right)^2 - n_0^2\right] u_{i,j}^m
$$

$$
\tag{3.142}
$$

The coefficients T and R are those defined in Eq. (3.115a,b). From this tridiagonal system, the coefficients are:

$$a_j = -\frac{1}{(\Delta y)^2} T_{i,j-1}^{m+1/2};$$

(3.143a)

$$b_j = \frac{4in_0k_0}{\Delta z} + \frac{2}{(\Delta y)^2} R_{i,j}^{m+1/2} - \frac{1}{2}k_0^2 \left[\left(n_{i,j}^{m+1/2} \right)^2 - n_0^2 \right];$$

(3.143b)

$$c_j = -\frac{1}{(\Delta y)^2} T_{i,j+1}^{m+1/2};$$

(3.143c)

$$r_j = \frac{T_{i,j-1}^m u_{i,j-1}^m + T_{i,j+1}^m u_{i,j+1}^m}{(\Delta y)^2} + \left\{ \frac{4in_0k_0}{\Delta z} - \frac{2}{(\Delta y)^2} R_{i,j}^m + \frac{k_0^2}{2} \left[\left(n_{i,j}^m \right)^2 - n_0^2 \right] \right\} u_{i,j}^m.$$

(3.143d)

The second component of the first step is performed by taking the finite-difference expression of Eq. (3.141), which yields:

$$\frac{4in_0k_0}{\Delta z} v_{i,j}^{m+1/2} - \frac{v_{i,j-1}^{m+1/2} - 2v_{i,j}^{m+1/2} + v_{i,j+1}^{m+1/2}}{(\Delta y)^2} - \frac{1}{2}k_0^2 \left[\left(n_{i,j}^{m+1/2} \right)^2 - n_0^2 \right] v_{i,j}^{m+1/2}$$

$$= \frac{4in_0k_0}{\Delta z} v_{i,j}^m + \frac{v_{i,j-1}^m - 2v_{i,j}^m + v_{i,j+1}^m}{(\Delta y)^2} + \frac{1}{2}k_0^2 \left[\left(n_{i,j}^m \right)^2 - n_0^2 \right] v_{i,j}^m$$

(3.144)

$$+ \frac{1}{4\Delta x \Delta y} \left(Z1_{i,j}^m u_{i+1,j+1}^m - Z2_{i,j}^m u_{i+1,j-1}^m - Z3_{i,j}^m u_{i-1,j+1}^m + Z4_{i,j}^m u_{i-1,j-1}^m \right)$$

$$+ \frac{1}{4\Delta x \Delta y} \left(Z1_{i,j}^{m+1/2} u_{i+1,j+1}^{m+1/2} - Z2_{i,j}^{m+1/2} u_{i+1,j-1}^{m+1/2} - Z3_{i,j}^{m+1/2} u_{i-1,j+1}^{m+1/2} + Z4_{i,j}^{m+1/2} u_{i-1,j-1}^{m+1/2} \right)$$

the $Z1$–$Z4$ coefficients being those defined in Eqs. (3.119eh). The Thomas algorithm's coefficients for that finite-difference equation forming the tridiagonal system are given by:

$$a_j = -\frac{1}{(\Delta y)^2};$$

(3.145a)

$$b_j = \frac{4in_0k_0}{\Delta z} + \frac{2}{(\Delta y)^2} - \frac{1}{2}k_0^2 \left[\left(n_{i,j}^m \right)^2 - n_0^2 \right];$$

(3.145b)

$$c_j = -\frac{1}{(\Delta y)^2};$$

(3.145c)

$$r_j = \frac{1}{(\Delta y)^2}\left[v_{i,j-1}^m + v_{i,j+1}^m\right] + \left\{\frac{4in_0k_0}{\Delta z} - \frac{2}{(\Delta y)^2} + \frac{1}{2}k_0^2\left[\left(n_{i,j}^m\right)^2 - n_0^2\right]\right\}v_{i,j}^m$$

$$+ \frac{1}{4\Delta x\Delta y}\left(Z1_{i,j}^m u_{i+1,j+1}^m - Z2_{i,j}^m u_{i+1,j-1}^m - Z3_{i,j}^m u_{i-1,j+1}^m + Z4_{i,j}^m u_{i-1,j-1}^m\right)$$

$$+ \frac{1}{4\Delta x\Delta y}\left(Z1_{i,j}^{m+1/2} u_{i+1,j+1}^{m+1/2} - Z2_{i,j}^{m+1/2} u_{i+1,j-1}^{m+1/2} - Z3_{i,j}^{m+1/2} u_{i-1,j+1}^{m+1/2} + Z4_{i,j}^{m+1/2} u_{i-1,j-1}^{m+1/2}\right)$$

$$(3.145\text{d})$$

3.3.2.2 Second Step

To complete the propagation along a distance Δz the second step must be performed by using Eq. (3.138b), which is the formal expression of the following equation:

$$\left(1 + \frac{i\Delta z}{4n_0k_0}\begin{bmatrix} A_x & C \\ 0 & B_x \end{bmatrix}\right)\cdot\begin{bmatrix} u \\ v \end{bmatrix}^{m+1} = \left(1 - \frac{i\Delta z}{4n_0k_0}\begin{bmatrix} A_x & C \\ 0 & B_x \end{bmatrix}\right)\cdot\begin{bmatrix} u \\ v \end{bmatrix}^{m+1/2}. \tag{3.146}$$

This vectorial equation can be now split into its two components, the first component being expressed as:

$$\left(1 + \frac{i\Delta z}{4n_0k_0}A_x\right)u^{m+1} + \frac{i\Delta z}{4n_0k_0}Cv^{m+1} = \left(1 - \frac{i\Delta z}{4n_0k_0}A_x\right)u^{m+1/2} - \frac{i\Delta z}{4n_0k_0}Cv^{m+1/2}, \tag{3.147}$$

while the second component is:

$$\left(1 + \frac{i\Delta z}{4n_0k_0}B_x\right)v^{m+1} = \left(1 - \frac{i\Delta z}{4n_0k_0}B_x\right)v^{m+1/2}. \tag{3.148}$$

The sequence to carry out this second step is drawn in Figure 3.24. Following that procedure, Eq. (3.148) is first used to calculate the y component of the magnetic field v^{m+1} from the known field $v^{m+1/2}$. Then, Eq. (3.147) allows us to obtain the x component of the magnetic field u^{m+1} from the known fields $u^{m+1/2}$, $v^{m+1/2}$ and v^{m+1}, thus finishing the whole propagation step of length Δz.

The numerical implementation of this second step starts by solving the finite-difference approximation of Eq. (3.148) (second component of the second step), expressed by:

$$\frac{4in_0k_0}{\Delta z}v_{i,j}^{m+1} - \frac{T_{i-1,j}^{m+1}v_{i-1,j}^{m+1} - 2R_{i,j}^{m+1}v_{i,j}^{m+1} + T_{i+1,j}^{m+1}v_{i+1,j}^{m+1}}{(\Delta x)^2} - \frac{1}{2}k_0^2\left[\left(n_{i,j}^{m+1}\right)^2 - n_0^2\right]v_{i,j}^{m+1}$$

$$= \frac{4in_0k_0}{\Delta z}v_{i,j}^{m+1/2} + \frac{T_{i-1,j}^{m+1/2}v_{i-1,j}^{m+1/2} - 2R_{i,j}^{m+1/2}v_{i,j}^{m+1/2} + T_{i+1,j}^{m+1/2}v_{i+1,j}^{m+1/2}}{(\Delta x)^2} + \frac{k_0^2}{2}\left[\left(n_{i,j}^{m+1/2}\right)^2 - n_0^2\right]v_{i,j}^{m+1/2}$$

$$(3.149)$$

where we are using the definition of T_{ij} and R_{ij} from Eq. (3.117a,b). The coefficients of the Thomas algorithm applied to this tridiagonal system are given by:

$$a_i = -\frac{1}{(\Delta x)^2} T_{i-1,j}^{m+1}; \tag{3.150a}$$

$$b_i = \frac{4in_0k_0}{\Delta z} + \frac{2}{(\Delta x)^2} R_{i,j}^{m+1} - \frac{1}{2}k_0^2\left[\left(n_{i,j}^{m+1}\right)^2 - n_0^2\right]; \tag{3.150b}$$

$$c_i = -\frac{1}{(\Delta x)^2} T_{i+1,j}^{m+1}; \tag{3.150c}$$

$$r_i = \frac{T_{i-1,j}^{m+1/2} v_{i-1,j}^{m+1/2} + T_{i+1,j}^{m+1/2} v_{i+1,j}^{m+1/2}}{(\Delta x)^2} + \left\{\frac{4in_0k_0}{\Delta z} - \frac{2R_{i,j}^{m+1/2}}{(\Delta x)^2} + \frac{k_0^2}{2}\left[\left(n_{i,j}^{m+1/2}\right)^2 - n_0^2\right]\right\} v_{i,j}^{m+1/2}. \tag{3.150d}$$

Finally, the propagation is completed after performing the first component of the second step (Eq. (3.147)), which has the following finite-difference expression:

$$\frac{4in_0k_0}{\Delta z} u_{i,j}^{m+1} - \frac{u_{i-1,j}^{m+1} - 2u_{i,j}^{m+1} + u_{i+1,j}^{m+1}}{(\Delta x)^2} - \frac{1}{2}k_0^2\left[\left(n_{i,j}^{m+1}\right)^2 - n_0^2\right] u_{i,j}^{m+1}$$

$$= \frac{4in_0k_0}{\Delta z} u_{i,j}^{m+1/2} + \frac{u_{i-1,j}^{m+1/2} - 2u_{i,j}^{m+1/2} + u_{i+1,j}^{m+1/2}}{(\Delta x)^2} + \frac{1}{2}k_0^2\left[\left(n_{i,j}^{m+1/2}\right)^2 - n_0^2\right] u_{i,j}^{m+1/2}$$

$$+ \frac{1}{4\Delta x\Delta y}\left(S1_{i,j}^{m+1/2} v_{i+1,j+1}^{m+1/2} - S2_{i,j}^{m+1/2} v_{i+1,j-1}^{m+1/2} - S3_{i,j}^{m+1/2} v_{i-1,j+1}^{m+1/2} + S4_{i,j}^{m+1/2} v_{i-1,j-1}^{m+1/2}\right)$$

$$+ \frac{1}{4\Delta x\Delta y}\left(S1_{i,j}^{m+1} v_{i+1,j+1}^{m+1} - S2_{i,j}^{m+1} v_{i+1,j-1}^{m+1} - S3_{i,j}^{m+1} v_{i-1,j+1}^{m+1} + S4_{i,j}^{m+1} v_{i-1,j-1}^{m+1}\right) \tag{3.151}$$

here, the $S1$–$S4$ coefficients are those defined in Eqs. (3.119a) to (3.119d). The coefficients of the tridiagonal linear system are:

$$a_i = -\frac{1}{(\Delta x)^2}; \tag{3.152a}$$

$$b_i = \frac{4in_0k_0}{\Delta z} + \frac{2}{(\Delta x)^2} - \frac{1}{2}k_0^2\left[\left(n_{i,j}^{m+1}\right)^2 - n_0^2\right]; \tag{3.152b}$$

$$c_i = -\frac{1}{(\Delta x)^2}; \tag{3.152c}$$

$$r_i = \frac{1}{(\Delta x)^2}\left[u_{i-1,j}^{m+1/2} + u_{i+1,j}^{m+1/2}\right] + \left\{\frac{4in_0k_0}{\Delta z} - \frac{2}{(\Delta x)^2} + \frac{1}{2}k_0^2\left[\left(n_{i,j}^{m+1/2}\right)^2 - n_0^2\right]\right\}u_{i,j}^{m+1/2}$$

$$+ \frac{1}{4\Delta x \Delta y}\left(S1_{i,j}^{m+1/2}v_{i+1,j+1}^{m+1/2} - S2_{i,j}^{m+1/2}v_{i+1,j-1}^{m+1/2} - S3_{i,j}^{m+1/2}v_{i-1,j+1}^{m+1/2} + S4_{i,j}^{m+1/2}v_{i-1,j-1}^{m+1/2}\right) \quad (3.152d)$$

$$+ \frac{1}{4\Delta x \Delta y}\left(S1_{i,j}^{m+1}v_{i+1,j+1}^{m+1} - S2_{i,j}^{m+1}v_{i+1,j-1}^{m+1} - S3_{i,j}^{m+1}v_{i-1,j+1}^{m+1} + S4_{i,j}^{m+1}v_{i-1,j-1}^{m+1}\right)$$

The FV-BPM algorithm based on the magnetic field transverse components is now used to obtain the quasi-TE and quasi-TM fundamental modes of the rib waveguide example shown in Figure 3.12. Transversal step sizes of $\Delta x = 0.02$ μm and $\Delta y = 0.1$ μm are used in the simulation and the computational window has 200×150 points in the vertical and horizontal directions, respectively. The quasi-TE polarized fundamental mode is obtained by propagating along the imaginary z-axis a H_x-polarized input Gaussian beam, by means of FV-BPM. Figure 3.34 presents the amplitude maps of the major and minor components (H_x and H_y, respectively), the ratio of maximum amplitudes being ~ 7.8, which is a clear indication of strong contrast index structure. Note also the continuity of the field across the index boundaries.

Additionally, the quasi-TM fundamental mode is obtained by launching a H_y-polarized input beam and propagating it by means of imaginary distance FV-BPM. Results of the simulation are given in Figure 3.35, where the ratio found between the major and minor components is ~ 10.4.

In order to see the accuracy of the different BPM approaches (scalar, semi-vectorial and FV-BPM), Figure 3.36 presents the normalized propagation constants b for quasi-TE and quasi-TM fundamental modes of the rib structure examined previously, defined as:

$$b = \frac{N_{eff}^2 - n_{subs}^2}{n_{core}^2 - n_{subs}^2}, \quad (3.153)$$

where $n_{subs} = 3.36$, $n_{core} = 3.44$ and N_{eff} is the effective index of the mode. Besides the data provided by the different BPM approaches, the values given by the effective index method [9] have

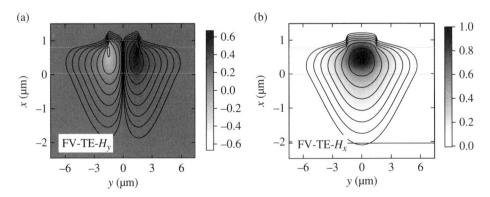

Figure 3.34 Maps of the magnetic field amplitude of the y component (a) and x component (b) of the quasi-TE fundamental mode of the rib waveguide shown in Figure 3.12

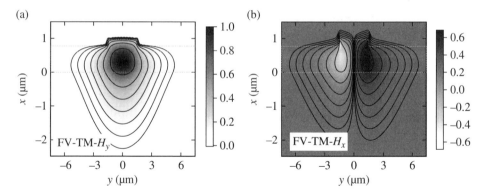

Figure 3.35 Maps of the magnetic field amplitude of the y component (a) and x component (b) of the quasi-TM fundamental mode of the rib waveguide shown in Figure 3.12

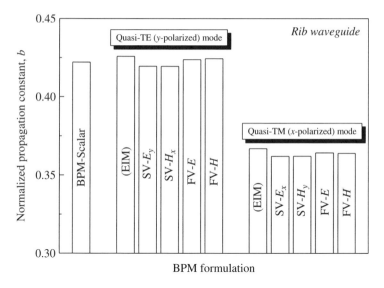

Figure 3.36 Comparison of the normalized propagation constants of the quasi-TE and quasi-TM fundamental modes of a strong index contrast rib waveguide, provided by the different BPM approaches. The values calculated by the effective index method have also been included

been included. A first salient feature from this graph is the high birefringence of the rib structure, and thus the scalar approach is insufficient. Also, it should be pointed out that the propagation constants provided by E-field and H-field formalisms are very close, both for semi-vectorial or full vectorial approaches. Let us note also that the values provided by FV-BPM differ from those obtained by SV-BPM, which indicates the non-negligible contribution of the cross-coupling terms in the wave equation. Therefore, FV-BPM has to be used to the correct description of strong contrast index structures.

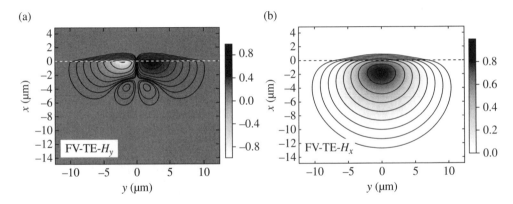

Figure 3.37 Maps of the magnetic field amplitude of the *y* component (a) and *x* component (b) of the quasi-TE fundamental mode of the diffused waveguide shown in Figure 3.18

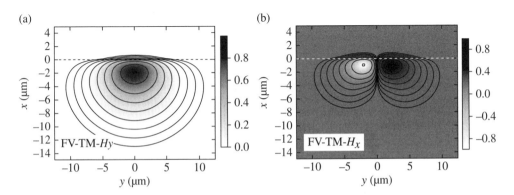

Figure 3.38 Maps of the magnetic field amplitude of the *y* component (a) and *x* component (b) of the quasi-TM fundamental mode of the diffused waveguide shown in Figure 3.18

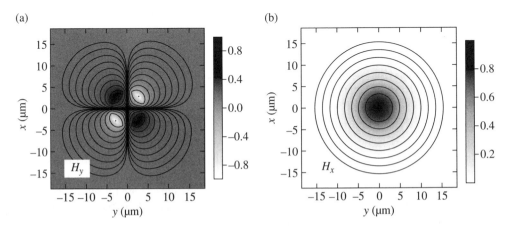

Figure 3.39 Maps of the magnetic field amplitude of the *y* component (a) and *x* component (b) of the HE$_{11}$ fundamental mode of the previously analysed step-index fibre

Figures 3.37 and 3.38 show the mode patterns of the fundamental quasi-TE and quasi-TM modes, respectively, for the transverse magnetic field components of the diffused waveguide previously depicted in Figure 3.18. The numerical simulation uses grid spacing of $\Delta x = \Delta y = 0.25\,\mu m$ and a computational window of 80×100 points along the x and y directions, respectively. The mode maps were obtained by imaginary-distance FV-BPM based on the magnetic field formulation, which gives similar effective indices for the modes to those previously calculated by E-field formulation. In this case, the ratios between the amplitudes for the major and minor magnetic field components are found to be ~ 140 and ~ 365 for quasi-TE and quasi-TM modes, respectively.

Finally, the single mode optical fibre previously presented in Figure 3.30 is now examined by means of FV-BPM based on the magnetic field formulation, with $\Delta x = \Delta y = 0.25\,\mu m$ and $N_x = N_y = 150$ grid points. Mode patterns of the HE_{11} mode obtained by imaginary-distance propagation are plotted in Figure 3.39, which gives a ratio between the maximum amplitudes of ~ 740 and is an indication of a low contrast index waveguide.

References

[1] Huang, W., Xu, C., Chu, S. and Chaudhuri, S.K. (1992) The finite-difference vector beam propagation method: analysis and assessment. *Journal of Lightwave Technology* **10**, 295–305.

[2] Huang, W.P., Xu, C.L., Chu, S.T. and Chaudhuri, S.K. (1991) A vector beam propagation method for guided-wave optics. *IEEE Photonics Technology Letters* **3**, 910–913.

[3] Huang, W.P., Xu, C.L. and Chaudhuri, S.K. (1991) A vector beam propagation method based on H fields. *IEEE Transactions Photonics Technology Letters* **3**, 1117–1120.

[4] Chung, Y. and Dagli, N. (1990) An assessment of finite difference beam propagation method. *IEEE Journal of Quantum Electronics* **26**, 1335–1339.

[5] Press, W.H., Teukolsky, S.A., Vetterling, W.T. and Flannery, B.P. (1996) *Numerical Recipes in Fortran 77: The Art of Scientific Computing*, Chapter 19. Cambridge University Press, New York.

[6] Hsuch, Y.L., Yang, M.C. and Chang, H.C. (1999) Three-dimensional noniterative full-vectorial beam propagation method based on the alternating direction implicit method. *Journal of Lightwave Technology* **17**, 2389–2397.

[7] Huang, W.P. and Xu, C.L. (1993) Simulation of three-dimensional optical waveguides by a full-vector beam propagation method. *Journal of Quantum Electronics* **29**, 2639–2649.

[8] Kunz, A., Zimulinda, F. and Heinlein, W.E. (1993) Fast three-dimensional split-step algorithm for vectorial wave propagation in integrated optics. *IEEE Photonics Technology Letters* **5**, 1073–1076.

[9] Lifante, G. (2003) *Integrated Photonics: Fundamentals*. John Wiley & Sons, Ltd, Chichester.

4

Special Topics on BPM

Introduction

In the last chapter we have shown the implementation of beam propagation method (BPM) techniques under three different approaches: scalar, semi-vectorial and full vectorial, based either on electric or magnetic transverse field components. Here, some extensions of the BPM based on finite-difference techniques will be presented. First, we present the wide-angle BPM, which relaxes the restriction of the application of BPM to paraxial waves and allows the simulation of light beams with large propagation angles with respect to the longitudinal direction. Reflection at a waveguide interface is discussed in the next section and some results of this problem are then to develop a BPM algorithm that can handle multiple reflections, known as bidirectional-BPM. Also, the simulation of light propagation in active media (with variable gain/losses), second-order non-linear media and anisotropic media are the topics covered in Sections 4.4–4.6, by developing the so called active-BPM, non-linear-BPM and anisotropic BPM. The last sections include the description of time-domain simulation techniques. These numerical techniques do not make any assumption of the temporal dependence of the optical field, and thus they can manage monochromatic waves as well as the propagation of optical pulses and can manage backward waves due to reflections at the discontinuities. Both the time-domain beam propagation method (TD-BPM) and finite-difference time-domain (FDTD) are explained in detail. Each section also provides selected examples to show the performance of the numerical methods and their range of applicability.

4.1 Wide-Angle Beam Propagation Method

The paraxiality restriction on the BPM, as well as the related restriction on index-contrast and multimode propagation, can be relaxed through the use of extensions referred as wide-angle BPM. The essential idea behind the various approaches is to relax the paraxial limitations

---

Beam Propagation Method for Design of Optical Waveguide Devices, First Edition. Ginés Lifante Pedrola.
© 2016 John Wiley & Sons, Ltd. Published 2016 by John Wiley & Sons, Ltd.

by incorporating the effect of the $\partial^2 u/\partial z^2$ term, which was neglected in the derivation of the basic BPM (Eq. (2.6)). The different approaches vary in the method and degree of approximation by which they accomplish this. The most popular formulation is referred to as the multi-step Padé-based wide-angle technique [1] and it is this approach that will be presented here.

4.1.1 Formalism of Wide-Angle-BPM Based on Padé Approximants

A simple approach to deriving a wide-angle BPM equation is to consider the scalar Helmholtz wave equation written in terms of the slowly varying field $u(x,y,z)$, but before making the slowly varying envelope approximation (SVEA) by neglecting the $\partial^2 u/\partial z^2$ term:

$$\frac{\partial^2 u}{\partial z^2} - 2in_0k_0\frac{\partial u}{\partial z} + \frac{\partial^2 u}{\partial x^2} + \frac{\partial^2 u}{\partial y^2} + k_0^2\left(n^2 - n_0^2\right)u = 0. \tag{4.1}$$

This last equation can be expressed as a function of two differential operators, $\mathcal{D}$ and $\mathcal{P}$:

$$\frac{1}{K^2}\mathcal{D}^2 u - \frac{2i}{K}\mathcal{D}u + \mathcal{P}u = 0, \tag{4.2}$$

where the $\mathcal{D}$ and $\mathcal{P}$ differential operators are defined as:

$$\mathcal{D} \equiv \frac{\partial}{\partial z}; \tag{4.3}$$

$$\mathcal{P} \equiv \frac{1}{K^2}\left[\frac{\partial^2}{\partial x^2} + \frac{\partial^2}{\partial y^2} + \left(k_0^2 n^2 - K^2\right)\right], \tag{4.4}$$

and K is the reference wavenumber ($k_0 n_0$). Also, the differential operator $\partial^2/\partial z^2$ is represented by $\mathcal{D}^2$. Putting aside the fact that $\mathcal{D}$ is a differential operator, Eq. (4.2) can be viewed as a quadratic equation to be solved for $\mathcal{D}$:

$$\frac{1}{K^2}\mathcal{D}^2 - \frac{2i}{K}\mathcal{D} + \mathcal{P} = 0, \tag{4.5}$$

yielding the following formal solution for a first-order equation in z:

$$\mathcal{D}u \equiv \frac{\partial u}{\partial z} = -iK\left(\sqrt{1+\mathcal{P}} - 1\right)u. \tag{4.6}$$

This equation is referred to as a one-way wave equation, since the first-order derivative only admits forward travelling waves (or backward waves if the signs are chosen appropriately, but not simultaneously). Although restricted to forward propagation, the equation is still exact in that no paraxiality approximation has been made. The difficulty is that before this equation can be integrated, the radical involving the differential operator $\mathcal{P}$ must be evaluated, as it cannot be explicitly solved due to the square root operator. One approach would be to use the Taylor expansion. First order, this leads to the standard paraxial BPM and to higher order it becomes more accurate and represents one approach to achieving a wide-angle scheme. However, expansion via Padé approximants is more accurate than the Taylor expansion for the same order of terms [2] and it is expressed as:

$$\left(\sqrt{1+\mathcal{P}}-1\right) \approx \frac{M_m(\mathcal{P})}{N_n(\mathcal{P})}, \tag{4.7}$$

where m and n are the highest degrees of $\mathcal{P}$ in the polynomials M and N, respectively. This approach leads to the following wide-angle equation:

$$\frac{\partial u}{\partial z} = -iK\frac{M_m(\mathcal{P})}{N_n(\mathcal{P})}u, \tag{4.8}$$

which contains the effect of (m,n) Padé approximants for the exact Helmholtz operator in Eq. (4.6).

To determine the Padé polynomials, let us rewrite the wave equation (4.1) in the following recursive way [1]:

$$\left.\frac{\partial}{\partial z}\right|_n = -iK\frac{\dfrac{\mathcal{P}}{2}}{1+\dfrac{i}{2K}\left.\dfrac{\partial}{\partial z}\right|_{n-1}}, \tag{4.9}$$

where it is assumed that the zero-order partial z-derivative is zero:

$$\left.\frac{\partial}{\partial z}\right|_0 = 0. \tag{4.10}$$

From the recursive relation (4.9), the first-order approximation (using the $(1, 0)$ Padé approximants) can be easily obtained:

$$\frac{\partial u}{\partial z} = -iK\frac{\mathcal{P}}{2}u, \tag{4.11}$$

which is the basic paraxial approximation for standard BPM. The next order approximation is obtained by substituting this last derivative in Eq. (4.9):

$$\left.\frac{\partial}{\partial z}\right|_2 = -iK\frac{\dfrac{\mathcal{P}}{2}}{1+\dfrac{i}{2K}\left(-\dfrac{iK\mathcal{P}}{2}\right)} = -iK\frac{\dfrac{\mathcal{P}}{2}}{1+\dfrac{\mathcal{P}}{4}}. \tag{4.12}$$

The next order approaches are the following:

$$\left.\frac{\partial}{\partial z}\right|_3 = -iK\frac{\frac{\mathcal{P}}{2}}{1+\frac{i}{2K}\left(-iK\frac{\frac{\mathcal{P}}{2}}{1+\frac{\mathcal{P}}{4}}\right)} = -iK\frac{\frac{\mathcal{P}}{2}+\frac{\mathcal{P}^2}{8}}{1+\frac{\mathcal{P}}{2}};\qquad(4.13)$$

$$\left.\frac{\partial}{\partial z}\right|_4 = -iK\frac{\frac{\mathcal{P}}{2}}{1+\frac{i}{2K}\left(-iK\frac{\frac{\mathcal{P}}{2}+\frac{\mathcal{P}^2}{8}}{1+\frac{\mathcal{P}}{2}}\right)} = -iK\frac{\frac{\mathcal{P}}{2}+\frac{\mathcal{P}^2}{4}}{1+\frac{3\mathcal{P}}{4}+\frac{\mathcal{P}^2}{16}};\qquad(4.14)$$

$$\left.\frac{\partial}{\partial z}\right|_5 = -iK\frac{\frac{\mathcal{P}}{2}}{1+\frac{i}{2K}\left(-iK\frac{\frac{\mathcal{P}}{2}+\frac{\mathcal{P}^2}{4}}{1+\frac{3\mathcal{P}}{4}+\frac{\mathcal{P}^2}{16}}\right)} = -iK\frac{\frac{\mathcal{P}}{2}+\frac{3\mathcal{P}^2}{8}+\frac{\mathcal{P}^3}{32}}{1+\mathcal{P}+\frac{3\mathcal{P}^2}{16}};\qquad(4.15)$$

$$\left.\frac{\partial}{\partial z}\right|_6 = -iK\frac{\frac{\mathcal{P}}{2}}{1+\frac{i}{2K}\left(-iK\frac{\frac{\mathcal{P}}{2}+\frac{3\mathcal{P}^2}{8}+\frac{\mathcal{P}^3}{32}}{1+\mathcal{P}+\frac{3\mathcal{P}^2}{16}}\right)} = -iK\frac{\frac{\mathcal{P}}{2}+\frac{\mathcal{P}^2}{2}+\frac{3\mathcal{P}^3}{32}}{1+\frac{5\mathcal{P}}{4}+\frac{3\mathcal{P}^2}{8}+\frac{\mathcal{P}^3}{64}}.\qquad(4.16)$$

This last one constitutes the (3,3) Padé order approximant. Table 4.1 summarizes four commonly used low-order Padé polynomials in terms of the operator $\mathcal{P}$ defined in Eq. (4.4).

4.1.2 Multi-step Method Applied to Wide-Angle BPM

Equation (4.8) can be written using the usual forward and backward finite differences as follows:

$$\frac{u^{m+1}-u^m}{\Delta z} = -iK\frac{M}{N}u^m;\qquad(4.17a)$$

$$\frac{u^{m+1}-u^m}{\Delta z} = -iK\frac{M}{N}u^{m+1}.\qquad(4.17b)$$

Table 4.1 Low order Padé polynomials of the operator (4.7) in terms of the differential operator $\mathcal{P}$ defined in Eq. (4.4)

Padé order	M_m	N_n
(1,0)	$\mathcal{P}/2$	1
(1,1)	$\mathcal{P}/2$	$1+\mathcal{P}/4$
(2,2)	$\mathcal{P}/2+\mathcal{P}^2/4$	$1+3\,\mathcal{P}/4+\mathcal{P}^2/16$
(3,3)	$\mathcal{P}/2+\mathcal{P}^2/2+3\,\mathcal{P}^3/32$	$1+5\,\mathcal{P}/4+3\mathcal{P}^2/8+\mathcal{P}^3/64$

By multiplying the first equation by $(1 - \alpha)$ and the second by α and adding the two equations, we obtain the Crank–Nicolson scheme:

$$u^{m+1} = \frac{N - iK\Delta z(1-\alpha)M}{N + iK\Delta z\alpha M} u^m. \tag{4.18}$$

Taking into account that M and N are polynomials of degree n, one can write [3]:

$$u^{m+1} = \frac{\displaystyle\sum_{i=0}^{n} \xi_i \mathcal{P}^i}{\displaystyle\sum_{i=0}^{n} \chi_i \mathcal{P}^i} u^m, \tag{4.19}$$

where $\xi_0 = \chi_0 = \mathcal{P}_0 = 1$ and the other ξ_i and χ_i are easily determined from the coefficients of the polynomials M and N.

Since a polynomial of degree n can always be factored in terms of its n roots, we may rewrite Eq. (4.19) as:

$$u^{m+1} = \frac{(1+\alpha_1\mathcal{P})(1+\alpha_2\mathcal{P})\ldots(1+\alpha_n\mathcal{P})}{(1+\beta_1\mathcal{P})(1+\beta_2\mathcal{P})\ldots(1+\beta_n\mathcal{P})} u^m. \tag{4.20}$$

Simple relationships exist between the sets of parameters ξ_i and α_i and between β_i and χ_i. For example, for $n = 2$ it yields:

$$\alpha_1 + \alpha_2 = \xi_1; \tag{4.21a}$$

$$\alpha_1\alpha_2 = \xi_2; \tag{4.21b}$$

$$\beta_1 + \beta_2 = \chi_1; \tag{4.21c}$$

$$\beta_1\beta_2 = \chi_2; \tag{4.21d}$$

And for $n = 3$ we have:

$$\alpha_1 + \alpha_2 + \alpha_3 = \xi_1; \tag{4.22a}$$

$$\alpha_1\alpha_2 + \alpha_2\alpha_3 + \alpha_1\alpha_3 = \xi_2; \tag{4.22b}$$

$$\alpha_1\alpha_2\alpha_3 = \xi_3; \tag{4.22c}$$

$$\beta_1 + \beta_2 + \beta_3 = \chi_1; \tag{4.22d}$$

$$\beta_1\beta_2 + \beta_2\beta_3 + \beta_1\beta_3 = \chi_2; \tag{4.22e}$$

$$\beta_1\beta_2\beta_3 = \chi_3; \tag{4.22f}$$

In general, the determination of the α_i and β_i coefficients requires the one-time solution of two nth-order algebraic equations.

Taking into account Eq. (4.20), it is apparent that an nth-order Padé propagator may be decomposed into an n-step algorithm for which the ith partial step takes the form [3]:

$$u^{m+\frac{i}{n}} = \frac{(1+\alpha_i \mathcal{P})}{(1+\beta_i \mathcal{P})} u^{m+\frac{i-1}{n}}, \ i = 1, 2, \ldots, n. \tag{4.23}$$

Each such partial step is unitary and tridiagonal for 2D propagation (and block tridiagonal for propagation in three dimensions). These two important properties imply that the resulting algorithm is fast and unconditionally stable. The run time for an nth-order propagator is obviously n times the paraxial run time. Because the latter is usually short, the resulting algorithm is capable of providing accurate wide-angle propagation with only a modest numerical penalty.

4.1.3   Numerical Implementation of Wide-Angle BPM

In order to show the application of the method, let us implement the wide angle formalism to scalar 2D propagation using the (2,2) Padé polynomials approach. In 2D problems, the scalar wave equation is given by:

$$\frac{\partial^2 u}{\partial z^2} - 2iK\frac{\partial u}{\partial z} + \frac{\partial^2 u}{\partial x^2} + \left(k^2 - K^2\right)u = 0, \tag{4.24}$$

where k denotes the position dependent wavenumber defined as $k = k_0 n$. The Padé approach for this equation reads:

$$\frac{\partial u}{\partial z} = -iK\frac{M_2(\mathcal{P})}{N_2(\mathcal{P})}u, \tag{4.25}$$

where the differential operator $\mathcal{P}$ reduces to:

$$\mathcal{P} \equiv \frac{1}{K^2}\left[\frac{\partial^2}{\partial x^2} + \left(k^2 - K^2\right)\right]. \tag{4.26}$$

The wave equation in terms of finite differences is in this case:

$$u^{m+1} = \frac{N - iK\Delta z(1-\alpha)M}{N + iK\Delta z\alpha M}u^m. \tag{4.27}$$

Using order 2 for the M and N polynomials, this equation is written as:

$$u^{m+1} = \frac{1 + \dfrac{3\mathcal{P}}{4} + \dfrac{\mathcal{P}^2}{16} - iK\Delta z(1-\alpha)\dfrac{\mathcal{P}}{2} - iK\Delta z(1-\alpha)\dfrac{\mathcal{P}^2}{4}}{1 + \dfrac{3\mathcal{P}}{4} + \dfrac{\mathcal{P}^2}{16} + iK\Delta z\alpha\dfrac{\mathcal{P}}{2} + iK\Delta z\alpha\dfrac{\mathcal{P}^2}{4}}u^m. \tag{4.28}$$

Reorganizing terms:

$$u^{m+1} = \frac{1 + P\left[\frac{3}{4} - \frac{iK\Delta z(1-\alpha)}{2}\right] + P^2\left[\frac{1}{16} - \frac{iK\Delta z(1-\alpha)}{4}\right]}{1 + P\left[\frac{3}{4} + \frac{iK\Delta z\alpha}{2}\right] + P^2\left[\frac{1}{16} + \frac{iK\Delta z\alpha}{4}\right]} u^m, \tag{4.29}$$

where the polynomial coefficients are given by:

$$\xi_1 = \frac{3}{4} - \frac{iK\Delta z(1-\alpha)}{2}; \tag{4.30a}$$

$$\xi_2 = \frac{1}{16} - \frac{iK\Delta z(1-\alpha)}{4}; \tag{4.30b}$$

$$\chi_1 = \frac{3}{4} + \frac{iK\Delta z\alpha}{2}; \tag{4.30c}$$

$$\chi_2 = \frac{1}{16} + \frac{iK\Delta z\alpha}{4}. \tag{4.30d}$$

Taking into account the relations in Eq. (4.21), the roots of the numerator and denominator polynomials can be obtained:

$$\alpha_1 = \frac{1}{2}\left[\xi_1 - \sqrt{\xi_1^2 - 4\xi_2}\right]; \tag{4.31a}$$

$$\alpha_2 = \frac{1}{2}\left[\xi_1 + \sqrt{\xi_1^2 - 4\xi_2}\right]; \tag{4.31b}$$

$$\beta_1 = \frac{1}{2}\left[\chi_1 - \sqrt{\chi_1^2 - 4\chi_2}\right]; \tag{4.31c}$$

$$\beta_2 = \frac{1}{2}\left[\chi_1 + \sqrt{\chi_1^2 - 4\chi_2}\right]. \tag{4.31d}$$

Now, the propagation of the optical field is split into two steps following the next scheme:

$$u^{m+\frac{1}{2}} = \frac{(1 + \alpha_1 P)}{(1 + \beta_1 P)} u^m; \tag{4.32a}$$

$$u^{m+1} = \frac{(1 + \alpha_2 P)}{(1 + \beta_2 P)} u^{m+\frac{1}{2}}. \tag{4.32b}$$

In terms of the coefficients α_j and β_j, the first step is performed by solving the next equation:

$$\left\{1+\frac{\beta_1}{K^2}\left[\frac{\partial^2}{\partial x^2}+\left(k^2-K^2\right)\right]\right\}u^{m+\frac{1}{2}}=\left\{1+\frac{\alpha_1}{K^2}\left[\frac{\partial^2}{\partial x^2}+\left(k^2-K^2\right)\right]\right\}u^m,\qquad(4.33)$$

which yields the following finite-difference equation:

$$u_j^{m+\frac{1}{2}}+\frac{\beta_1}{K^2}\frac{u_{j-1}^{m+\frac{1}{2}}-2u_j^{m+\frac{1}{2}}+u_{j+1}^{m+\frac{1}{2}}}{\Delta x^2}+\frac{\beta_1}{K^2}\left(k^2-K^2\right)u_j^{m+\frac{1}{2}}$$

$$=u_j^m+\frac{\alpha_1}{K^2}\frac{u_{j-1}^m-2u_j^m+u_{j+1}^m}{\Delta x^2}+\frac{\alpha_1}{K^2}\left(k^2-K^2\right)u_j^m \qquad(4.34)$$

This equation forms a tridiagonal system, from where its coefficients a_j, b_j, c_j and r_j are given by:

$$a_j=\frac{\beta_1}{K^2\Delta x^2};\qquad(4.35a)$$

$$b_j=1-\frac{2\beta_1}{K^2\Delta x^2}+\frac{\beta_1}{K^2}\left(k^2-K^2\right);\qquad(4.35b)$$

$$c_j=\frac{\beta_1}{K^2\Delta x^2};\qquad(4.35c)$$

$$r_j=u_j^m+\frac{\alpha_1}{K^2}\frac{u_{j-1}^m-2u_j^m+u_{j+1}^m}{\Delta x^2}+\frac{\alpha_1}{K^2}\left(k^2-K^2\right)u_j^m.\qquad(4.35d)$$

This tridiagonal system can be solved efficiently by using the Thomas algorithm (Appendix B).

Once the first step has been performed, the calculated new field $u^{m+1/2}$ is used to perform the second step to obtain finally the field u^{m+1}. The procedure is identical to that exposed here but now using coefficients α_2 and β_2 (instead of the α_1 and β_1 coefficients).

The applicability, accuracy and effectiveness of the wide-angle BPM are now shown through some examples. First, the simulation of a 2D Gaussian beam propagating in free space with a 30° tilt with respect to the z-axis is performed by both paraxial and wide-angle BPM. For small angles, where paraxiality is fulfilled, conventional BPM gives accurate results. Nevertheless, when the propagation angle (with respect to the z-direction) increases, paraxiality is lost and the numerical result using classical BPM degrades its accuracy. In these cases, wide-angle BPM is needed. Figure 4.1 shows the simulation of the free space propagation of a 30° tilted Gaussian beam, with a width of 4 μm and at $\lambda = 1.55$ μm. The numerical simulations are performed using transversal grid spacing of $\Delta x = 0.04$ μm and longitudinal steps of $\Delta z = 0.05$ μm. As the beam propagates, it suffers a shift along the transversal direction and also broadens because of diffraction. The input at $z = 0$ and the outputs at $z = 20$ μm are plotted in the same simulation window in order to facilitate the comparison of their relative positions and intensities. As shown, the wide-angle BPM result (using Padé (2,2) approximants) has a closer intensity to the exact solution with a slight shift in position, while the paraxial BPM calculation propagates at the wrong angle and incorrectly preserves the symmetric Gaussian shape, rather than the spread out, asymmetric profile.

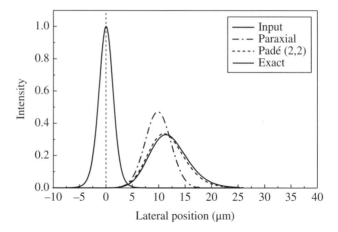

Figure 4.1 Intensity profiles of the input and outputs resulting from the 2D propagation of a 30° phase tilt Gaussian beam along a distance of 20 μm through free space. Continuous line: input profile ($z = 0$). Dashed-dotted line: profile of the Gaussian beam obtained by paraxial BPM. Dashed line: profile computed by wide-angle BPM, using Padé (2,2) approximants. Thin continuous line: exact profile

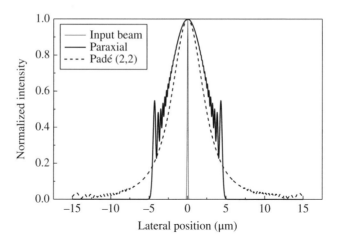

Figure 4.2 Intensity profile of a 100 nm width Gaussian beam after 6.4 μm propagation in a homogeneous medium with refractive index $= 2.2$. Thin continuous line: input profile ($z = 0$). Continuous line: profile of the Gaussian beam after 6.4 μm propagation in the z direction, obtained by paraxial BPM. Dashed line: profile at 6.4 μm computed by wide-angle BPM, using Padé (2,2) approximants

The propagation a highly divergent beam through a homogeneous medium provides a second example to illustrate the improvement of wide-angle BPM with respect to conventional BPM. Figure 4.2 represents the intensity profile of a 100 nm width Gaussian beam with $\lambda = 532$ nm after a propagation of 6.5 μm in a homogeneous medium with refractive index $n = 2.2$, computed using paraxial BPM (continuous line) and wide-angle BPM (dashed line). Grid spacing of $\Delta x = 0.01$ μm and longitudinal steps of $\Delta z = 0.1$ μm are used. It is clear that paraxial BPM cannot correctly simulate the propagation of the highly divergent beam due to the presence of a large amount of high spatial frequencies. By contrast, Padé (2,2) is able to model the broadening of the narrow Gaussian beam quite correctly, although some spurious noise at the tail of the profile is observed. Of course, higher order Padé approximants can be used to obtain better accuracy of the propagation.

A last example is provided to demonstrate that when using wide-angle BPM schemes the choice of reference index can be more relaxed to obtain high accuracy in optical propagation. For such a purpose, the coupling length for TE-propagation of two parallel planar waveguides is examined. The coupler consists of two identical step-index planar waveguides of 1.6 μm width, separated by 2 μm, having a core refractive index of 1.46 and surrounded by a medium of unity refractive index (Figure 4.3).

Figure 4.4 shows the percentage in the relative error of the computed coupling length of the device at a working wavelength of $\lambda = 1.55$ μm as a function of the reference index used in the simulation, for different BPM approaches using a propagation step of $\Delta z = 0.2$ μm. The paraxial simulation (Padé (1,0), dashed line) requires us to perform the simulation with a reference

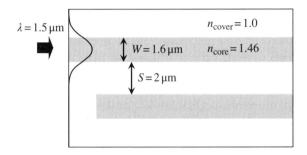

Figure 4.3   Coupler made of two identical planar waveguides

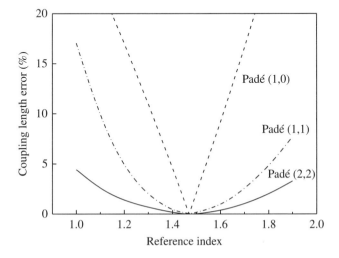

Figure 4.4   Influence of the reference index chosen in the BPM algorithm on the coupling length accuracy of a parallel coupler for different BPM approaches (paraxial and wide-angle)

index close to 1.47 to obtain an acceptable degree of accuracy. By contrast, using wide angle-BPM with Padé (2,2) (continuous line), a very wide range on the reference index (from 1.0 to 1.9) can be used preserving the accuracy of the propagation with errors lower than 5% in the coupling length. In this sense, wide-angle BPM also gives improved accuracy in the simulation for multimode propagation, where a large range of modal effective indices can be present.

The wide-angle BPM based on Padé approximant operators and the multi-step method here described developed by Hadley [1, 3] is the most commonly employed technique to improve the numerical accuracy for wide-angle simulations of 2D structures in the scalar approach. Extension of the algorithm to a 2D-vectorial BPM has been also demonstrated [4]. When it is applied to 3D structures, large sparse matrices are involved where the matrix algebraic equation is no longer tridiagonal and needs to be solved using iterative methods. However, alternative approaches have been proposed and semi-vectorial 3D wide-angle BPM numerical schemes, without an iteration procedure, have been developed [5, 6].

4.2   Treatment of Discontinuities in BPM

Conventional BPMs here described so far are useful only if the refractive index of the structure varies sufficiently slowly along the propagation direction so that the reflections at each step, or the accumulated reflections along a certain distance, are negligible. Nevertheless, many practical integrated optical devices involve junctions of different waveguides (having different materials and/or different cross sections), laser facets, gratings, multiple dielectric layers and so on, which give rise to non-negligible reflected fields. Thus, for an accurate modelling of the optical fields propagation in such structures, which should include reflection and/or transmission, it is necessary to develop special algorithms that consider coupling of the forward and backward waves [7]. In this section, the problem of reflection at a waveguide interface is discussed, such as the problem of the reflection at a waveguide laser facet. Some results of this problem are then used in the Section 4.3 to develop a BPM algorithm that can handle multiple reflections, known as bidirectional-BPM.

4.2.1   Reflection and Transmission at an Interface

Let us consider the 2D Helmholtz wave equation for the transverse component ψ of a monochromatic TE (transverse electric) or TM (transverse magnetic) electromagnetic field:

$$\frac{\partial^2 \psi}{\partial x^2} + \frac{\partial^2 \psi}{\partial z^2} + k_0^2 n^2(x,z)\psi = 0; \quad \text{TE} \qquad (4.36a)$$

$$\frac{\partial}{\partial x}\left(\frac{1}{n^2(x,z)}\frac{\partial \psi}{\partial x}\right) + \frac{\partial^2 \psi}{\partial z^2} + k_0^2 n^2(x,z)\psi = 0. \quad \text{TM} \qquad (4.36b)$$

For TE waves ψ represents the component E_y of the electric field, while ψ is the component H_y of the magnetic field for TM waves and $n(x,z)$ is the refractive index of the structure. We now deal with the case of an incident field propagating in the $+z$ direction from a region A with refractive index $n_A(x)$ into a region B with refractive index $n_B(x)$. The refractive index of the structure is therefore:

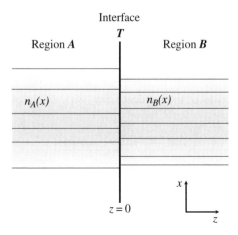

Figure 4.5   Diagram to illustrate the reflectivity problem. Two layered media, A and B, are separated by an interface T situated at $z = 0$. If an incident wave from region A impinges the interface T, then a reflected wave and a transmitted wave are generated

$$n(x,z) = \begin{cases} n_A(x) & z \leq 0 \\ n_B(x) & z > 0 \end{cases} \tag{4.37}$$

where the regions A and B are composed of layered medium, as shown in Figure 4.5.

The fields in region A and B can be expressed in terms of the fields at the interface T (located at $z = 0$) as [8]:

$$\psi_A = \exp\left[-i\sqrt{\mathcal{L}_A}z\right] \psi_A^+ + \exp\left[+i\sqrt{\mathcal{L}_A}z\right] \psi_A^-; \tag{4.38a}$$

$$\psi_B = \exp\left[-i\sqrt{\mathcal{L}_B}z\right] \psi_B^+ + \exp\left[+i\sqrt{\mathcal{L}_B}z\right] \psi_B^-, \tag{4.38b}$$

where the $+$ and $-$ superscripts denote forward and backward fields, respectively. In these formulae, $\mathcal{L}_A$ and $\mathcal{L}_B$ denote pseudo-differential operators, defined by:

$$\mathcal{L}_j \equiv \begin{cases} \dfrac{\partial^2}{\partial x^2} + k_0^2 n_j^2(x) & \text{TE} \\ \\ \dfrac{\partial}{\partial x}\left(\dfrac{1}{n_j^2(x)}\dfrac{\partial}{\partial x}\right) + k_0^2 n_j^2(x) & \text{TM} \end{cases} \quad (j = A, B) \tag{4.39}$$

If an incident field ψ_A^+ is launched onto the interface T, then reflected ψ_A^- and transmitted ψ_B^+ fields are induced.

For TE incidence ($\psi \equiv E_y$) the boundary conditions imply that both ψ and $d\psi/dz$ must be continuous at the interface:

$$\psi_A^+ + \psi_A^- = \psi_B^+ + \psi_B^-; \tag{4.40a}$$

$$\sqrt{\mathcal{L}_A}\psi_A^+ - \sqrt{\mathcal{L}_A}\psi_A^- = \sqrt{\mathcal{L}_B}\psi_B^+ - \sqrt{\mathcal{L}_B}\psi_B^-. \tag{4.40b}$$

In region A the electric field is composed of a forward travelling incident field and a backward travelling reflected field, while only a forward propagating transmitted field exists in region B and therefore $\psi_B^- = 0$. From these two relations, we can obtain the expressions for the reflected and transmitted fields in terms of the incident field:

$$\psi_A^- = \frac{\sqrt{\mathcal{L}_A} - \sqrt{\mathcal{L}_B}}{\sqrt{\mathcal{L}_A} + \sqrt{\mathcal{L}_B}}\psi_A^+; \tag{4.41a}$$

$$\psi_B^+ = \frac{2\sqrt{\mathcal{L}_A}}{\sqrt{\mathcal{L}_A} + \sqrt{\mathcal{L}_B}}\psi_A^+, \tag{4.41b}$$

In a similar way, for TM fields the component of the field to be considered is H_y and both ψ and $(1/n^2)d\psi/dz$ should be continuous at the interface T, yielding:

$$\psi_A^- = \frac{\frac{1}{n_A^2}\sqrt{\mathcal{L}_A} - \frac{1}{n_B^2}\sqrt{\mathcal{L}_B}}{\frac{1}{n_A^2}\sqrt{\mathcal{L}_A} + \frac{1}{n_B^2}\sqrt{\mathcal{L}_B}}\psi_A^+; \tag{4.42a}$$

$$\psi_B^+ = \frac{\frac{2}{n_A^2}\sqrt{\mathcal{L}_A}}{\frac{1}{n_A^2}\sqrt{\mathcal{L}_A} + \frac{1}{n_B^2}\sqrt{\mathcal{L}_B}}\psi_A^+. \tag{4.42b}$$

where ψ_A^+, ψ_A^- and ψ_B^+ denote the incident, reflected and transmitted magnetic fields at T, respectively.

The difficulty encountered in evaluating the reflected and transmitted fields using equations in Eq. (4.41) (or (4.42)) is associated with the determination of the square root of the characteristic operators, $\mathcal{L}_A$ and $\mathcal{L}_B$. One method of simplification is achieved by approximating the square root of the operators by the Taylor expansion or by a rational function of the operators using Padé approximants [8]. For doing that, the $\mathcal{L}_j$ operator is rewritten in the form:

$$\mathcal{L}_j \equiv K_j^2 \left(1 + X_j\right), \tag{4.43}$$

where the dimensionless operator X_j has been defined as:

$$X_j \equiv \begin{cases} \dfrac{1}{K_j^2}\left[\dfrac{\partial^2}{\partial x^2} + k_0^2 n_j^2(x) - K_j^2\right] & \text{TE} \\[3mm] \dfrac{1}{K_j^2}\left[n_j^2(x)\left[\dfrac{\partial}{\partial x}\left(\dfrac{1}{n_j^2(x)}\dfrac{\partial}{\partial x}\right)\right] + k_0^2 n_j^2(x) - K_j^2\right] & \text{TM} \end{cases} \tag{4.44}$$

and the reference wavenumber K_j has been introduced. As we will see, this reference wavenumber must be adequately chosen to minimize the value of the X_j operator and so to obtain a good approximation, which should be exact for plane waves in homogeneous medium, choosing K_j as the wavenumber of the medium.

The square root operators in Eq. (4.41) (or Eq. (4.42)) is usually rationalized by using Padé approximants of order (m,m) [9]:

$$\sqrt{L_j} \equiv K_j \sqrt{1 + X_j} \approx K_j \left[1 + \sum_{p=1}^{m} \frac{a_p X_j}{1 + b_p X_j} \right]. \qquad (4.45)$$

These coefficients are given by the following expressions [10]:

$$a_p = \frac{2}{2n+1} \sin^2 \left(\frac{p\pi}{2n+1} \right); \qquad (4.46a)$$

$$b_p = \cos^2 \left(\frac{p\pi}{2n+1} \right). \qquad (4.46b)$$

If the reference wavenumber K_j and the Padé coefficients a_p and b_p are all real-valued, then the approximation of the squared root of the operator L_j corresponds to a mapping of the real axis onto itself. Therefore, although the propagating modes, corresponding to real-to-real mapping, are accurate to the order of the approximation, the evanescent modes, which should be mapped onto the positive imaginary axis, only retain zeroth-order accuracy for any m [11]. Consequently, the evanescent fields generated at the interface are treated as propagating fields and cannot decay or grow as they should. This inappropriate treatment of evanescent modes leads to computational instability, because the evanescent modes are assigned real eigenvalues that can result in a vanishing denominator in evaluation of the reflection operator [9]. Even if the simulation remains stable, calculation of reflection and transmission coefficients can be grossly inaccurate as an evanescent field propagates through the system rather than being localized at the interface [10].

The solution of these inaccuracy and instability problems is to find a better propagator that does not map to the real axis. Two options include choosing a complex-valued reference wavenumber K_j or using a complex representation of the Padé approximation. In the former approach [12], a phase φ is added to the reference wavenumber:

$$\bar{K}_j \equiv K_j e^{i\varphi}, \qquad (4.47)$$

where K_j is the original real-valued reference wavenumber. Its imaginary part accommodates evanescent waves that can be generated when guided modes on the high index side become evanescent on the low index side. Nevertheless, in high contrast-index cases, a single choice of the reference wavenumber does not adequately match either the guided or the evanescent mode behaviour because of the spread of wavenumbers generated at the interface.

A second approach consists of rotating the branch cut of the square-root operator in making the Padé approximants [10]. That is, when approximating the square root of the operator, L_j, the branch cut of the square root function is rotated away from the original negative real axis:

$$\sqrt{L_j} = K_j e^{i\alpha/2} \sqrt{1 + \left[(1 + X_j) e^{-i\alpha} - 1 \right]} \approx K_j e^{i\alpha/2} \left[1 + \sum_{p=1}^{m} \frac{a_p \left[(1 + X_j) e^{-i\alpha} - 1 \right]}{1 + b_p \left[(1 + X_j) e^{-i\alpha} - 1 \right]} \right], \qquad (4.48)$$

where α is the rotation angle and K_j can be either real-valued or complex-valued. It has been shown that the complex wavenumber method and the rotated branch cut method with real-valued wavenumber are equivalent if $\alpha = 2\varphi$ [11].

4.2.2 Implementation Using First-Order Approximation to the Square Root

Assuming TE waves, we approximate the square root of the operator L_j to its first-order polynomial expansion:

$$\sqrt{L_j} = K_j \sqrt{1 + X_j} \approx K_j \left[1 + \frac{X_j}{2}\right] = \frac{1}{2K_j} \left(\frac{\partial^2}{\partial x^2} + k_0^2 n_j^2 + K_j^2\right). \tag{4.49}$$

Using this approximation, the reflected field for TE incidence is:

$$\psi_A^- \approx \frac{\dfrac{K_B}{K_A}\left(\dfrac{\partial^2}{\partial x^2} + k_0^2 n_A^2 + K_A^2\right) - \left(\dfrac{\partial^2}{\partial x^2} + k_0^2 n_B^2 + K_B^2\right)}{\dfrac{K_B}{K_A}\left(\dfrac{\partial^2}{\partial x^2} + k_0^2 n_A^2 + K_A^2\right) + \left(\dfrac{\partial^2}{\partial x^2} + k_0^2 n_B^2 + K_B^2\right)} \psi_A^+, \tag{4.50}$$

which is the formal representation of:

$$\left[\left(\frac{K_B}{K_A} + 1\right)\frac{\partial^2}{\partial x^2} + \frac{K_B}{K_A}k_0^2 n_A^2 + k_0^2 n_B^2 + K_A K_B + K_B^2\right]\psi_A^-$$

$$= \left[\left(\frac{K_B}{K_A} - 1\right)\frac{\partial^2}{\partial x^2} + \frac{K_B}{K_A}k_0^2 n_A^2 - k_0^2 n_B^2 + K_A K_B - K_B^2\right]\psi_A^+ \tag{4.51}$$

Now approximating the operators by the usual finite difference schemes, it results in:

$$\left(\frac{K_B}{K_A} + 1\right)\frac{u_{j-1}^- - 2u_j^- + u_{j+1}^-}{(\Delta x^2)} + \left(\frac{K_B}{K_A}k_0^2 n_A^2 + k_0^2 n_B^2 + K_A K_B + K_B^2\right)u_j^-$$

$$= \left(\frac{K_B}{K_A} - 1\right)\frac{u_{j-1}^+ - 2u_j^+ + u_{j+1}^+}{(\Delta x^2)} + \left(\frac{K_B}{K_A}k_0^2 n_A^2 - k_0^2 n_B^2 + K_A K_B - K_B^2\right)u_j^+ \tag{4.52}$$

where the incident field ψ_A^+ at the position x_j has been represented by u_j^+ and the reflected field in the region A ψ_A^- is denoted by u_j^-. This set of equations forms a tridiagonal system that can be solved very efficiently by using the Thomas algorithm, as was seen in Chapter 2. The coefficients for such system are the following:

$$a_j = \left(\frac{K_B}{K_A} + 1\right)/(\Delta x)^2; \tag{4.53a}$$

$$b_j = -2\left(\frac{K_B}{K_A} + 1\right) / (\Delta x)^2 + \frac{K_B}{K_A} k_0^2 n_A^2 - k_0^2 n_B^2 + K_A K_B + K_B^2; \tag{4.53b}$$

$$c_j = \left(\frac{K_B}{K_A} + 1\right) / (\Delta x)^2; \tag{4.53c}$$

$$r_j = \left[-2\left(\frac{K_B}{K_A} + 1\right) / (\Delta x)^2 + \frac{K_B}{K_A} k_0^2 n_A^2 - k_0^2 n_B^2 + K_A K_B - K_B^2\right] u_j^+ \\ + \left(\frac{K_B}{K_A} + 1\right)\left(u_{j-1}^+ + u_{j+1}^+\right) / (\Delta x)^2 \tag{4.53d}$$

Proceeding in a similar way for the transmitted field ψ_B^+, it gives:

$$\psi_B^+ \approx \frac{2\frac{K_B}{K_A}\left(\frac{\partial^2}{\partial x^2} + k_0^2 n_A^2 + K_A^2\right)}{\frac{K_B}{K_A}\left(\frac{\partial^2}{\partial x^2} + k_0^2 n_A^2 + K_A^2\right) + \left(\frac{\partial^2}{\partial x^2} + k_0^2 n_B^2 + K_B^2\right)} \psi_A^+ \tag{4.54}$$

which is the formal representation of:

$$\left[\left(\frac{K_B}{K_A} + 1\right)\frac{\partial^2}{\partial x^2} + \frac{K_B}{K_A} k_0^2 n_A^2 + k_0^2 n_B^2 + K_A K_B + K_B^2\right]\psi_B^+ \\ = \left[2\frac{K_B}{K_A}\frac{\partial^2}{\partial x^2} + 2\frac{K_B}{K_A} k_0^2 n_A^2 + 2 K_A K_B\right]\psi_A^+ \tag{4.55}$$

Now approximating the operators by the usual finite-difference schemes, it yields:

$$\left(\frac{K_B}{K_A} + 1\right)\frac{w_{j-1}^+ - 2w_j^+ + w_{j+1}^+}{(\Delta x)^2} + \left(\frac{K_B}{K_A} k_0^2 n_A^2 + k_0^2 n_B^2 + K_A K_B + K_B^2\right)w_j^+ \\ = 2\frac{K_B}{K_A}\frac{u_{j-1}^+ - 2u_j^+ + u_{j+1}^+}{(\Delta x)^2} + \left(2\frac{K_B}{K_A} k_0^2 n_A^2 + 2 K_A K_B\right)u_j^+ \tag{4.56}$$

where the forward field at the region B ψ_B^+ has been denoted by w_j^+. This set of equations forms again a tridiagonal system, which can be solved to obtain the transmitted field. Nevertheless, once the reflected field ψ_A^- is calculated, the transmitted field ψ_B^+ can be obtained directly by using the relation (4.40a), recalling that $\psi_B^- = 0$.

Figure 4.6 shows one example of application of the reflectivity calculations in waveguides. The input beam (left) corresponds to the fundamental mode of the planar waveguide. When the incident beam reaches the waveguide/air interface, the radiation splits in a reflected and a transmitted beam. In addition, only a fraction of the reflected field corresponds to the fundamental waveguide mode (see Appendix C). Following the geometry shown in that figure, the reflectivity of the fundamental mode incident on the waveguide facet as a function of the core width, for TE and TM incidences and for two different index contrasts, is presented in Figure 4.7. The

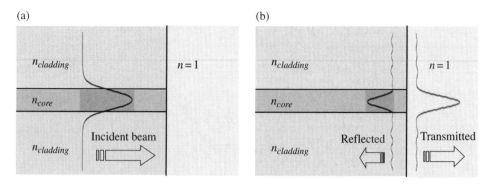

Figure 4.6 Geometry of the waveguide–air discontinuity. (a) Waveguide mode propagating to the waveguide-end. (b) Reflected and transmitted beams generated at the discontinuity

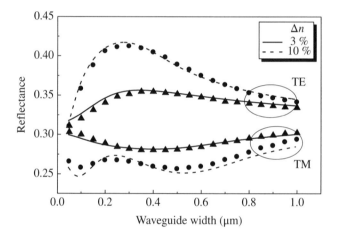

Figure 4.7 TE and TM reflectivity of the fundamental mode for a step-index planar waveguide as a function of the core width, at a wavelength of 0.86 μm, for two different cladding refractive indices, being the core refractive index 3.60. Dashed lines: calculations based on Padé (1,0) for an index contrast of 10%; continuous lines: calculations using Padé (1,0) for an index contrast of 3%. Symbols represent the exact values

example simulates the case of a semiconductor laser, where the waveguide air provides the necessary feedback to maintain the laser oscillation. The core has a refractive index of 3.60 (GaAs), while for the refractive index of the cladding two values have been considered: 3.24 for high index contrast ($\Delta n = 10\%$) and 3.492 for medium index contrast ($\Delta n = 3\%$). The working wavelength in the numerical simulation is $\lambda = 0.86$ μm.

The reflectance curves show maxima and minima, which can be understood qualitatively considering the mode size and the angular spread of the fundamental TE-mode of the symmetric planar waveguide as a function of the core width (Figure 4.8), besides the reflectance of planar waves at the semiconductor–air interface (Figure 4.9) [13, 14]. The Brewster angle and the critical angle of an incident plane wave into air from GaAs ($n = 3.60$) are $\sim15.5°$ and $\sim16.1°$, respectively. For large core widths, the Fourier spectrum (spatial frequencies) in the lateral direction is quite narrow. This results in angles smaller than the Brewster angle for the incident waves. As the core width is decreased, initially the lateral modes become more confined and their spectrum broadens. As a result, the reflectivity, which follows the TM behavior in the lateral direction, decreases as well. This, of course, contradicts the TE reflectivity. If the field of the mode is decomposed into plane waves with propagation vectors having a large incident angle extending over 16°, the reflectivity of the TE wave is larger and that of the TM wave is small, in comparison with that of the normal incident plane wave. The field distribution is most

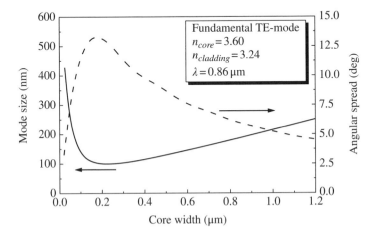

Figure 4.8   Continuous line: mode size of the fundamental TE-mode of the symmetric planar waveguide as a function of the core width, at a wavelength of $\lambda = 0.86\,\mu m$, showing a minimum at $d \approx 0.22\,\mu m$. Dashed line: angular spread of the spatial frequency spectrum of the fundamental guided mode

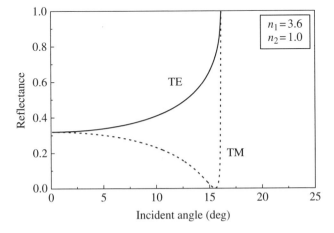

Figure 4.9   TE and TM reflectances for planar waves at the semiconductor $(n_1 = 3.6)$ – air $(n_2 = 1.0)$ interface. The Brewster angle is situated at 15.3° and the angle for total internal reflection is located at 16.1°

concentrated in the waveguide core at the thickness where the reflectivity shows the maxima for TE waves.

In general, the reflectivity of the TE wave is larger that of the TM wave and the difference between them is large enough to select TE oscillation in usual double-heterostructure injection lasers [13].

4.3   Bidirectional BPM

Among several proposals that deal with bidirectional BPM, one of the most popular and efficient is a bidirectional BMP algorithm based on an iterative method [12]. The iterative bidirectional BPM approach presented here can handle problems involving an arbitrary number of dielectric interfaces and the algorithm has significantly reduced time and memory requirements compared to other multi-interface techniques.

4.3.1   Formulation of Iterative Bi-BPM

The approach of the iterative bidirectional BPM [12] starts by assuming that the reflected field is known at the input of the structure. Then a BPM-based transfer matrix approach is applied to relate the forward and backward fields at the output to those at the input. Based upon this relation and the boundary conditions, the reflected field is found in an iterative manner. This approach is simple to program and has the advantage that only one pair of fields must be stored.

For two-dimensional structures, the fields at any fixed position in the propagation direction (z-direction) are represented by a two-component vector $\boldsymbol{\psi}(x)$ consisting of the forward (ψ^{+}) and backward (ψ^{-}) fields:

$$\boldsymbol{\psi}(x) = \begin{pmatrix} \psi^{+}(x) \\ \psi^{-}(x) \end{pmatrix}. \tag{4.57}$$

The light propagation through the structure is modelled via two kinds of matrices: a propagation matrix ($\mathcal{P}$), which transfers $\boldsymbol{\psi}$ from one end of a z homogeneous (or quasi-homogeneous) region to the other end and an interface matrix ($\mathcal{T}$), which transfers $\boldsymbol{\psi}$ from one side of an interface to the other side. Thus, given an arbitrary dielectric structure, the light propagation can be described using alternating products of a series of $\mathcal{P}$s and $\mathcal{T}$s (Figure 4.10). The overall transfer matrix M, which relates the field at the output region $\boldsymbol{\psi}_{out}$ to the field at the input region $\boldsymbol{\psi}_{in}$, is then built as:

$$\mathcal{M} \equiv \mathcal{P}_n \mathcal{T}_{n-1,n} \mathcal{P}_{n-1} \mathcal{T}_{n-2,n-1} \cdots \cdots \mathcal{P}_3 \mathcal{T}_{2,3} \mathcal{P}_2 \mathcal{T}_{1,2} \mathcal{P}_1. \tag{4.58}$$

and therefore:

$$\begin{pmatrix} \psi_{out}^{+} \\ \psi_{out}^{-} \end{pmatrix} = \mathcal{M} \begin{pmatrix} \psi_{in}^{+} \\ \psi_{in}^{-} \end{pmatrix}. \tag{4.59}$$

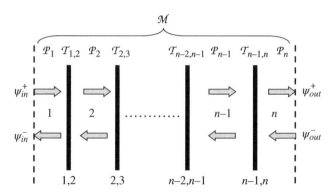

Figure 4.10 Schematic representation for the bidirectional BPM calculation, where the structure is divided on n longitudinal invariant sections, separated by $(n-1)$ discrete discontinuities

The propagation matrix $\mathcal{P}_j$ propagates the forward and backward fields independently, using conventional BPM for ψ^+ and 'time-reversed' conventional BPM for ψ^-:

$$
\mathcal{P}_j = \begin{pmatrix} e^{-i\int L_j dz} & 0 \\ 0 & e^{i\int L_j dz} \end{pmatrix},
\tag{4.60}
$$

where $L_j \equiv \frac{\partial^2}{\partial x^2} + k_0^2 n_j^2(x)$, being $n_j(x)$ the refractive index profile in region j and k_0 is the wavenumber in vacuum. Paraxial BPM is sufficient for many cases, but for problems with large index variation, a wide-angle BPM algorithm must be employed to yield the correct phase.

On the other hand, the transition matrix $\mathcal{T}_{AB}$ for the interface between region A (at position j) and next region B (at position $j+1$) is given by:

$$
\mathcal{T}_{AB} = \frac{1}{2} \begin{pmatrix} 1 + g\sqrt{L_B^{-1}}\sqrt{L_A} & 1 - g\sqrt{L_B^{-1}}\sqrt{L_A} \\ 1 - g\sqrt{L_B^{-1}}\sqrt{L_A} & 1 + g\sqrt{L_B^{-1}}\sqrt{L_A} \end{pmatrix},
\tag{4.61}
$$

with $g = 1$ for TE case and $g = (n_B/n_A)^2$ for TM propagation. This means that the fields in region B are related with the fields in region A by the following expression:

$$
\begin{pmatrix} \psi_B^+ \\ \psi_B^- \end{pmatrix} = \frac{1}{2} \begin{pmatrix} 1 + g\sqrt{L_B^{-1}}\sqrt{L_A} & 1 - g\sqrt{L_B^{-1}}\sqrt{L_A} \\ 1 - g\sqrt{L_B^{-1}}\sqrt{L_A} & 1 + g\sqrt{L_B^{-1}}\sqrt{L_A} \end{pmatrix} \begin{pmatrix} \psi_A^+ \\ \psi_A^- \end{pmatrix}.
\tag{4.62}
$$

As consequence, the forward field in region **B** is:

$$
\psi_B^+ = \frac{1}{2}\left(1 + g\sqrt{L_B^{-1}}\sqrt{L_A}\right)\psi_A^+ + \frac{1}{2}\left(1 - g\sqrt{L_B^{-1}}\sqrt{L_A}\right)\psi_A^-,
\tag{4.63}
$$

which can be more conveniently written as:

$$
\psi_B^+ = \left(\frac{\sqrt{L_B} + g\sqrt{L_A}}{2\sqrt{L_B}} \right) \psi_A^+ + \left(\frac{\sqrt{L_B} - g\sqrt{L_A}}{2\sqrt{L_B}} \right) \psi_A^-.
\tag{4.64}
$$

Similarly, the backward field in region B is derived as:

$$
\psi_B^- = \left(\frac{\sqrt{L_B} - g\sqrt{L_A}}{2\sqrt{L_B}} \right) \psi_A^+ + \left(\frac{\sqrt{L_B} + g\sqrt{L_A}}{2\sqrt{L_B}} \right) \psi_A^-.
\tag{4.65}
$$

The difficulty encountered in evaluating the reflected and transmitted fields in Eqs. (4.64) and (4.65) is associated with the determination of the square root of the characteristic operators. One way of simplification is achieved by approximating the square root of the operators by a polynomial expansion, or by a rational function of the operators using Padé approximants. For doing that, first the L_j operator is rewritten in the form:

$$
L_j \equiv \frac{\partial^2}{\partial x^2} + k_0^2 n_j^2(x) = K_j^2 \left[1 + \frac{\frac{\partial^2}{\partial x^2} + k_0^2 n_j^2(x) - K_j^2}{K_j^2} \right] = K_j^2(1 + X_j),
\tag{4.66}
$$

where we have defined the dimensionless operator X_j as:

$$
X_j \equiv \frac{\frac{\partial^2}{\partial x^2} + k_0^2 n_j^2(x) - K_j^2}{K_j^2},
\tag{4.67}
$$

and we have introduced the reference wavenumber K_j. As we will see later, this reference wavenumber must be properly chosen to obtain a good approximation. In many practical applications, the square root operator is usually rationalized by its first-order polynomial expansion:

$$
\sqrt{L_j} \equiv K_j \sqrt{1 + X_j} \approx K_j \left[1 + \frac{1}{2} X_j \right] = \frac{1}{2K_j} \left[\frac{\partial^2}{\partial x^2} + k_0^2 n_j^2(x) + K_j^2 \right].
\tag{4.68}
$$

To obtain stable propagation, the reference wavenumber should be chosen to be complex, as shown in the previous section. For low index contrast structures a small imaginary part (small rotation angles, $\varphi \sim 0.1°$) of the reference wavenumber is enough to reach stability, but higher rotation angles are necessary to maintain stable bidirectional propagation in high index contrast structures. Also, if the portion of evanescent fields created at the interfaces are significant, then one should use a complex reference wavenumber for modelling the propagation operator as well [11].

4.3.2　Finite-Difference Approach of the Bi-BPM

Using the first-order polynomial approximation of the square root operators let us derivate the finite-difference equations that govern the transmitted field into region **B**. From Eq. (4.64) and using the approximation (4.68), we obtain:

$$
2\psi_B^+ = \left(\frac{\left[\dfrac{\partial^2}{\partial x^2} + k_0^2 n_B^2(x) + K_B^2 \right] + g\dfrac{K_B}{K_A}\left[\dfrac{\partial^2}{\partial x^2} + k_0^2 n_A^2(x) + K_A^2 \right]}{\dfrac{\partial^2}{\partial x^2} + k_0^2 n_B^2(x) + K_B^2} \right) \psi_A^+
$$

$$
+ \left(\frac{\left[\dfrac{\partial^2}{\partial x^2} + k_0^2 n_B^2(x) + K_B^2 \right] - g\dfrac{K_B}{K_A}\left[\dfrac{\partial^2}{\partial x^2} + k_0^2 n_A^2(x) + K_A^2 \right]}{\dfrac{\partial^2}{\partial x^2} + k_0^2 n_B^2(x) + K_B^2} \right) \psi_A^-
$$

(4.69)

which is equivalent to:

$$
2\left[\frac{\partial^2}{\partial x^2} + k_0^2 n_B^2 + K_B^2 \right] \psi_B^+ = \left(\left[\frac{\partial^2}{\partial x^2} + k_0^2 n_B^2 + K_B^2 \right] + g\frac{K_B}{K_A}\left[\frac{\partial^2}{\partial x^2} + k_0^2 n_A^2 + K_A^2 \right] \right) \psi_A^+
$$

$$
+ \left(\left[\frac{\partial^2}{\partial x^2} + k_0^2 n_B^2 + K_B^2 \right] - g\frac{K_B}{K_A}\left[\frac{\partial^2}{\partial x^2} + k_0^2 n_A^2 + K_A^2 \right] \right) \psi_A^-
$$

(4.70)

Now, from Eq. (4.65), the corresponding backward field in region **B** yields:

$$
2\left[\frac{\partial^2}{\partial x^2} + k_0^2 n_B^2 + K_B^2 \right] \psi_B^- = \left(\left[\frac{\partial^2}{\partial x^2} + k_0^2 n_B^2 + K_B^2 \right] + g\frac{K_B}{K_A}\left[\frac{\partial^2}{\partial x^2} + k_0^2 n_A^2 + K_A^2 \right] \right) \psi_A^+
$$

$$
+ \left(\left[\frac{\partial^2}{\partial x^2} + k_0^2 n_B^2 + K_B^2 \right] + g\frac{K_B}{K_A}\left[\frac{\partial^2}{\partial x^2} + k_0^2 n_A^2 + K_A^2 \right] \right) \psi_A^-
$$

(4.71)

The finite difference expression of the forward (transmitted) field in region **B** governed by Eq. (4.70) is given by:

$$
2\left[-\frac{2}{(\Delta x)^2} + k_0^2 n_B^2 + K_B^2 \right] \psi_{B,j}^+ + \frac{2}{(\Delta x)^2}\left(\psi_{B,j-1}^+ + \psi_{B,j+1}^+ \right)
$$

$$
= \left[-\frac{2\left(1+g\frac{K_B}{K_A}\right)}{(\Delta x)^2} + k_0^2 n_B^2 + g\frac{K_B}{K_A}k_0^2 n_A^2 + K_B^2 + gK_BK_A \right] \psi_{A,j}^+ + \frac{\left(1+g\frac{K_B}{K_A}\right)}{(\Delta x)^2}\left(\psi_{A,j-1}^+ + \psi_{A,j+1}^+\right)
$$

$$
+ \left[-\frac{2\left(1-g\frac{K_B}{K_A}\right)}{(\Delta x)^2} + k_0^2 n_B^2 - g\frac{K_B}{K_A}k_0^2 n_A^2 + K_B^2 - gK_BK_A \right] \psi_{A,j}^+ + \frac{\left(1-g\frac{K_B}{K_A}\right)}{(\Delta x)^2}\left(\psi_{A,j-1}^- + \psi_{A,j+1}^-\right)
$$

$$
\tag{4.72}
$$

This equation represents a tridiagonal system, with the following coefficients:

$$
a_j = \frac{2}{(\Delta x)^2}; \tag{4.73a}
$$

$$
b_j = -\frac{4}{(\Delta x)^2} + 2k_0^2 n_B^2 + 2K_B^2; \tag{4.73b}
$$

$$
c_j = \frac{2}{(\Delta x)^2}; \tag{4.73c}
$$

$$
r_j = \left[-\frac{2\left(1+g\frac{K_B}{K_A}\right)}{(\Delta x)^2} + k_0^2 n_B^2 + g\frac{K_B}{K_A}k_0^2 n_A^2 + K_B^2 + gK_BK_A \right] \psi_{A,j}^+ + \frac{\left(1+g\frac{K_B}{K_A}\right)}{(\Delta x)^2}\left(\psi_{A,j-1}^+ + \psi_{A,j+1}^+\right)
$$

$$
+ \left[-\frac{2\left(1+g\frac{K_B}{K_A}\right)}{(\Delta x)^2} + k_0^2 n_B^2 - g\frac{K_B}{K_A}k_0^2 n_A^2 + K_B^2 - gK_BK_A \right] \psi_{A,j}^- + \frac{\left(1-g\frac{K_B}{K_A}\right)}{(\Delta x)^2}\left(\psi_{A,j-1}^- + \psi_{A,j+1}^-\right)
$$

$$
\tag{4.73d}
$$

Proceeding in a very similar way, the coefficients of the tridiagonal system corresponding to the backward (reflected) field in region **B** are expressed as:

$$
a_j = \frac{2}{(\Delta x)^2}; \tag{4.74a}
$$

$$
b_j = -\frac{4}{(\Delta x)^2} + 2k_0^2 n_B^2 + 2K_B^2; \tag{4.74b}
$$

$$
c_j = \frac{2}{(\Delta x)^2}; \tag{4.74c}
$$

$$
r_j = \left[-\frac{2\left(1-g\dfrac{K_B}{K_A}\right)}{(\Delta x)^2} + k_0^2 n_B^2 - g\frac{K_B}{K_A}k_0^2 n_A^2 + K_B^2 - gK_B K_A \right] \psi_{A,j}^+ + \frac{\left(1+g\dfrac{K_B}{K_A}\right)}{(\Delta x)^2}\left(\psi_{A,j-1}^+ + \psi_{A,j+1}^+\right)
$$

$$
+ \left[-\frac{2\left(1+g\dfrac{K_B}{K_A}\right)}{(\Delta x)^2} + k_0^2 n_B^2 - g\frac{K_B}{K_A}k_0^2 n_A^2 + K_B^2 - gK_B K_A \right] \psi_{A,j}^- + \frac{\left(1-g\dfrac{K_B}{K_A}\right)}{(\Delta x)^2}\left(\psi_{A,j-1}^- + \psi_{A,j+1}^-\right)
$$

$$(4.74d)$$

This formulation allows us to evaluate ψ_{out}, given ψ_{in}, with the only limitation of the approximation introduced in the square root of the differential operator $\mathcal{L}_j$. However, in real situations we are given ψ_{in}^+ and must determine ψ_{in}^- such that the boundary condition that no backward field be present at the output, that is, $\psi_{out}^- = 0$, is satisfied. The problem thus requires the solution of:

$$
\begin{pmatrix} \psi_{out}^+ \\ 0 \end{pmatrix} = M \cdot \begin{pmatrix} \psi_{in}^+ \\ \psi_{in}^- \end{pmatrix}.
\tag{4.75}
$$

To solve this implicit equation for ψ_{in}, the following iterative scheme can be used:

1. ψ_{in}^+ is given; ψ_{in}^- is guessed (e.g., set to 0).
2. For $n = 0, 1, 2, \ldots$ do

   a. $\begin{pmatrix} \psi_{out}^+ \\ 0 \end{pmatrix} = \mathcal{M} \cdot \begin{pmatrix} \psi_{in}^+ \\ \psi_{in}^- \end{pmatrix}$;

   b. if $\left|\psi_{out}^-\right| < tolerance$, stop; otherwise,

   c. $\begin{pmatrix} * \\ \psi_{in}^- \end{pmatrix} = \mathcal{N} \cdot \begin{pmatrix} \psi_{out}^+ \\ \gamma \psi_{out}^- \end{pmatrix}$, where $N = M^{-1}$, $0 < \gamma < 1$ and $*$ indicates an unused component.

In brief, at each iteration (n), the current guess for $\psi_{in}^{-(n)}$ is propagated along with $\psi_{in}^{+(n)}$ forward through the system. Then, $\psi_{out}^{-(n)}$, which should be zero when we reach a solution, is dampened toward zero. The fields $\psi_{out}^{+(n)}$ and $\psi_{out}^{-(n)}$ are now propagated backward through the structure to obtain a new guess for $\psi_{in}^{-(n+1)}$. This scheme is repeated until the power carried by the field $\psi_{out}^{-(n)}$ is lower than a chosen value. The flow chart of this algorithm is shown in Figure 4.11. The damping factor, γ, which can be viewed as a form of an under-relaxation approach, is selected to optimize convergence. This scheme is straightforward and works well in many problems, however, it does not always converge for structures that are complex or have a high index contrast.

Let us note that the forward and backward fields in Eq. (4.59) and the fields used to calculate the reflected and transmitted fields using the interface operator $\mathcal{T}$ are the *complete* fields, while the fields that we have used to make the description of the conventional BPM were the slowly varying envelopes (SVEs) of the fields. Therefore, after propagation of the SVE field using

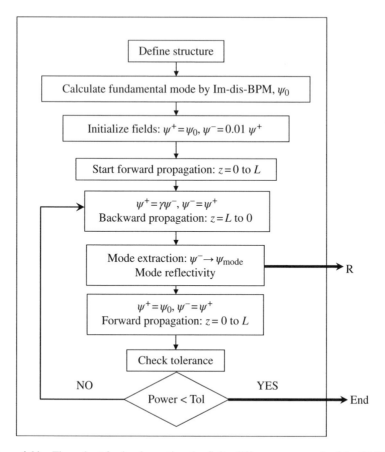

Figure 4.11 Flow chart for implementing the finite-difference approach of the Bi-BPM

BPM along a longitudinal invariant section, one must construct the complete field to compute the reflected and transmitted field using:

$$\psi(x,z) = u(x,z)e^{-iKz}, \qquad (4.76)$$

where $\psi(x,z)$ is the complete field, $u(x,z)$ is the SVE field and K is the reference wavenumber used in the conventional BPM.

4.3.3 Example of Bidirectional BPM: Index Modulation Waveguide Grating

As an example of bidirectional BPM, let us calculate the mode reflectivity of a waveguide index grating. In this kind of structures, refractive index discontinuities exist along the direction of propagation of the optical field; hence a bidirectional technique that considers coupling of the forward and backward waves is needed for modelling the reflection and/or transmission, as a non-neglected fraction of light is back-scattered along the propagation direction. First, let us consider a graded index waveguide, where a longitudinal smooth index modulation has been induced, giving rise to a waveguide index grating with low contrast (Figure 4.12).

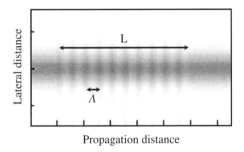

Figure 4.12 Two-dimensional waveguide index grating, fabricated by inducing a longitudinal index modulation on a graded index waveguide

The unperturbed two-dimensional waveguide has a symmetric Gaussian profile, with a substrate index of $n = 1.5$ and a maximum index increase of $\Delta n = 0.03$, having a width of 2 μm. The grating consists of a longitudinal index modulation of sinusoidal type with an index modulation depth of $\delta n = 0.01$. At the operating wavelength of $\lambda = 1.52$ μm, the effective index of the fundamental TE mode of the unperturbed waveguide is $N = 1.51984$. Thus, the maximum reflectivity at this wavelength is achieved for a grating period of $\Lambda = \lambda/2\,N = 0.5$ μm.

In this example, the reference wavenumber used in the bi-BPM calculation is the same as that used in the normal BPM propagation and is taken as $K_{ref} = k_0 N$. It is worth mentioning that a real wavenumber could give rise to instabilities in the bidirectional algorithm and thus a complex wavenumber should be used. Nevertheless, in this small index contrast grating, is it enough to use a very small rotation angle to reach stability and we have used $\varphi = 0.1°$. Regarding the damping factor γ used in the propagation scheme, as a rule of thumb, a recommended choice is to use a damping factor close to the expected reflectivity. Thus, for low reflectivities, a good election for γ is ∼0.1. In these cases, 2–5 iterations are usually enough to reach convergence. On the other hand, the convergence is poorer for high values of the reflectivity. For example, at 1.52 μm (maximum reflectivity), the damping factor should be close to ∼0.8 and thus 10–20 iterations are required at least to obtain an accurate solution.

Figure 4.13 shows the reflectivity of the fundamental TE mode for a waveguide grating of length $L = 50$ μm as a function of the operating wavelength, where a maximum reflectivity of ∼83.4% is obtained at $\lambda = 1.52$ μm. As both the waveguide and the index grating present smooth and low index contrast, this waveguide grating has very low polarization dependence and thus the TM mode reflectivity spectrum is very similar to that shown in Figure 4.13. Figure 4.14 shows the numerical results of the TE mode reflectivity using the bidirectional BPM in the waveguide index grating as a function of the grating length. The numerical data can be fit quite well to the theoretical dependence following a squared hyperbolic tangent function [15]:

$$R = \tanh^2(\kappa L), \qquad (4.77)$$

with a coupling coefficient of $\kappa = 3.05 \times 10^{-2}$ μm^{-1}.

The effect of the propagation step and the transversal grid discretization on the simulation accuracy is now examined. Figure 4.15 shows the mode reflectivity of the waveguide grating at $\lambda = 1.535$ μm as a function of the propagation step. Although for normal BPM (and reversal BPM) large propagation steps are possible while maintaining accuracy, for waveguide grating simulations the discretization of the longitudinal structure should be fine enough to resolve the

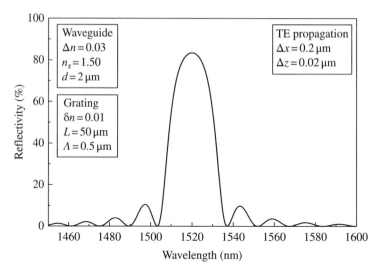

Figure 4.13 TE-mode reflectivity of a sinusoidal index waveguide grating as a function of the wavelength. Parameters of the waveguide grating and parameters of the simulation are given in the insets

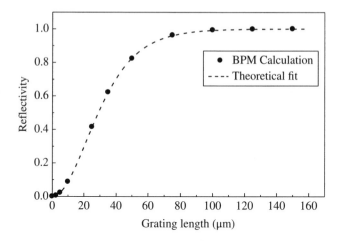

Figure 4.14 Maximum reflectivity ($\lambda = 1.52\,\mu\text{m}$) as a function of the waveguide grating length

salient features of the structure, in particular the sinusoidal variation on the refractive index modulation. In general, for sinusoidal index grating with period Λ, good results are obtained for propagation steps lower than $\sim\Lambda/20$. In our case, this means propagation step Δz shorter than $\sim 0.025\,\mu\text{m}$. The influence of the transversal step Δx on the reflectivity provided by the bi-BPM is plotted in Figure 4.16, indicating that reasonable results are obtained for transversal discretizations lower than $0.2\,\mu\text{m}$ in the smooth graded index waveguide example.

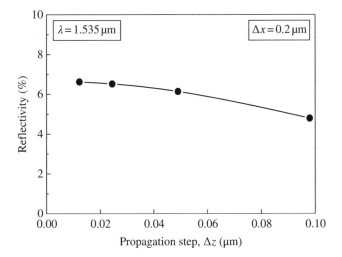

Figure 4.15 Calculated mode reflectivity of the waveguide index grating as a function of the propagation step

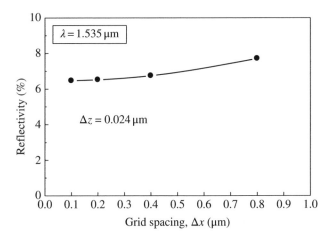

Figure 4.16 Effect of the transversal discretization on the calculated reflectivity

4.4 Active Waveguides

The performance of waveguide optical amplifiers and waveguide lasers operating in continuous wave regime (CW) based on rare-earth (RE) doped materials (active media) can be modelled by using the Overlap Integral (OI) method [16], which is applicable to straight waveguides and assumes invariant transverse profiles of the modes (see Appendix N). The model involves two types of differential equations: those describing the RE^{3+} population dynamics and those describing the forward propagation of the optical fields (and also backward propagation of the fields in the case of laser devices) [17]. For the more general case of active z-variant structures [18], the BPM algorithm provides a route of modelling the propagating fields in active media, with the possibility of energy exchange between the different monochromatic waves. Within a semi-classical approach, the rate equation formalism is used to describe the population dynamics of the different RE^{3+} levels, while complex refractive indices are allowed in the BPM algorithm to take into account gain/losses of the beams [19, 20].

4.4.1 Rate Equations in a Three-Level System

In order to evaluate the performance of the optical amplifier, it is necessary in the first place to describe the rate equations that govern the population dynamics of the levels involved in the amplification process. In general, the rate equations should include absorption, spontaneous emission and stimulated emission, besides other processes such as cross-relaxation or up-conversion processes. Here, we will analyse the equations that govern the population dynamics in a three-level system, which can be applied to the case of Erbium ions for amplification purposes in the range of telecom wavelengths.

Figure 4.17 sketches the main processes involved in an optical amplifier at around 1.5 μm based on Er^{3+} ions, under pumping at 0.98 μm. In this three-level system the amplification (or laser oscillation) at the signal wavelength λ_s takes place between the levels (2) and (1), whereas the absorption at the wavelength pump λ_p populates the level (3) from the ground level (1). In the case of Er^{3+} ions, levels (1), (2) and (3) corresponds to the multiplets $^4I_{15/2}$, $^4I_{13/2}$ and $^4I_{11/2}$, respectively. In what follows, the Amplified Spontaneous Emission (ASE) is neglected, which is a good approximation for modelling optical amplifiers with gain lower than 20–25 dB.

The (1) → (3) absorption process is regulated by the pump rate R_{13}. The transition (3) → (1) can take place by stimulated emission with a rate R_{31}, or by spontaneous emission with probability A_{31}. Level (3) can also decay to the level (2) by radiative and non-radiative channels, with probabilities A_{32} and W_{32}^{NR}, respectively. In addition, transitions form level (2) to the ground level (1) can be either by spontaneous de-excitation (including radiative and non-radiative processes), with total probability A_{2m}, or by stimulated emission with a rate of W_{12}. The value of A_{2m} is the inverse of the measured lifetime of the level (2). Finally, in presence of photons at

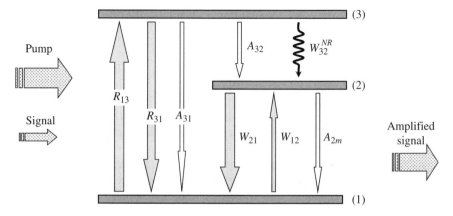

Figure 4.17   Relevant transitions occurring in a three-level system

wavelength of λ_s, induced transition $(1) \rightarrow (2)$ can also take place at a rate W_{21}. As the magnitudes R_{ij}, W_{ij} and A_{ij} represent number of transitions per units of time, they are expressed in units of s^{-1}.

Following the several processes that occur between the levels, the rate equations describing their population dynamics are:

$$\frac{dN_1}{dt} = (R_{31} + A_{31})N_3 + (A_{2m} + W_{21})N_2 - (R_{13} + W_{12})N_1; \tag{4.78a}$$

$$\frac{dN_2}{dt} = \left(A_{32} + W_{32}^{NR}\right)N_3 - (A_{2m} + W_{21})N_2 + W_{12}N_1; \tag{4.78b}$$

$$\frac{dN_3}{dt} = R_{13}N_1 - \left(R_{31} + A_{31} + A_{32} + W_{32}^{NR}\right)N_3; \tag{4.78c}$$

$$N = N_1 + N_2 + N_3, \tag{4.78d}$$

where N_i represents the population density (in units of ions/m^3) in the i-th level ($i = 1, 2, 3$). N_1 corresponds to the ground state, N_3 to the upper level that is pumped at λ_p and N_2 is the population of the upper level where the amplification of the signal at λ_s takes place. The concentration of active ions is denoted by N (Eq. (4.78d)). At the steady state, all the populations remain constant and thus all the derivatives in Eqs. (4.78a–d), are equal to zero.

The system of Eqs. (4.78a–d) can be solved analytically in the steady state, where the populations of the levels are expressed as:

$$N_3 = \frac{R_{13}(A_{2m} + W_{21})}{R_{13}\left(A_{32} + W_{32}^{NR}\right) + W_{12}(R_{31} + A_{3m}) + (A_{2m} + W_{21})(A_{3m} + R_{13} + R_{31})} N; \tag{4.79a}$$

$$N_1 = \frac{A_{2m} + W_{21}}{A_{2m} + W_{21} + W_{12}} N - \frac{A_{32} + W_{32}^{NR} + A_{2m} + W_{21}}{A_{2m} + W_{21} + W_{12}} N_3; \tag{4.79b}$$

$$N_2 = \frac{W_{12}}{A_{2m} + W_{21}} N_1 + \frac{A_{32} + W_{32}^{NR}}{A_{2m} + W_{21}} N_3. \tag{4.79c}$$

In Eq. (4.79a) we have introduced the parameter A_{3m}, which is the inverse of the measured lifetime of the level (3) and which is related to other spectroscopic parameters by:

$$A_{3m} \equiv \frac{1}{\tau_3} = A_{31} + A_{32} + W_{32}^{NR}. \tag{4.80}$$

On the other hand, the pump and absorption rates at the pump wavelength λ_p are related to the pump intensity I_p (units of W/m^2) through:

$$R_{13}(x, y) = \frac{\sigma_{13}}{h_c/\lambda_p} I_p(x, y); \tag{4.81a}$$

$$R_{31}(x,y) = \frac{\sigma_{31}}{h_c/\lambda_p} I_p(x,y), \qquad (4.81b)$$

where σ_{13} and σ_{31} are the absorption and emission cross-sections (units of m^2), respectively, h is the Planck's constant and c is the speed of light in free space. Similarly, the induced transition rates W_{12} and W_{21} due to the presence of photons at the signal wavelength λ_s are given by the relations:

$$W_{12}(x,y) = \frac{\sigma_{12}}{h_c/\lambda_s} I_s(x,y); \qquad (4.82a)$$

$$W_{21}(x,y) = \frac{\sigma_{21}}{h_c/\lambda_s} I_s(x,y). \qquad (4.82b)$$

σ_{12} and σ_{21} being the absorption and emission cross-sections, respectively, corresponding to the transitions between the levels (1) and (2) and I_s is the radiation intensity at the signal wavelength λ_s.

4.4.2 Optical Attenuation/Amplification

In order to simulate the optical attenuation/amplification in waveguide configuration using BPM equations, we must simultaneously perform the propagation of two monochromatic beams: one of them corresponding to the pump and a second beam from the signal wavelength. On the other hand, the energy variation experimented by each beam is controlled by considering a complex refractive index:

$$\widetilde{n}(x,y) = n(x,y) + i\frac{c}{2\omega}\alpha(x,y) = n(x,y) + i\kappa(x,y). \qquad (4.83)$$

The imaginary part of the refractive index distribution at a particular transversal plane of the propagation, corresponding to the signal wavelength, depends on the absorption and emission cross sections of the transitions between the levels (1) and (2) and their populations and is given by:

$$\kappa_s(x,y) = \frac{\lambda_s}{4\pi}[\sigma_{21}N_2(x,y) - \sigma_{12}N_1(x,y) - \widetilde{\alpha}_s], \qquad (4.84)$$

where we have included the parameter $\widetilde{\alpha}_s$ (units of m^{-1}), which takes into account the intrinsic propagation losses of the passive waveguide at the signal wavelength. In a similar way, the imaginary part of the refractive index for the pump wavelength is given by:

$$\kappa_p(x,y) = \frac{\lambda_s}{4\pi}[\sigma_{31}N_3(x,y) - \sigma_{13}N_1(x,y) - \widetilde{\alpha}_p]. \qquad (4.85)$$

In order to proceed with the simulation in an optical amplifier, we follow the next sequence:

1. The initial pump and signal fields are defined, with the correct amplitudes to take account the absolute pump and signal powers (in Watts).
2. The induced transition rates $R_{ij}(x,y)$ and $W_{ij}(x,y)$, at a particular transversal plane, z, are calculated in terms of the intensity of the beams both for the pump and signal using Eqs. (4.81) and (4.82).
3. From Eqs. (4.79a–c), the populations of each level are calculated.
4. The imaginary part of the refractive index distribution, for both pump and signal wavelengths, are calculated using Eqs. (4.84) and (4.85).
5. Using BPM, the pump and signal are propagated simultaneously from the transversal plane z to the next plane $z + \Delta z$, considering complex refractive indices in the wave equations.

The optical fields used to simulate the propagation can not be scaled arbitrarily, but must be set at their correct amplitudes to take into account the specific power carried by the pump and the signal beams, as the amplification/attenuation depends directly on the absolute values of the intensities. Also, at the input the fundamental modes of the channel waveguide are usually chosen, both for the pump and the signal. For doing that, initially a propagation is performed along the imaginary z-axis, obtaining a field $\psi(x,y)$, which gives the transversal distribution of the fundamental mode (in arbitrary units). On the other hand and in the scalar approach, the power P (in Watts) carried by the input beam as a function of the discretized intensity $I(x,y)$ (in units of W/m^2) is given by:

$$P = \iint I(x,y)dxdy = \iint C^2 |\psi(x,y)|^2 dxdy = C^2 \iint |\psi(x,y)|^2 dxdy, \qquad (4.86)$$

where C is a constant. Thus, the optical field that must be chosen for initializing the propagation is given by:

$$u(x,y) = \sqrt{\frac{P}{\iint |\psi(x,y)|^2 dxdy}} \, \psi(x,y), \qquad (4.87)$$

where P is the injected pump power in the channel waveguide.

Now, the rates $R_{ij}(x,y)$ and $W_{ij}(x,y)$ can be calculated in terms of the intensity of the beams, both for the pump and signal, using Eqs. (4.81) and (4.82) and from Eqs. (4.79a–c) the populations of each level are determined. This allows us to obtain the imaginary part of the refractive index distribution, for both pump and signal wavelengths, using Eqs. (4.84) and (4.85).

4.4.3   Channel Waveguide Optical Amplifier

As an example of active-BPM, we will present the study of an optical amplifier around 1.5 μm based on channel waveguides fabricated in Er^{3+}-doped phosphate glass. The BPM-scalar approximation is used here as it is a good approach to simulate optical propagation in low-contrast waveguides and usually the optical amplification takes place at a particular light polarization.

Sodium-alumino-phosphate glass is a chemically durable material developed for use in active and passive waveguide devices. The channels waveguides are fabricated by ion exchange in a molten KNO_3 or $AgNO_3$ salt bath, which gives rise to graded index waveguides with maximum index increases ranging from 0.005 to 0.02, depending on the fabrication parameters [21]. On the other hand, this glass can be doped with Er_2O_3, at concentration levels selected anywhere from 0 to 10 wt%, to provide an active medium for amplification at 1.53 μm [22].

Figure 4.18a shows the refractive index map of a channel waveguide made in a doped Er^{3+}-doped glass substrate. The refractive index is assumed to have a profile corresponding to the formula given in Eq. (3.84), with $n_s = 1.513$, $\Delta n = 0.012$, $d = 4$ μm and $W = 2$ μm. The cover is assumed to be air ($n_c = 1.000$). This waveguide is designed to be monomode at 1.534 μm and supports three modes at the pump wavelength (0.98 μm).

Using this index profile, the intensity profiles of the fundamental modes, corresponding to the diffused channel waveguide at the pump and signal wavelengths, are obtained by scalar Im-Dis-BPM (Figure 4.18b,c). The overlap between theses modes is calculated to be $\Gamma = 0.801$. The fundamental modes at 0.98 and at 1.534 μm are used to start the propagation in the

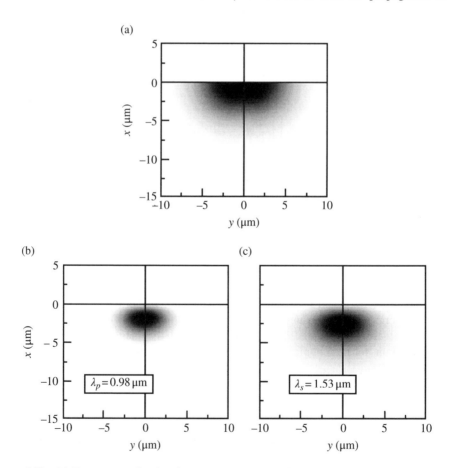

Figure 4.18   (a) Transverse refractive index map of the active channel waveguide. (b, c) Intensity profiles of the fundamental modes at the pump and signal wavelengths

Table 4.2 Spectroscopic parameters used in the simulation of the optical amplifier

Parameter	Symbol	Value
Wavelength pump	λ_p	0.98 μm
Wavelength signal	λ_s	1.534 μm
Absorption cross section $1 \rightarrow 3$	σ_{13}	0.258×10^{-20} cm^2
Emission cross section $3 \rightarrow 1$	σ_{31}	0.258×10^{-20} cm^2
Absorption cross section $1 \rightarrow 2$	σ_{12}	0.726×10^{-20} cm^2
Emission cross section $2 \rightarrow 1$	σ_{21}	0.672×10^{-20} cm^2
Lifetime level (2)	$\tau_2 = 1/A_{2m}$	9.18 ms
Lifetime level (3)	$\tau_3 = 1/A_{3m}$	2 μs
Radiative probability $3 \rightarrow 2$	A_{32}	60 s^{-1}
Radiative probability $3 \rightarrow 1$	A_{31}	300 s^{-1}
Er^{3+} concentration	N	1.89×10^{20} cm^{-3}
Intrinsic losses at λ_s	$\widetilde{\alpha}_s$	4.6×10^{-6} μm^{-1}
Intrinsic losses at λ_p	$\widetilde{\alpha}_p$	6.9×10^{-6} μm^{-1}

simulations, where their powers at the input should be selected depending on the parameter performance of the amplifier to be studied.

Table 4.2 presents the relevant spectroscopic parameters of the Er^{3+} ions in phosphate glass needed to perform the simulation of the optical amplifier [22, 23]. It is assumed that the substrate is homogeneously doped with Er^{+3} ions at a concentration of 1.89×20 ions/cm^3. At this concentration level and due to the weak interaction among the rare ions in phosphate glass, the cooperative upconversion process can be ignored. The spontaneous decay from level (2) and the ground level is essentially radiative, while the spontaneous decay from level (3) is predominantly non-radiative since the energy gap between levels $^4I_{13/2}$ and $^4I_{11/2}$ of the Er^{3+} ions is small compared to the maximum phonon energy of the phosphate glass matrix. The intrinsic losses are $\widetilde{\alpha}_s = 0.2$ dB/cm and $\widetilde{\alpha}_p = 0.3$ dB/cm at the signal and pump wavelengths, respectively.

First, we analyse the influence of the computational window discretization on the simulation results. The computed gain for a waveguide amplifier of 1 cm length and pumped with 100 mW of input power is plotted in Figure 4.19 as a function of the grid step along the horizontal direction used to perform the BPM calculations. The power of the signal beam is set to 1 μW (small-signal gain regime). The figure indicates that transverse mesh sizes lower than 0.5 μm are enough to obtain reliable results. On the other hand, to provide accurate numerical simulation the propagation steps should be shorter than 2 μm, as can be seen from Figure 4.20.

Now, BPM analysis is to be used to determine the characteristics of the waveguide amplifier. The gain as a function of the pump power in dBm units[1], in the small signal gain regime, is plotted in Figure 4.21. In order to get net gain for a 1-cm long waveguide amplifier it must be pumped with powers higher than $\sim$10 dBm (10 mW). Lower pump powers produce attenuation of the signal beam. At a pump power of 20 dBm (100 mW), the gain is close to 3 dB and for higher powers the amplifier starts to saturate.

Figure 4.22 shows the gain of the amplifier as a function of the signal power level. As can be seen, the amplifier has a gain close to 2.5 dB at low signal powers (small signal gain regime) and it drops for a higher signal power. Even at a certain signal power (around 100 mW), the

---

[1] $P(\text{dBm}) = 10\log_{10}[P(\text{mW})]$

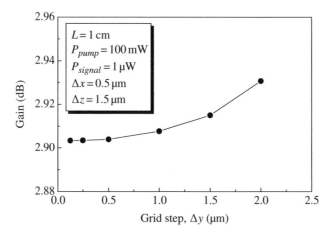

Figure 4.19 Gain in a waveguide amplifier based on Er^{3+}-doped phosphate glass, as a function of the transversal size step along the y direction, in the small signal regime

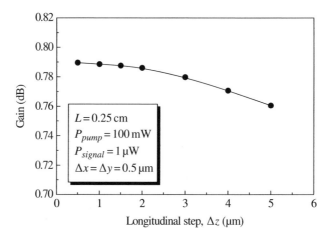

Figure 4.20 Gain in an Er^{3+}-doped phosphate glass waveguide amplifier for a pump power of 100 mW, as a function of the longitudinal step.

amplifier does not produce net gain due to the absorption of the Er ions at the signal wavelength and to the intrinsic propagation losses.

Finally, the performance of the amplifier as a function of the waveguide length is numerically simulated and this is presented in Figure 4.23 for different pump powers. It is seen that, for a particular pump power, an optimum length of the device for reaching maximum gains of the amplifier exists. Also, this optimum length increases as the pump power steps up.

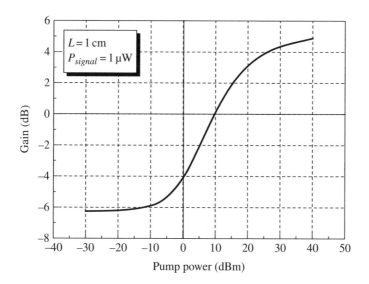

Figure 4.21 Gain of the waveguide amplifier as a function of the pump power for a small input signal

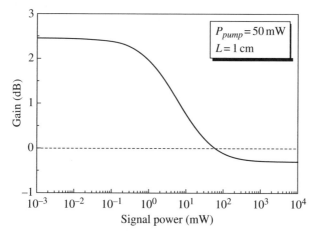

Figure 4.22 Gain of the waveguide amplifier as a function of the signal input power, for a pump power of 10 mW. Saturation is evident for signal input power beyond 0.1 mW

4.5 Second-Order Non-Linear Beam Propagation Techniques

Due to the versatile and efficient frequency conversion ability, second-order non-linear effects in waveguide structures, especially those implemented with quasi-phase matching techniques, have found applications in many areas. To understand the function of these devices, theoretical modelling for the non-linear effects is important. Although analytical treatments have been done to provide a quick insight into non-linear effects, many approximations are required. Thus, when a large or irregular geometrical variation exists and the depletion of the pump wave is not

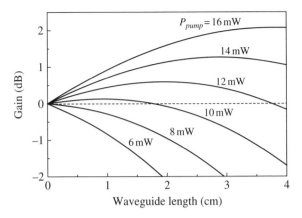

Figure 4.23   Gain of the optical amplifier as a function of the waveguide length, for different pump powers

negligible, precise analytical modelling of these second-order non-linear devices becomes quite difficult. Therefore, for a general and accurate analysis of light propagation in second-order non-linear structures, numerical methods are required [24, 25]. In particular, the BPM is a suitable numerical tool that can be extended to simulate second-order non-linear effects. Non-linear BPM analysis can provide an invaluable insight into the understanding of device behaviour and wave–device interactions. It also enables the construction of accurate models incorporating material characteristics as well as wave propagation effects. In particular, the method is able to deal with the depletion of the fundamental wave and with problems involving geometrical variation of the waveguide structure.

4.5.1   Paraxial Approximation of Second-Order Non-Linear Wave Equations

Here we will develop the paraxial approach of the second-order non-linear equation using the slowly varying approximation, starting from the non-linear equations in terms of the transverse fields obtained in Section 1.3.2. To do that, first the fields are expressed as a function of slowly varying amplitudes and fast oscillating phases on the form of paraxial waves:

$$E_t(r) = u_t e^{-iKz}, \tag{4.88}$$

where u_t denotes the SVE of the transverse electric field.

If the waves propagate mainly along the z-direction, the slowly varying amplitude for paraxial propagation (neglecting second-order derivatives with respect to z) is applied:

$$\frac{\partial^2 u}{\partial z^2} \ll \frac{\partial u}{\partial z} K, \tag{4.89}$$

and the first terms in Eq. (1.95) can be approximated by:

$$\nabla^2 E_t(r) \approx \left[\nabla_\perp^2 u_t(r) - K^2 u_t(r) - 2iK \frac{\partial u_t(r)}{\partial z} \right] e^{-iKz}. \tag{4.90}$$

Substituting this expression in the non-linear wave equations (1.95) one obtains:

$$2iK_1\frac{\partial \boldsymbol{u}_{1t}}{\partial z} = \nabla_\perp^2 \boldsymbol{u}_{1t} + \nabla_t\left[\frac{\nabla_t n_1^2}{n_1^2}\cdot \boldsymbol{u}_{1t}\right] + \left(k_{01}^2 n_1^2 - K_1^2\right)\boldsymbol{u}_{1t} + \left[k_{01}^2 \chi_{ijk} u_{3j} u_{2k}^* e^{-i(K_3-K_2-K_1)z}\right]_t ; \quad (4.91a)$$

$$2iK_2\frac{\partial \boldsymbol{u}_{2t}}{\partial z} = \nabla_\perp^2 \boldsymbol{u}_{2t} + \nabla_t\left[\frac{\nabla_t n_1^2}{n_2^2}\cdot \boldsymbol{u}_{2t}\right] + \left(k_{02}^2 n_2^2 - K_2^2\right)\boldsymbol{u}_{2t} + \left[k_{02}^2 \chi_{ijk} u_{3j} u_{1k}^* e^{-i(K_3-K_2-K_1)z}\right]_t ; \quad (4.91b)$$

$$2iK_3\frac{\partial \boldsymbol{u}_{3t}}{\partial z} = \nabla_\perp^2 \boldsymbol{u}_{3t} + \nabla_t\left[\frac{\nabla_t n_3^2}{n_3^2}\cdot \boldsymbol{u}_{3t}\right] + \left(k_{03}^2 n_3^2 - K_3^2\right)\boldsymbol{u}_{3t} + \left[k_{03}^2 \chi_{ijk} u_{1j} u_{2k} e^{+i(K_3-K_2-K_1)z}\right]_t , \quad (4.91c)$$

where $\boldsymbol{u}_{1t}$, $\boldsymbol{u}_{2t}$ and $\boldsymbol{u}_{3t}$ are the transverse fields at angular frequencies ω_1, ω_2 and ω_3, respectively, where $\omega_3 = \omega_1 + \omega_2$ holds for energy conservation. In addition, k_{01}, k_{02} and k_{03} are their respective wavenumbers in vacuum. Also, K_1, K_2 and K_3 denote their respective reference wavenumbers, defined by:

$$K_1 = n_{01} k_{01}, K_2 = n_{02} k_{02}, K_3 = n_{03} k_{03}. \quad (4.92)$$

Here, n_{01}, n_{02} and n_{03} are the reference indexes, which are usually set to the effective indexes of the waveguide modes of interest.

Here, we will assume that only one transverse component is enough to describe each monochromatic beam and that the second term on the right hand side in Eqs. (4.91a–c) can be neglected. These two approximations lead to the scalar approach of the non-linear paraxial wave equations, which reduces to:

$$2iK_{01}n_{01}\frac{\partial u_1}{\partial z} = \frac{\partial^2 u_1}{\partial x^2} + k_{01}^2\left(n_1^2 - n_{01}^2\right)u_1 + k_{01}^2\chi_1^{(2)} e^{-i\Delta Kz} u_3 u_2^*; \quad (4.93a)$$

$$2iK_{02}n_{02}\frac{\partial u_2}{\partial z} = \frac{\partial^2 u_2}{\partial x^2} + k_{02}^2\left(n_2^2 - n_{02}^2\right)u_2 + k_{02}^2\chi_2^{(2)} e^{-i\Delta Kz} u_3 u_1^*; \quad (4.93b)$$

$$2iK_{03}n_{03}\frac{\partial u_3}{\partial z} = \frac{\partial^2 u_3}{\partial x^2} + k_{03}^2\left(n_3^2 - n_{03}^2\right)u_3 + k_{03}^2\chi_3^{(2)} e^{+i\Delta Kz} u_1 u_2, \quad (4.93c)$$

where the mismatch parameter Δk is defined as:

$$\Delta k \equiv k_{03} n_{03} - k_{01} n_{01} - k_{02} n_{02}, \quad (4.94)$$

and the second-order susceptibilities are:

$$\chi_1^{(2)} \equiv \chi^{(2)}(\omega_1 : \omega_3, -\omega_2); \quad (4.95a)$$

$$\chi_2^{(2)} \equiv \chi^{(2)}(\omega_2 : \omega_3, -\omega_1); \quad (4.95b)$$

$$\chi_3^{(2)} \equiv \chi^{(2)}(\omega_3 : \omega_1, -\omega_2). \quad (4.95c)$$

To derive the finite-difference scheme of the coupled parabolic equations, we define the following finite-difference operators and variables:

$$\mathcal{L}_x u_i^{m,l} = \frac{1}{(\Delta x)^2}\left(u_i^{m-1,l} - 2u_i^{m,l} + u_i^{m+1,l}\right);$$ (4.96a)

$$\mathcal{L}_{0i}^{m,l} = k_{0i}^2\left[\left(n_i^{m,l}\right)^2 - n_{0i}^2\right] \quad i = 1, 2, 3;$$ (4.96b)

$$\mathcal{F}_i^{m,l} = k_{0i}^2 \chi_i^{(2)m,l} e^{-i\Delta k l \Delta z} \quad i = 1, 2;$$ (4.96c)

$$\mathcal{F}_3^{m,l} = k_{03}^2 \chi_3^{(2)m,l} e^{+i\Delta k l \Delta z}.$$ (4.96d)

The step sizes along the x and the z directions are denoted by Δx and Δz, with m and l representing the indexes along these two directions, respectively. In this way, the variable $u^{m,l}$ represents the electric field at the point $(x_m, z_l) = (m\Delta x, l\Delta z)$.

In previous chapters, we have used the implicit second-order-accurate Crank–Nicolson scheme to solve parabolic partial differential equations. In the linear case, it resulted in a tridiagonal system of linear equations, which can be solved with good efficiency. However, in the presence of non-linearities, the non-linear coupling terms cannot be split exactly to yield a linear system of equations. A simple way to handle this problem is to 'lag' part of the non-linear terms, which are approximated by the known fields in the previous step. This approach gives rise to a scheme that is only first-order accurate in z. If we want to have a scheme that is consistently second-order accurate, the non-linear source term should be integrated by other methods. The resulting difference equations then involve undetermined non-linear source terms in the next step where several iterative schemes can be used to solve this problem. Among them, the fixed-point iteration is chosen for the iterative-finite-difference non-linear beam propagation method (IFD-NL-BPM) [26, 27]. It has the advantage of great simplicity and requires minimal modification from the standard linear BPM.

The IFD-NL-BPM is described as follows. First, one set of solutions is obtained by assuming that the new non-linear terms is equal to that of the previous step. This zero order solution is denoted by $u^{m,l(0)}$, which is the initial guess of the electric field in the following iteration step. The iterative algorithm can be written as:

$$\frac{2ik_{01}n_{01}}{\Delta z}\left(u_1^{m,l+1^{(t)}} - u_1^{m,l}\right) = \frac{1}{2}\left(\mathcal{L}_x + \mathcal{L}_{01}^{m,l+1/2}\right)\left(u_1^{m,l+1^{(t)}} + u_1^{m,l}\right)$$

$$+ \frac{1}{2}\left(\mathcal{F}_1^{m,l} u_3^{m,l} u_2^{m,l*} + \mathcal{F}_1^{m,l+1} u_3^{m,l+1^{(t-1)}} u_2^{m,l+1^{(t-1)*}}\right)$$ (4.97a)

$$\frac{2ik_{02}n_{02}}{\Delta z}\left(u_2^{m,l+1^{(t)}} - u_2^{m,l}\right) = \frac{1}{2}\left(\mathcal{L}_x + \mathcal{L}_{02}^{m,l+1/2}\right)\left(u_2^{m,l+1^{(t)}} + u_2^{m,l}\right)$$

$$+ \frac{1}{2}\left(\mathcal{F}_2^{m,l} u_3^{m,l} u_1^{m,l*} + \mathcal{F}_2^{m,l+1} u_3^{m,l+1^{(t-1)}} u_1^{m,l+1^{(t-1)*}}\right)$$ (4.97b)

$$\frac{2ik_{03}n_{03}}{\Delta z}\left(u_3^{m,l+1^{(t)}}-u_3^{m,l}\right)=\frac{1}{2}\left(\mathcal{L}_x+\mathcal{L}_{03}^{m,l+1/2}\right)\left(u_3^{m,l+1^{(t)}}+u_3^{m,l}\right)$$
$$+\frac{1}{2}\left(\mathcal{F}_3^{m,l}u_1^{m,l}u_2^{m,l}+\mathcal{F}_3^{m,l+1}u_1^{m,l+1^{(t-1)}}u_2^{m,l+1^{(t-1)}}\right) \tag{4.97c}$$

where the superscript t is the iteration count and $u^{m,l+1^{(t)}}$ represents the tth iteration field. Now, these equations can be solved readily because there are no undetermined non-linear terms. This procedure is executed repeatedly until the difference between successive iterations is smaller than a given tolerance value.

To evaluate the power exchanged between the fields, we use the power per unit length (in the y direction) carried by the wave i ($i=1$–3) at a distance $z=l\Delta z$, which is given by:

$$P_i^l/L_y=\frac{\Delta x n_{0i}}{2\eta_0}\sum_{m=1}^{M}\left|u_i^{m,l}\right|^2, \tag{4.98}$$

where M is the number of discretization points in the transversal direction and η_0 is the free space impedance, defined by:

$$\eta_0\equiv\sqrt{\frac{\mu_0}{\varepsilon_0}}=120\pi\ \Omega. \tag{4.99}$$

4.5.2   Second-Harmonic Generation in Waveguide Structures

The most common case in second-order processes is the second-harmonic (SH) generation, which is a degenerate situation encountered when two frequencies are equal ($\omega_1=\omega_2$). In this case, we have that $\omega_s=2\omega_f$, where the subscript s stands for the second harmonic wave and f indicates the fundamental wave. The paraxial equations that govern the two field envelopes u_f and u_s are given by [28]:

$$2ik_{0f}n_{0f}\frac{\partial u_f}{\partial z}=\frac{\partial^2 u_f}{\partial x^2}+k_{0f}^2\left(n_f^2-n_{0f}^2\right)u_f+k_{0f}^2\chi^{(2)}e^{i\Delta kz}u_su_f^*; \tag{4.100a}$$

$$2ik_{0s}n_{0s}\frac{\partial u_s}{\partial z}=\frac{\partial^2 u_s}{\partial x^2}+k_{0s}^2\left(n_s^2-n_{0s}^2\right)u_s+k_{0s}^2\frac{\chi^{(2)}}{2}e^{-i\Delta kz}u_fu_f, \tag{4.100b}$$

where the mismatch parameter Δk is now given by:

$$\Delta k\equiv 2k_{0f}n_{0f}-k_{0s}n_{0s}. \tag{4.101}$$

To obtain the finite-difference formulation of Eqs. (4.100a) and (4.100b), we define the following operators:

$$\mathcal{L}_xu_i^{m,l}=\frac{1}{(\Delta x)^2}\left(u_i^{m-1,l}-2u_i^{m,l}+u_i^{m+1,l}\right); \tag{4.102a}$$

$$\mathcal{L}_{0i}^{m,l} = k_{0i}^2 \left[\left(n_i^{m,l} \right)^2 - n_{0i}^2 \right] \quad i = f, s; \tag{4.102b}$$

$$\mathcal{F}_f^{m,l} = k_{0f}^2 \chi^{(2)m,l} e^{+i\Delta kl\Delta z}; \tag{4.102c}$$

$$\mathcal{F}_s^{m,l} = \frac{1}{2} k_{0s}^2 \chi^{(2)m,l} e^{-i\Delta kl\Delta z}. \tag{4.102d}$$

With these definitions, the IFD-NL-BPM algorithm applied to the parabolic equations (4.100a) and (4.100b) gives rise to [27]:

$$
\frac{2ik_{0f}n_{0f}}{\Delta z} \left(u_f^{m,l+1^{(t)}} - u_f^{m,l} \right) = \frac{1}{2} \left(\mathcal{L}_x + \mathcal{L}_{0f}^{m,l+1/2} \right) \left(u_f^{m,l+1^{(t)}} + u_f^{m,l} \right)
$$
$$
+ \left(\mathcal{F}_f^{m,l} u_s^{m,l} u_f^{m,l*} + \mathcal{F}_f^{m,l+1} u_s^{m,l+1^{(t-1)}} u_f^{m,l+1^{(t-1)*}} \right) \tag{4.103a}
$$

$$
\frac{2ik_{0s}n_{0s}}{\Delta z} \left(u_s^{m,l+1^{(t)}} - u_s^{m,l} \right) = \frac{1}{2} \left(\mathcal{L}_x + \mathcal{L}_{0s}^{m,l+1/2} \right) \left(u_s^{m,l+1^{(t)}} + u_s^{m,l} \right)
$$
$$
+ \left(\mathcal{F}_s^{m,l} u_f^{m,l} u_f^{m,l} + \mathcal{F}_s^{m,l+1} u_f^{m,l+1^{(t-1)}} u_f^{m,l+1^{(t-1)}} \right) \tag{4.103b}
$$

In most cases, two iterations are enough to achieve convergence. On the other hand, the propagation step should be carefully chosen, as the accuracy of the results depends on the non-linear susceptibility value as well as the field amplitudes involved in the propagation.

To show the applicability of this formalism, let us apply it to the propagation of the first guided TE mode of the fundamental wave at $\lambda_f = 1.55\,\mu m$ along a phase-matched planar waveguide, where the core of the waveguide (film region) presents second-order non-linearity (Figure 4.24). The waveguide consists of a $0.44\,\mu m$ thick layer with $n_{film} = 3.60555$, surrounded by air ($n_{cover} = 1.0$) and the substrate with $n_{subs} = 3.1$. To fulfil the phase-matching condition

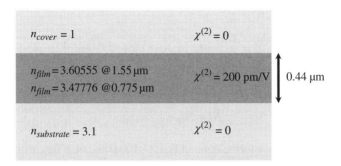

Figure 4.24 Asymmetric planar waveguide showing second-order non-linearity in the core. The refractive indices of the structure have been chosen to fulfil perfect phase-matching condition for TE modes at the fundamental wavelength of $1.55\,\mu m$

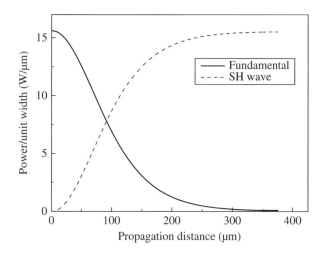

Figure 4.25 Evolution powers of the fundamental and second harmonic waves in a phase-matched planar waveguide

($\Delta k = 0$), the fundamental field and the first guided mode of the second-harmonic wave should have the same effective refractive index. This requirement is met by choosing a value of $n_{film} = 3.47776$ at $\lambda_s = 0.775$ μm inside the guiding layer. The non-linear susceptibility has a value of $\chi^{(2)} = 200$ pm/V, corresponding to the non-vanishing element of the non-linear susceptibility tensor of the bulk GaAs crystal [29].

Figure 4.25 shows the power per unit length (in the y-direction) of the fundamental and the second harmonic waves after excitation with the first guided mode of the fundamental wave with a peak amplitude of 100 V/μm, previously calculated by imaginary-distance BPM. This amplitude value corresponds to a power per unit width of 15.6 W/μm. The simulation results have been obtained by IFD-NL-BPM using a propagation step of $\Delta z = 0.2$ μm, with only two iterations and using a reference refractive index of $N_{ref} = 3.41$ for both fields.

As the phase calculations of the fields are critical for the correct propagation of the two waves, the reference indices for both the fundamental and the second harmonic waves should be chosen as close as possible to the effective indices of the propagating modes, which should be calculated by imaginary-distance BPM. As the reference index used for the propagation deviates from the effective index of the waves involved in the propagation, the propagation step needed to obtain reliable results must be shortened. In Figure 4.26 the maximum efficiency conversion in the phase-matched waveguide is plotted as a function of the propagation step, for two different reference indices. If the reference index is close to the effective index of the two fields ($N_{ref} = 3.41$, $N_{eff} = 3.40778$), convergence results are obtained for propagation steps close to 1 μm (circles). Nevertheless, if the reference index differs to the modes effective index ($N_{ref} = 3.1$), propagation steps shorter than 0.01 μm are needed to obtain a correct solution (squares).

The input field (at $z = 0$), which corresponds to the first-guided mode of the fundamental wave with amplitude 100 V/μm, is plotted in Figure 4.27 (continuous line), corresponding with a power of 15.6 W/μm. The second harmonic wave is set close to zero at this point. After a propagation of ~375 μm along the phase-matched planar waveguide, the fundamental wave is almost completely depleted and its energy is transferred to the SH wave. The field profile

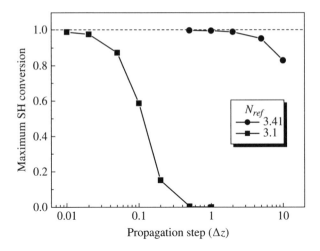

Figure 4.26 Dependence of the calculated maximum efficiency conversion in a phase-matched waveguide as a function of the propagation step, for two different reference refractive indices. The effective index of the phase-matched fundamental and second-harmonic modes is 3.40 780

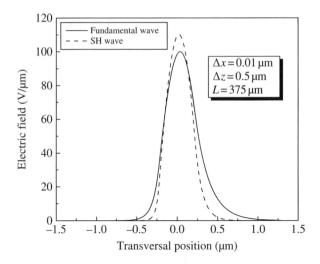

Figure 4.27 Electric field amplitude of the input fundamental wave (continuous line) and the second harmonic wave computed at a distance of 375 μm (dashed line), in a phase-matched planar waveguide

of the generated SH wave is plotted in Figure 4.27 (dashed line), which corresponds to the first guided mode of the waveguide at $\lambda_s = 0.775$ μm (which is bimodal at this wavelength). The peak amplitude of the SH field is 110.2 V/μm and the computed power carried by this field using Eq. (4.98) is 15.6 W/μm, indicating energy conservation in the frequency conversion process.

4.6   BPM in Anisotropic Waveguides

For light propagation in linear and dielectrically anisotropic media, the displacement and the electric field vectors are related through the $D = \varepsilon E$ where ε is the permittivity tensor. If the material is non-active and lossless, the permittivity tensor becomes real and symmetric [30]. Furthermore, if the crystal orientation is properly chosen, the matrix is diagonal. Under this situation, off-diagonal elements can appear if some external fields are applied, for example, if an electric field is applied in electro-optic crystals [31]. Here, we consider transverse anisotropic materials, constrained so that one of the material principal axis points in the main propagation direction of the waveguide. Under this assumption, the permittivity tensor takes the form [32]:

$$\varepsilon = \begin{pmatrix} \varepsilon_{xx} & \varepsilon_{xy} & 0 \\ \varepsilon_{yx} & \varepsilon_{yy} & 0 \\ 0 & 0 & \varepsilon_{zz} \end{pmatrix}. \tag{4.104}$$

To obtain the BPM equations in anisotropic media [33], we start with the wave equation (1.92) in terms of the transverse fields. As in the case of isotropic media, if the transverse components of an electromagnetic field are known, then the longitudinal component can be obtained by using the first Maxwell equation $\nabla(\varepsilon E) = 0$, thus allowing us to describe the full vectorial properties of the electromagnetic field using its transverse components [34]. As we proceed in previous sections, the electric transverse fields in Eq. (1.92) are expressed in terms of the transverse electric field envelope Ψ_t:

$$E_t = \Psi_t e^{-in_0 k_0 z} \text{ or } \begin{bmatrix} E_x \\ E_y \end{bmatrix} = \begin{bmatrix} \Psi_x \\ \Psi_y \end{bmatrix} e^{-in_0 k_0 z}, \tag{4.105}$$

where n_0 is a reference refractive index. Making use of the SVEA:

$$\left| \frac{\partial^2 \Psi_t}{\partial z^2} \right| \ll 2n_0 k_0 \left| \frac{\partial \Psi_t}{\partial z} \right|. \tag{4.106}$$

we obtain the paraxial vectorial wave equation:

$$2in_0 k_0 \frac{\partial}{\partial z} \begin{bmatrix} \Psi_x \\ \Psi_y \end{bmatrix} = \begin{bmatrix} P_{xx} & P_{xy} \\ P_{yx} & P_{yy} \end{bmatrix} \begin{bmatrix} \Psi_x \\ \Psi_y \end{bmatrix}, \tag{4.107}$$

where the differential operators, P_{ij}, are defined in this case as:

$$P_{xx}\Psi_x = \frac{\partial}{\partial x}\left(\frac{1}{\varepsilon_{zz}} \frac{\partial}{\partial x}(\varepsilon_{xx}\Psi_x) \right) + \frac{\partial}{\partial x}\left(\frac{1}{\varepsilon_{zz}} \frac{\partial}{\partial y}(\varepsilon_{yx}\Psi_x) \right) + \frac{\partial^2 \Psi_x}{\partial y^2} + k_0^2\left(\frac{\varepsilon_{xx}}{\varepsilon_0} - n_0^2 \right)\Psi_x; \tag{4.108a}$$

$$P_{yy}\Psi_y = \frac{\partial^2 \Psi_y}{\partial x^2} + \frac{\partial}{\partial y}\left(\frac{1}{\varepsilon_{zz}} \frac{\partial}{\partial y}(\varepsilon_{yy}\Psi_y) \right) + \frac{\partial}{\partial y}\left(\frac{1}{\varepsilon_{zz}} \frac{\partial}{\partial x}(\varepsilon_{xy}\Psi_y) \right) + k_0^2\left(\frac{\varepsilon_{yy}}{\varepsilon_0} - n_0^2 \right)\Psi_y; \tag{4.108b}$$

$$P_{xy}\Psi_y = \frac{\partial}{\partial x}\left(\frac{1}{\varepsilon_{zz}}\frac{\partial}{\partial y}\left(\varepsilon_{yy}\Psi_y\right)\right) + \frac{\partial}{\partial x}\left(\frac{1}{\varepsilon_{zz}}\frac{\partial}{\partial x}\left(\varepsilon_{xy}\Psi_y\right)\right) - \frac{\partial^2\Psi_y}{\partial x\partial y} + \frac{\varepsilon_{xy}}{\varepsilon_0}k_0^2\Psi_y; \tag{4.108c}$$

$$P_{yx}\Psi_x = \frac{\partial}{\partial y}\left(\frac{1}{\varepsilon_{zz}}\frac{\partial}{\partial x}\left(\varepsilon_{xx}\Psi_x\right)\right) + \frac{\partial}{\partial y}\left(\frac{1}{\varepsilon_{zz}}\frac{\partial}{\partial y}\left(\varepsilon_{yx}\Psi_x\right)\right) - \frac{\partial^2\Psi_x}{\partial y\partial x} + \frac{\varepsilon_{yx}}{\varepsilon_0}k_0^2\Psi_x. \tag{4.108d}$$

It should be noted that in these equations both the material and the geometrical properties of the waveguides contribute to the polarization dependence, that is, $P_{xx} \neq P_{yy}$ and coupling, that is, $P_{xy} \neq 0$ and $P_{yx} \neq 0$. If the material anisotropy and the geometric polarization effect are weak, then the terms $\frac{\partial}{\partial x}\left(\frac{1}{\varepsilon_{yx}}\frac{\partial}{\partial y}\left(\varepsilon_{xx}\Psi_x\right)\right)$, $\frac{\partial}{\partial y}\left(\frac{1}{\varepsilon_{zz}}\frac{\partial}{\partial x}\left(\varepsilon_{xy}\Psi_y\right)\right)$, $\frac{\partial}{\partial x}\left(\frac{1}{\varepsilon_{zz}}\frac{\partial}{\partial y}\left(\varepsilon_{xy}\Psi_y\right)\right)$ and $\frac{\partial}{\partial y}\left(\frac{1}{\varepsilon_{zz}}\frac{\partial}{\partial y}\left(\varepsilon_{yx}\Psi_x\right)\right)$ in Eq. (4.108) may be neglected. These terms, which contain an off-diagonal coefficient and at least one transverse differentiation at the same time, are the 'product' of the geometric and material effects and are expected to be very small for practical cases [33, 34]. Neglecting these terms, the operators defined in Eq. (4.108) can then be approximated by:

$$P_{xx}\Psi_x = \frac{\partial}{\partial x}\left(\frac{1}{\varepsilon_{zz}}\frac{\partial}{\partial x}\left(\varepsilon_{xx}\Psi_x\right)\right) + \frac{\partial^2\Psi_x}{\partial y^2} + k_0^2\left(\frac{\varepsilon_{xx}}{\varepsilon_0} - n_0^2\right)\Psi_x; \tag{4.109a}$$

$$P_{yy}\Psi_y = \frac{\partial^2\Psi_y}{\partial x^2} + \frac{\partial}{\partial y}\left(\frac{1}{\varepsilon_{zz}}\frac{\partial}{\partial y}\left(\varepsilon_{yy}\Psi_y\right)\right) + k_0^2\left(\frac{\varepsilon_{yy}}{\varepsilon_0} - n_0^2\right)\Psi_y; \tag{4.109b}$$

$$P_{xy}\Psi_y = \frac{\partial}{\partial x}\left(\frac{1}{\varepsilon_{zz}}\frac{\partial}{\partial y}\left(\varepsilon_{yy}\Psi_y\right)\right) - \frac{\partial^2\Psi_y}{\partial x\partial y} + \frac{\varepsilon_{xy}}{\varepsilon_0}k_0^2\Psi_y; \tag{4.109c}$$

$$P_{yx}\Psi_x = \frac{\partial}{\partial y}\left(\frac{1}{\varepsilon_{zz}}\frac{\partial}{\partial x}\left(\varepsilon_{xx}\Psi_x\right)\right) - \frac{\partial^2\Psi_x}{\partial y\partial x} + \frac{\varepsilon_{yx}}{\varepsilon_0}k_0^2\Psi_x. \tag{4.109d}$$

Equation (4.107), besides the operators defined in Eq. (4.109), is the basic BPM equation that governs the electric transverse field envelopes in anisotropic media under the conditions previously pointed out. The numerical techniques to solve them using finite difference schemes are similar to those previously presented in Chapter 3.

For two-dimensional structures (planar waveguides) the terms that contain derivatives with respect to the y-coordinate vanish. In such cases, the operators defined in Eq. (4.109) simplify to:

$$P_{xx}\Psi_x = \frac{\partial}{\partial x}\left(\frac{1}{\varepsilon_{zz}}\frac{\partial}{\partial x}\left(\varepsilon_{xx}\Psi_x\right)\right) + k_0^2\left(\frac{\varepsilon_{xx}}{\varepsilon_0} - n_0^2\right)\Psi_x; \tag{4.110a}$$

$$P_{yy}\Psi_y = \frac{\partial^2\Psi_y}{\partial x^2} + k_0^2\left(\frac{\varepsilon_{yy}}{\varepsilon_0} - n_0^2\right)\Psi_y; \tag{4.110b}$$

$$P_{xy}\Psi_y = \frac{\varepsilon_{xy}}{\varepsilon_0}k_0^2\Psi_y; \tag{4.110c}$$

$$P_{yx}\Psi_x = \frac{\varepsilon_{yx}}{\varepsilon_0}k_0^2\Psi_x. \tag{4.110d}$$

Let us observe that, even in planar structures, energy transfer between TE and TM modes can exist if the waveguide is anisotropic, due to the mixing terms (4.110c) and (4.110d); the off-diagonal elements, ε_{xy} and ε_{yx} being responsible for coupling.

The wave equation (4.107) for the transversal electric field envelopes in two-dimensional structures is then given by:

$$2in_0k_0\frac{\partial \Psi_x}{\partial z} = \frac{\partial}{\partial x}\left(\frac{1}{\varepsilon_{zz}}\frac{\partial}{\partial x}(\varepsilon_{xx}\Psi_x)\right) + k_0^2\left(\frac{\varepsilon_{xx}}{\varepsilon_0} - n_0^2\right)\Psi_x + \frac{\varepsilon_{xy}}{\varepsilon_0}k_0^2\Psi_y; \tag{4.111a}$$

$$2in_0k_0\frac{\partial \Psi_y}{\partial z} = \frac{\partial^2 \Psi_y}{\partial x^2} + k_0^2\left(\frac{\varepsilon_{yy}}{\varepsilon_0} - n_0^2\right)\Psi_y + \frac{\varepsilon_{yx}}{\varepsilon_0}k_0^2\Psi_x. \tag{4.111b}$$

These coupled equations can be solved using finite difference schemes. The problem arises from the unknown field component Ψ_y at the longitudinal step $(m + 1)$, which is necessary to solve Eq. (4.111a). A good approach consists of taking the field $\Psi_y^{m+1} = \Psi_y^m$ in Eq. (4.111a) and then solving it for the field Ψ_x^{m+1} using the conventional Thomas algorithm. Then Eq. (4.111b) is used to calculate the field Ψ_y^{m+1} by once again applying the Thomas algorithm. The numerical procedure finishes by solving Eq. (4.111a), now with the corrected field Ψ_y^{m+1} previously calculated. The algorithm works well in most practical cases, even in structures with longitudinal modulated permittivity tensor, provided that the longitudinal step is sufficiently small compared with the modulation period.

4.6.1   TE ↔ TM Mode Conversion

One example of light propagation in anisotropic waveguides is provided by a double hetero-structure GaAs pn-junction waveguide electro-optic modulator [35]. The GaAs is an isotropic crystal with refractive index of $n = 3.6$ at $\lambda = 1$ µm. Therefore, its permittivity tensor is:

$$\boldsymbol{\varepsilon} = \varepsilon_0\begin{pmatrix} n^2 & 0 & 0 \\ 0 & n^2 & 0 \\ 0 & 0 & n^2 \end{pmatrix} = \varepsilon_0\begin{pmatrix} 12.96 & 0 & 0 \\ 0 & 12.96 & 0 \\ 0 & 0 & 12.96 \end{pmatrix}. \tag{4.112}$$

On the other hand, the GaAs crystal exhibits linear electro-optic effect (see Appendices F and G) and its electro-optic matrix is given by [36]:

$$r_{ij} = \begin{pmatrix} 0 & 0 & 0 \\ 0 & 0 & 0 \\ 0 & 0 & 0 \\ 1.1 & 0 & 0 \\ 0 & 1.1 & 0 \\ 0 & 0 & 1.1 \end{pmatrix} \times 10^{-12} \ m/V. \tag{4.113}$$

Figure 4.28 shows a schematic of this GaAs-Al$_x$Ga$_{1-x}$As [110] pn-junction waveguide [35]. As the GaAs crystal is electro-optic, the isotropic waveguide medium becomes anisotropic by the

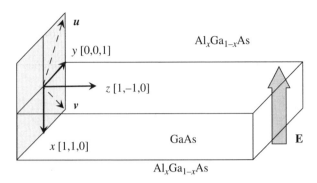

Figure 4.28   The application of an external field along the [1,1,0] direction modifies the dielectric tensor of the isotropic medium via the electro-optic effect, inducing the appearance of off-diagonal elements in the (x,y,z) coordinate system, which makes the polarization conversion possible

application of an electric field. If the external electric field is applied along the [110] direction (see Appendix G), the new principal axes of the perturbed medium are along the u, v and z directions, as indicated in Figure 4.28. In this coordinate system, the dielectric tensor is diagonal, but in the waveguide coordinate system (x,y,z), the dielectric tensor has non-null non-diagonal elements.

Via the linear electro-optic effect, the application of an electric field of 50 V/μm along the x-axis ([110] crystallographic direction) gives rise to the following modified permittivity tensor (Appendix G):

$$\boldsymbol{\varepsilon} = \varepsilon_0 \begin{pmatrix} 12.960006 & -0.0092 & 0 \\ -0.0092 & 12.960006 & 0 \\ 0 & 0 & 12.96 \end{pmatrix}. \tag{4.114}$$

The new principal axes are along x', y' and z as drawn in Figure 4.28. Let us note that two of the elements in the diagonal are slightly modified, although the most salient feature is the apparition of two off-diagonal elements in the matrix. This last circumstance provides the possibility of mode conversion between orthogonal polarizations via the non-diagonal elements, ε_{xy} and ε_{yx}.

To simulate the light propagation in the planar waveguide we assume an index profile of the form [35]:

$$n(x) = n_s + \Delta n / \cosh^2(2x/d), \tag{4.115}$$

with $n_s = 3.6$, $\Delta n = 0.001$ and $d = 5.6$ μm and x indicates the in-depth coordinate. This structure is monomode for both TE and TM propagation at $\lambda = 1$ μm.

Figure 4.29 shows the power of the TE and TM guided modes in the waveguide as a function of the propagation distance using anisotropic BPM at the operation wavelength of 1 μm, after launching the fundamental TE mode. A complete power transfer from the TE mode to the TM mode is achieved after a propagation distance of 200 μm. This value is in close agreement with results from coupled-wave theory [35].

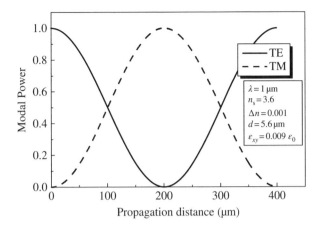

Figure 4.29 Optical power of the TE mode (continuous line) and TM mode (dashed line) for an electro-optic modulator based on GaAs pn-junction waveguide, calculated by using anisotropic BPM with a step size of $\Delta z = 0.5 \, \mu m$

4.7 Time Domain BPM

So far, the optical propagation described up to now has been performed in the frequency domain; that is, assuming monochromatic light of a very well-defined frequency, ω_0, whose temporal dependence on the fields was of the form of $\exp(i\omega_0 t)$. This assumption has allowed us to simplify the problem where we have implemented quite robust and efficient BPM algorithms for the simulation of light propagation in complex structures.

Although the conventional BPM has been extended to account for reflection due longitudinal discontinuities, in the so-called bidirectional BPM, an iterative procedure has to be carried out until a self-consistent solution is obtained. An alternative approach to managing backward waves due to reflections at the discontinuities is the use of time-domain simulation techniques. In addition, as these techniques do not make any assumption of the temporal dependence of the optical field, they can manage monochromatic waves as well as the propagation of optical pulses.

If the optical pulse contains several optical cycles, one can use the slow-wave approximation to simplify the problem of optical propagation. This approach, called time-domain beam propagation method, is of practical application for simulating light propagation in 1D, 2D and 3D optical structures [37, 38]. Also, it is possible to extend the method to treat narrow temporal pulses in the so-called broad-band approach, although its implementation for 2D optical structures requires iteration procedures [38].

On the other hand, powerful numerical methods have been established for treating full vectorial waves using time-domain techniques. These methods are known as finite-difference time-domain approaches and are the right choices for a rigorous treatment of light propagation in complex structures [39]. The only drawback of the technique relies on the enormous computational resources required for the simulation of large optical structures, especially for light propagation in 2D and 3D optical structures and the maximum time-step allowed, imposed by the Courant–Friedrichs–Levy (CFL) stability criterion [40]. This numerical method will be presented in Section 4.8.

4.7.1   Time-Domain Beam Propagation Method (TD-BPM)

Time-domain analysis of light propagation in optical structures can be developed efficiently by using TD-BPM's on the basis of the SVEA [41]. In the conventional TD-BPM, the SVEA is applied only to the time term, which allows us to simulate light propagation in complex structures, including reflection and diffraction at any angle. If the SVEA is also applied to the z coordinate, the numerical method is more efficient, although the light propagation analysis precludes the description of reflection, as paraxial propagation is considered [42]. With the implicit scheme for discretization in time used in TD-BPM approach, a time step is chosen to be larger than that in the FD-TD because it is not limited by the CFL condition.

The TD-BPM technique for the simulation of optical pulses in inhomogeneous structures (optical waveguides) starts with the wave equation for the electric field:

$$\nabla \times \nabla \times \mathcal{E} + \frac{n^2}{c^2}\frac{\partial^2 \mathcal{E}}{\partial t^2} = 0, \tag{4.116}$$

where $n = n(x,y,z)$ is the refractive index of the waveguide structure and c is the speed of light in free space. If the optical pulse is considered to be centred at a carrier angular frequency, ω_0, this fast frequency is extracted to identify a time-SVE of the wave as follows:

$$\mathcal{E}(r,t) = u(r,t)e^{i\omega_0 t}, \tag{4.117}$$

where $u(r,t)$ represents the temporal envelop of the electric field. Substituting this expression in Eq. (4.116) leads to:

$$\frac{n^2}{c^2}\frac{\partial^2 u}{\partial t^2} + 2i\frac{\omega_0 n^2}{c^2}\frac{\partial u}{\partial t} = -\nabla \times \nabla \times u + \frac{\omega_0^2 n^2}{c^2}u. \tag{4.118}$$

If the modulation frequency is much lower than the optical carrier frequency, the slow-wave approximation (SVEA):

$$\left|\frac{\partial^2 u}{\partial t^2}\right| \ll 2\omega_0 \left|\frac{\partial u}{\partial t}\right| \tag{4.119}$$

can be applied and the equation for the electric field envelope is then reduced to the following first-order equation in time:

$$2i\frac{\omega_0 n^2}{c^2}\frac{\partial u}{\partial t} = -\nabla \times \nabla \times u + \frac{\omega_0^2 n^2}{c^2}u. \tag{4.120}$$

Here, we start developing the conventional narrow-band TD-BPM applied to waves in 1D structures, which is suitable for modelling the optical propagation of wide temporal pulses. Then, extension of the algorithm to broad-band analysis is analysed, which is required to treat short temporal pulses. Finally, TD-BPM in 2D optical waveguides is presented, making use of the alternating direction implicit (ADI) technique in the narrow-band approach.

4.7.2   Narrow-Band 1D-TD-BPM

For one-dimensional problems, the wave equation (4.120) simplifies to:

$$2i\frac{\omega_0 n^2}{c^2}\frac{\partial u}{\partial t} = \frac{\partial^2 u}{\partial z^2} + \frac{\omega_0^2 n^2}{c^2}u. \tag{4.121}$$

To attain second-order accuracy on time steps, the standard Crank–Nicolson scheme is implemented to the wave equation, yielding:

$$2i\frac{\omega_0 n^2}{c^2}\frac{u^{m+1}-u^m}{\Delta t} = \left(\frac{\partial^2}{\partial z^2} + \frac{\omega_0^2 n^2}{c^2}\right)\frac{u^{m+1}-u^m}{2}, \tag{4.122}$$

where u^m represents the field value at time $t = m\Delta t$, Δt being the temporal discretization step. The finite-difference approach applied to the spatial derivative results in:

$$\begin{aligned}
&\frac{4i\omega_0 n_j^2}{c^2\Delta t}u_j^{m+1} - \frac{u_{j-1}^{m+1}-2u_j^{m+1}+u_{j+1}^{m+1}}{(\Delta z)^2} - \frac{\omega_0^2 n_j^2}{c^2}u_j^{m+1} \\
&= \frac{4i\omega_0 n_j^2}{c^2\Delta t}u_j^m + \frac{u_{j-1}^m-2u_j^m+u_{j+1}^m}{(\Delta z)^2} + \frac{\omega_0^2 n_j^2}{c^2}u_j^m
\end{aligned} \tag{4.123}$$

where the subscript j denote the spatial discretization and Δz is the spatial increments. This finite difference equation forms a tridiagonal system, which is solved by the standard Thomas method. The coefficients of the linear system are:

$$a_j = -\frac{1}{(\Delta z)^2}; \tag{4.124a}$$

$$b_j = \frac{4i\omega_0 n_j^2}{c^2\Delta t} + \frac{2}{(\Delta z)^2} - \frac{\omega_0^2 n_j^2}{c^2}; \tag{4.124b}$$

$$c_j = -\frac{1}{(\Delta z)^2}; \tag{4.124c}$$

$$r_j = \frac{1}{(\Delta z)^2}\left[u_{j-1}^m + u_{j+1}^m\right] + \left\{\frac{4i\omega_0 n_j^2}{c^2\Delta t} - \frac{2}{(\Delta z)^2} + \frac{\omega_0^2 n_j^2}{c^2}\right\}u_j^m; \tag{4.124d}$$

Let us examine the propagation of a pulse in 1D using TD-BPM. The pulse initially propagates in the vacuum ($n_1 = 1.00$) and reaches a medium with refractive index of $n_2 = 3.00$ (Figure 4.30a). The initial pulse (at $t = 0$) has a Gaussian shape with the following expression:

$$u(z, t = 0) = \exp\left[-\left(\frac{z-z_0}{\sigma_z}\right)^2\right]\cdot e^{-ik(z-z_0)}, \tag{4.125}$$

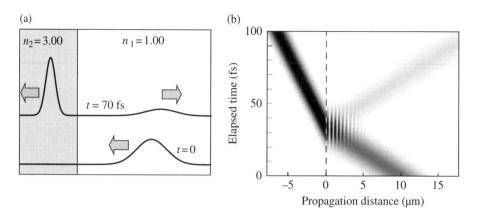

Figure 4.30 Temporal evolution of a pulse propagating in vacuum, using 1D TD-BPM. The pulse reaches the interface with a medium with refractive index $n_2 = 3.00$ located at $z = 0$, when it suffers reflection and transmission. (a) Incident pulse at t = 0, and reflected and transmitted pulses at t = 70 fs. (b) Spatial distribution of the light intensity as a function of the time

where k is the wavevector in the medium and is given by $k = 2\pi n_1/\lambda$, with $\lambda = 1\ \mu m$ and the width of the Gaussian is $\sigma_z = 4\ \mu m$. Initially, the pulse is located at $z_0 = 10\ \mu m$. This pulse contains several cycles and thus the assumption of the narrow-band is satisfied.

When the pulse reaches the interface, it splits on a reflected pulse and a transmitted pulse. Figure 4.30b shows the evolution of the pulse as a function of time, simulated by using 1D TD-BPM with a carrier frequency $\omega_0 = 2\pi c/\lambda$. At the edges of the simulation window, transparent boundary conditions (TBCs) are imposed [43], which allows the radiation to pass free through the window limits.

The calculated pulse velocity in the vacuum as a function of the longitudinal step Δz (for time steps of $\Delta t = 0.1$ fs) is plotted in Figure 4.31. It indicates that the correct speed is attained for longitudinal steps shorter than $\Delta z = 0.05\ \mu m$. As a rule of thumb, reasonable results are achieved for $\Delta z < \lambda/20\ n$.

Also, it is important to know the effect of the time step used in the propagation on the numerical results. Figure 4.32 plots such numerical results, indicating that a very large time step can be used in the propagation, as large as ~ 2 fs, which is a direct consequence of the implicit scheme used in the TD-BPM.

Finally, the ability of TD-BPM in computing the reflected power at the interface between vacuum and the medium with $n_2 = 3.00$ is examined. Figure 4.33 shows the calculated reflectivity as a function of the longitudinal step. To achieve reasonable results ($R_{theo} = 0.25$, dashed line), the longitudinal step Δz must be shorter than 0.02 μm. In fact the commented rule $\Delta z < \lambda/20\ n$ gives a value of 0.015 μm, using $n = 3.00$ (the highest refractive index involved in the reflection process).

4.7.3   Wide-Band 1D-TD-BPM

The restriction of TD-BPM to simulate wide spatial pulses (narrow-band approach) can be relaxed by improving the accuracy of the temporal derivatives in quite a similar way as we

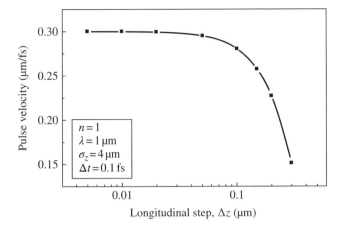

Figure 4.31 Group velocity of a Gaussian pulse propagating in the vacuum calculated by TD-BPM as a function of the longitudinal propagation step

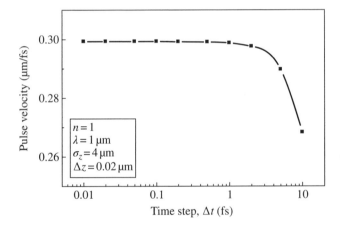

Figure 4.32 Group velocity of a Gaussian pulse propagating in the vacuum calculated by TD-BPM as a function of the time propagation step

did when we introduced wide-angle BPM to overcome paraxial propagation in conventional BPM. Here, we present the analysis of wide-band TD-BPM, which allows the simulation of optical pulses with very few optical cycles [44].

Let us consider the scalar version of the wave equation (4.116) and introduce an operator $\mathcal{P}$ defined as:

$$\mathcal{P} \equiv \frac{\dfrac{c^2}{n^2}\nabla^2 + \omega_0^2}{\omega_0^2},$$ (4.126)

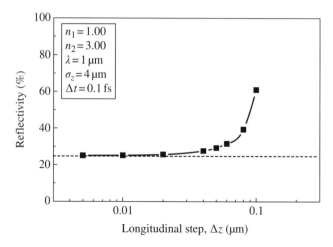

Figure 4.33 Reflectivity at the interface of Figure 4.30 calculated by TD-BPM as a function of the longitudinal propagation step

where ω_0 is the central frequency. With this definition, the scalar wave equation can be factorized into the expression [44]:

$$\left(\frac{\partial}{\partial t} + i\omega_0\sqrt{1-\mathcal{P}}\right)\left(\frac{\partial}{\partial t} - i\omega_0\sqrt{1-\mathcal{P}}\right)E = 0. \qquad (4.127)$$

For fields propagating forward in time, only the term in the second bracket in Eq. (4.127) needs to be considered. Therefore, the forward in time equation reads as:

$$\frac{\partial E}{\partial t} = \left(i\omega_0\sqrt{1-\mathcal{P}}\right)E. \qquad (4.128)$$

Now substituting the SVE field $u(\mathbf{r},t)$ defined in Eq. (4.117) into Eq. (4.128), the following differential equation is obtained:

$$\frac{\partial u}{\partial t} = i\omega_0\left(1-\sqrt{1-\mathcal{P}}\right)u. \qquad (4.129)$$

Now defining the operator Q as:

$$Q \equiv -i\omega_0\left(1-\sqrt{1-\mathcal{P}}\right), \qquad (4.130)$$

the formal solution of the discretized version of Eq. (4.129) is given by:

$$u(t+\Delta t) = e^{Q\Delta t}u(t). \qquad (4.131)$$

The discretization of Eq. (4.129) using of the standard Crank–Nicolson implicit scheme gives the general formula to the time advance of the field:

$$\left[1 - \frac{1}{2}Q\ \Delta t\right]u(t + \Delta t) = \left[1 + \frac{1}{2}Q\ \Delta t\right]u(t). \tag{4.132}$$

It is worth noting that this solution can be alternatively obtained by using the Caley's form of the exponential function in Eq. (4.131).

The approximation of the Q operator for the conventional TD-BPM (narrow-band) is obtained by the first-order approximation of Eq. (4.130) using a Taylor series expansion of the square-root operator and is given by:

$$Q \approx -\frac{i}{2\omega_0}\left(\omega_0^2 + \frac{c^2}{n^2}\nabla^2\right). \tag{4.133}$$

This narrow-band approximation has been the approach that has been used to provide the example shown previously.

Putting aside the scalar assumption for the field, the wave equation (4.129) established for the field envelope is exact and has not approximation in it. Starting from this equation, it is therefore possible to obtain an accurate description of the time marching of the fields using the field envelope. For doing that, the Q operator is expressed as a rational polynomial in $\mathcal{P}$ by applying the Padé approximation to the square-root operator, which is more accurate than the Taylor expansion for the same order of terms. The Padé approximants applied to the operator in the right hand side of Eq. (4.129) is expressed as a quotient of two polynomials M_m and N_n:

$$\left(1 - \sqrt{1 - \mathcal{P}}\right) \approx \frac{M_m(\mathcal{P})}{N_n(\mathcal{P})}. \tag{4.134}$$

Using this approximation, Eq. (4.129) reads as:

$$\frac{\partial u}{\partial t} = -i\omega_0 \frac{M_m(\mathcal{P})}{N_n(\mathcal{P})}u. \tag{4.135}$$

This equation contains the effect of an (m,n) Padé approximants for the exact operator in Eq. (4.129), where m and n are the highest degrees of $\mathcal{P}$ in the polynomials M and N, respectively. Appendix I shows the recursive way to obtain the Padé polynomial and Table 4.3

Table 4.3 Polynomials of the Padé approximant for the operator defined in Eq. (4.125)

Padé order	M_m	N_n
(1,0)	$\mathcal{P}/2$	1
(1,1)	$\mathcal{P}/2$	$1 - \mathcal{P}/4$
(2,2)	$\mathcal{P}/2 - \mathcal{P}^2/4$	$1 - 3\ \mathcal{P}/4 + \mathcal{P}^2/16$
(3,3)	$\mathcal{P}/2 - \mathcal{P}^2/2 + 3\ \mathcal{P}^3/32$	$1 - 5\ \mathcal{P}/4 + 3\ \mathcal{P}^2/8 - \mathcal{P}^3/64$

summarizes the most commonly used Padé polynomials of the operator (4.134) in terms of the operator $\mathcal{P}$ defined in Eq. (4.126).

By factorizing the resultant polynomials on the right- and left-hand sides of Eq. (4.132), the multi-step method can be then conveniently applied to solve it [44]:

$$\prod_{j=0}^{m}\left(1+c_j\mathcal{P}\right)u^{n+1} = \prod_{j=0}^{m}\left(1+c_j^*\mathcal{P}\right)u^{n}, \tag{4.136}$$

where m is the order of the Padé (m,m) approximant. The equation for a single time step is then split into m steps, with the jth step taking the following form:

$$\left(1+c_j\mathcal{P}\right)u^{n+(j+1)/m} = \left(1+c_j^*\mathcal{P}\right)u^{n+j/m}, \tag{4.137}$$

To show the improved accuracy gained by using higher order Padé approximants, let us simulate the propagation of a 1D narrow optical pulse in vacuum. The initial pulse has the same functional form of formula (4.124), with $\sigma_z = 1\,\mu m$ and at $t=0$ is located at $z_0 = -6\,\mu m$ (Figure 4.34, thick line). After a propagation time of 40 fs, the pulse has advanced 12 µm and should be located at $z = +6\,\mu m$. Although the propagation using Padé (1,0) approximant gives the correct position of the pulse (thin line), it does not reproduce the correct shape. Better approximation is gained with Padé (1,1) approximant (dashed line), but is necessary to perform the simulation by using Padé (2,2) approximant to obtain good results (dotted line).

In order to analyse the origin of the accuracy gained by higher-order Padé polynomials, let us examine the phase velocity obtained in the simulation by the different Padé orders. To do that,

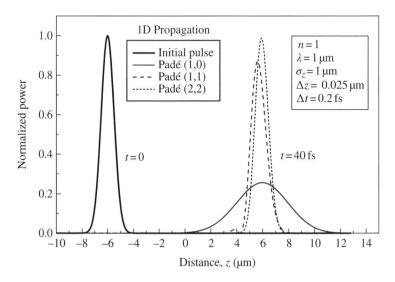

Figure 4.34 Temporal evolution of a narrow pulse propagating in vacuum, using different approaches of 1D TD-BPM. The initial pulse is centred at $z_0 = -6\,\mu m$

first the spatial frequencies $w(k,t)$ are obtained by fast Fourier transform (FFT) of the spatial profile of the optical pulse:

$$w(k,t) = \frac{1}{\sqrt{2\pi}} \int\limits_{-\infty}^{+\infty} u(z,t)e^{-ikz}dz. \tag{4.138}$$

Then, it is assumed that the spatial frequencies $w(k,t)$ are related through:

$$w(k,t+\Delta t) = w(k,t)e^{i\omega\Delta t}, \tag{4.139}$$

that can be used to calculate the relation between the components of the angular frequency ω of the optical pulse and their wavenumber k:

$$\omega = -\frac{i}{\Delta t}\ln\left[\frac{w(k,t+\Delta t)}{w(k,t)}\right], \tag{4.140}$$

which gives the relation dispersion $\omega(k)$ for the pulse. From this formula the phase velocity can be then easily calculated:

$$v_{phase} \equiv \frac{\omega(k)}{k}. \tag{4.141}$$

For a pulse through a homogeneous medium with refractive index n, the phase velocity is equal to the group velocity and its value is just c/n. Figure 4.35 shows the spatial frequency spectrum of the initial pulse in Figure 4.34 ($\sigma_z = 1\ \mu\mathrm{m}$), centred at $k = 2\pi/\lambda = 6.28\ \mu\mathrm{m}^{-1}$ (thick line), besides the phase velocity numerically calculated by performing propagation using different Padé orders. The phase velocity of the optical pulse obtained by propagation using Padé (1,0) approach shows a strong variation with the wavenumber, although at $k = 6.28\ \mu\mathrm{m}^{-1}$

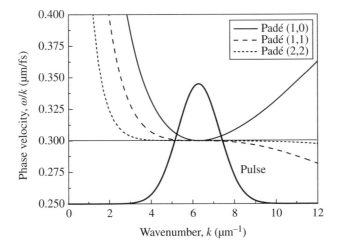

Figure 4.35 Computed phase velocity of a pulse obtained by different Padé orders

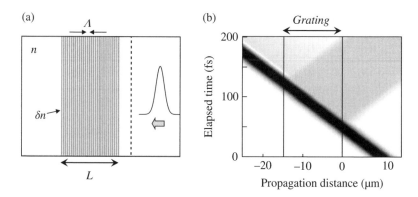

Figure 4.36 (a) Scheme of an optical pulse incident on a Bragg grating; vertical dashed line indicates the position for computing the reflected field. (b) Temporal evolution of the pulse

the calculated speed gives the correct value of 0.3 μm/fs. This implies that different spatial frequency components of the pulse will experience different phase velocities and thus the pulse will broadens as it propagates, as was observed in the plot of Figure 4.34. The Padé (1,1) algorithm shows a flat response of the phase velocity around 6.28 μm^{-1}, but it soon varies as the wavenumber differs from the central peak of the spatial frequency spectrum. Therefore, the pulse will experience deformation as it advances in the propagation. By contrast, the simulation of optical propagation using Padé (2,2) algorithm shows a flat response in the whole range of the spatial frequency spectrum. Thus, the pulse correctly preserves its shape with the propagation, as was shown in Figure 4.34.

Finally, the reflectivity spectral response of a sinusoidal Bragg grating is studied by 1D-TD-BPM (Figure 4.36a), with $n = 1.5$, $\delta n = 0.02$, $\Lambda = 0.5$ μm and $L = 15$ μm. For doing that, an optical pulse is launched at $z_0 = 0$ and the reflected field at a selected point $z_1 = 2$ μm is recorded at each time step. Figure 4.36b presents the TD-BPM propagation of the pulse during the first 200 fs. The reflectivity of the grating is calculated by the ratio between the Fourier transforms of the reflected field and the incident field:

$$R(\Delta\omega) = \frac{\left| \int u_{reflected}(z_1,t)e^{i\Delta\omega t}dt \right|}{\left| \int u_{input}(z_1,t)e^{i\Delta\omega t}dt \right|}. \tag{4.142}$$

Figure 4.37 shows the spectral response of the 1D Bragg grating computed by TD-BPM using Padé (1,0) (full circles) and (2,2) (open circles) orders. For comparison, the exact reflectivity is included (continuous line). For the calculations, an input pulse following the expression (4.125) is used, with a $\lambda = 1.5$ μm, $n = 1.5$ and $\sigma_z = 2$ μm. In the numerical simulation, a time step of $\Delta t = 0.5$ fs is chosen, for a total time of $T = 2048$ fs. This gives a spectral sampling ($\Delta\omega = 2\pi/T$) of $\Delta\lambda \approx 4$ nm at around $\lambda = 1.5$ μm, which is small enough to reproduce the main features of the reflectivity spectrum. The results are in close agreement with the exact solution, although the reflectivity at the main peak is slightly higher than the exact values due to the discretization of the structure. Also, the response for wavelength larger than 1.65 μm deviates from the exact solution for Padé (1,0), due to the fact that the temporal second derivatives is omitted in this approach.

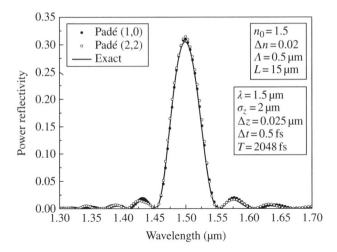

Figure 4.37   Spectral response of a sinusoidal index grating, simulated by TD-BPM

4.7.4   Narrow-Band 2D-TD-BPM

Let us develop now the algorithm to simulate the light propagation in 2D optical structures by narrow-band TD-BPM, based on the ADI method [45]. Wide-band 2D-TD-BPM algorithms have been also developed, but they require the use of iteration techniques [38]. Here we assume that the transverse and propagation directions are along the x- and z-axis and there is no variation in the y-direction. In this case, the 2D wave equation for the time envelope field $u(x,z,t)$ can be written as:

$$2i\frac{\omega_0 n^2}{c^2}\frac{\partial u}{\partial t} = \frac{\partial^2 u}{\partial x^2} + \frac{\partial^2 u}{\partial z^2} + \frac{\omega_0^2 n^2}{c^2}u, \quad \text{TE wave}; \tag{4.143a}$$

$$2i\frac{\omega_0}{c^2}\frac{\partial u}{\partial t} = \frac{\partial}{\partial x}\left(\frac{1}{n^2}\frac{\partial u}{\partial x}\right) + \frac{\partial}{\partial z}\left(\frac{1}{n^2}\frac{\partial u}{\partial z}\right) + \frac{\omega_0^2}{c^2}u, \quad \text{TM wave}, \tag{4.143b}$$

where u refers to the y component of the SVE electric field for TE propagation, or to the y-component of the SVE magnetic field for TM polarization. We now restrict ourselves to the analysis of TE propagation in what follows, where the TM case can be similarly treated.

The wave equation (4.143a) that governs the SVE field for TE propagation can be expressed as:

$$2i\frac{\omega_0 n^2}{c^2}\frac{\partial u}{\partial t} = \mathcal{H}u, \tag{4.144}$$

where the differential operator, $\mathcal{H}$, is defined as:

$$\mathcal{H}u \equiv \frac{\partial^2 u}{\partial x^2} + \frac{\partial^2 u}{\partial z^2} + \frac{\omega_0^2 n^2}{c^2}u. \tag{4.145}$$

In order to solve this differential equation using finite-difference approximations, the ADI technique is implemented where the operator $\mathcal{H}$ is split in two terms:

$$\mathcal{H} = \mathcal{H}_x + \mathcal{H}_y, \tag{4.146}$$

defined as follows:

$$\mathcal{H}_x u \equiv \frac{\partial^2 u}{\partial x^2} + \frac{1}{2}\frac{\omega_0^2 n^2}{c^2} u; \tag{4.147a}$$

$$\mathcal{H}_z u \equiv \frac{\partial^2 u}{\partial z^2} + \frac{1}{2}\frac{\omega_0^2 n^2}{c^2} u. \tag{4.147b}$$

The implementation of the standard Crank–Nicolson scheme to Eq. (4.144) has the form:

$$2i\frac{\omega_0 n^2}{c^2}\frac{u^{m+1} - u^m}{\Delta t} = (\mathcal{H}_x + \mathcal{H}_z)\frac{u^{m+1} + u^m}{2}. \tag{4.148}$$

The ADI method allows us to factorize the operator as:

$$\left(\frac{4i\omega_0 n^2}{c^2 \Delta t} - \mathcal{H}_x\right)\left(\frac{4i\omega_0 n^2}{c^2 \Delta t} - \mathcal{H}_z\right)u^{m+1} = \left(\frac{4i\omega_0 n^2}{c^2 \Delta t} + \mathcal{H}_x\right)\left(\frac{4i\omega_0 n^2}{c^2 \Delta t} + \mathcal{H}_z\right)u^m. \tag{4.149}$$

To perform the propagation, this equation is split in two sub-steps. First, the optical field at the time step m is used to calculate the intermediate field at the fictitious time position $m + 1/2$, following the scheme:

$$\left(\frac{4i\omega_0 n^2}{c^2 \Delta t} - \mathcal{H}_z\right)u^{m+1/2} = \left(\frac{4i\omega_0 n^2}{c^2 \Delta t} + \mathcal{H}_z\right)u^m. \tag{4.150}$$

Then, to complete the propagation, a second step is performed, using the field previously calculated $u^{m+1/2}$ to obtain finally the field u^{m+1}:

$$\left(\frac{4i\omega_0 n^2}{c^2 \Delta t} - \mathcal{H}_x\right)u^{m+1} = \left(\frac{4i\omega_0 n^2}{c^2 \Delta t} + \mathcal{H}_x\right)u^{m+1/2}. \tag{4.151}$$

The finite difference approach of the first sub-step (Eq. (4.150)) following the standard Crank–Nicolson scheme can be written as:

$$\frac{4i\omega_0 n^2}{c^2 \Delta t}u_{i,j}^{m+1/2} - \frac{u_{i,j-1}^{m+1/2} - 2u_{i,j}^{m+1/2} + u_{i,j+1}^{m+1/2}}{(\Delta z)^2} - \frac{1}{2}\frac{\omega_0^2 n^2}{c^2}u_{i,j}^{m+1/2}$$
$$= \frac{4i\omega_0 n^2}{c^2 \Delta t}u_{i,j}^m + \frac{u_{i,j-1}^m - 2u_{i,j}^m + u_{i,j+1}^m}{(\Delta z)^2} + \frac{1}{2}\frac{\omega_0^2 n^2}{c^2}u_{i,j}^m \tag{4.152}$$

where $u_{i,j}^m$ denotes the field at a discreted time $t = m\Delta t$ and at the spatial position $x = i\Delta x$ and $z = j\Delta z$. The coefficients of the tridiagonal system are:

$$a_j = -\frac{1}{(\Delta z)^2};\qquad\qquad(4.153a)$$

$$b_j = \frac{4i\omega_0 n^2}{c^2\Delta t} + \frac{2}{(\Delta z)^2} - \frac{1}{2}\frac{\omega_0^2 n^2}{c^2};\qquad\qquad(4.153b)$$

$$c_j = -\frac{1}{(\Delta z)^2};\qquad\qquad(4.153c)$$

$$r_j = \frac{1}{(\Delta z)^2}\left[u_{i,j-1}^m + u_{i,j+1}^m\right] + \left\{\frac{4i\omega_0 n^2}{c^2\Delta t} - \frac{2}{(\Delta z)^2} + \frac{1}{2}\frac{\omega_0^2 n^2}{c^2}\right\}u_{i,j}^m.\qquad(4.153d)$$

In a very similar way, we proceed with the second sub-step (Eq. (4.151)) derived from the application of the ADI method:

$$\frac{4i\omega_0 n^2}{c^2\Delta t}u_{i,j}^{m+1} - \frac{u_{i-1,j}^{m+1} - 2u_{i,j}^{m+1} + u_{i+1,j}^{m+1}}{(\Delta x)^2} - \frac{1}{2}\frac{\omega_0^2 n^2}{c^2}u_{i,j}^{m+1}$$

$$= \frac{4i\omega_0 n^2}{c^2\Delta t}u_{i,j}^{m+1/2} + \frac{u_{i-1,j}^{m+1/2} - 2u_{i,j}^{m+1/2} + u_{i+1,j}^{m+1/2}}{(\Delta x)^2} + \frac{1}{2}\frac{\omega_0^2 n^2}{c^2}u_{i,j}^{m+1/2}$$

$$(4.154)$$

The coefficients of the tridiagonal system are now given by:

$$a_i = -\frac{1}{(\Delta x)^2};\qquad\qquad(4.155a)$$

$$b_i = \frac{4i\omega_0 n^2}{c^2\Delta t} + \frac{2}{(\Delta x)^2} - \frac{1}{2}\frac{\omega_0^2 n^2}{c^2};\qquad\qquad(4.155b)$$

$$c_i = \frac{1}{(\Delta x)^2};\qquad\qquad(4.155c)$$

$$r_i = \frac{1}{(\Delta x)^2}\left[u_{i-1,j}^{m+1/2} + u_{i+1,j}^{m+1/2}\right] + \left\{\frac{4i\omega_0 n^2}{c^2\Delta t} - \frac{2}{(\Delta x)^2} + \frac{1}{2}\frac{\omega_0^2 n^2}{c^2}\right\}u_{i,j}^{m+1/2}.\qquad(4.155d)$$

The stability of this numerical scheme is unconditionally stable and non-dissipative for any values of Δt, so the non-physical power loss is eliminated. The convergence occurs with a discretization error $O[a(c\Delta t)^2 + (\Delta x)^2 + (\Delta z)^2]$.

To show the applicability of the 2D-TD-BPM, let us simulate the propagation of a TE pulse travelling along a step index waveguide, with $n_{core} = 3.6$, $n_{cladding} = 3.564$ and $d = 1.4\,\mu\text{m}$

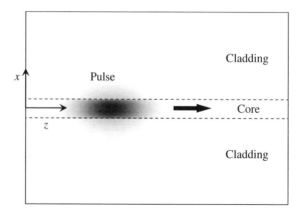

Figure 4.38   Scheme of a pulse travelling along a step-index waveguide

(Figure 4.38). The input pulse is excited at $t = 0$, which consists of the fundamental TE-eigenmode field in the x-direction and a Gaussian profile in the z-direction, that is:

$$u(x, z, t = 0) = f(x) e^{-\left(\frac{z - z_0}{\sigma_z}\right)^2} e^{-i\beta(z - z_0)}, \tag{4.156}$$

where $f(x)$ is the transversal profile corresponding to the fundamental TE mode of the planar waveguide at $\lambda = 1.55 \ \mu m$, $\sigma_z = 1 \ \mu m$ is the width of the pulse along the z-direction, $\beta = 14.54 \ \mu m^{-1}$ is the propagation constant of the waveguide mode and $z_0 = 2 \ \mu m$ is the central position of the input optical pulse.

First, we examine the computed speed of the pulse as a function of the longitudinal step Δz used in the simulation (Figure 4.39), using a transversal grid spacing of $\Delta x = 0.04 \ \mu m$ and a propagation step of $\Delta t = 0.1 \ fs$. The converged value of the speed is given by $v = c/N_{eff}$, with N_{eff} the effective index of the mode, which gives a pulse speed of $0.0836 \ \mu m/fs$ ($N_{eff} = 3.587$) that is represented by a dotted line in the figure. The simulations indicate that longitudinal step shorter than $0.02 \ \mu m$ provides pulse speed values close to the exact value.

At the edges of the computational region, TBCs can be implemented [46]. This algorithm is suitable for situations in which the fields near the boundary can be locally approximated by a single outgoing plane wave [43]. Nevertheless, some small errors from this assumption can create spurious incoming waves and given enough time this noise from the spurious waves can cause the fields at the boundary to exponentially grow and significantly depart from the correct propagation. When long propagation times are required in the simulation, for instance when the spectral response of grating is analysed through FFT of the reflected field, the implementation of perfectly matched layers (PMLs) is more suitable at the boundaries [40], in spite of the fact that some extra regions for the PML should be added in the computational window. The wave equations incorporating PML layers are given by [47]:

$$2i \frac{n^2 \omega_0}{c^2} \frac{\partial u}{\partial t} = \frac{s_x}{s} \frac{\partial}{\partial x} \left(\frac{s_x}{s} \frac{\partial u}{\partial x} \right) + \frac{s_z}{s} \frac{\partial}{\partial z} \left(\frac{s_z}{s} \frac{\partial u}{\partial z} \right) + \frac{n^2 \omega_0^2}{c^2} u, \quad \text{TE waves}; \tag{4.157a}$$

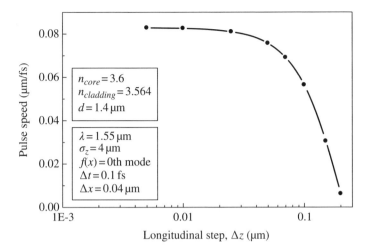

Figure 4.39 Speed of a pulse travelling along a step-index waveguide as a function of the longitudinal step, simulated by 2D-TD-BPM

$$2i\frac{\omega_0}{c^2}\frac{\partial u}{\partial t} = \frac{s_x}{s}\frac{\partial}{\partial x}\left(\frac{1}{n^2}\frac{s_x}{s}\frac{\partial u}{\partial x}\right) + \frac{s_z}{s}\frac{\partial}{\partial z}\left(\frac{1}{n^2}\frac{s_z}{s}\frac{\partial u}{\partial z}\right) + \frac{\omega_0^2}{c^2}u, \quad \text{TM waves}, \qquad (4.157b)$$

where $u(x,z,t)$ stands for the SVE of E_y for TE waves or the SVE of H_y for TM waves. The PML parameter s is given by:

$$s = 1 + i\frac{3c}{4\omega_0 nd}\left(\frac{\rho}{d_{PML}}\right)\ln R, \quad \text{in PML region}; \qquad (4.158a)$$

$$s = 1, \quad \text{in non-PML region}, \qquad (4.158b)$$

where ρ the distance inside the PML region from the beginning of PML, d_{PML} is the PML thickness and R is the theoretical reflectivity from the PML region. For the PML regions perpendicular to the x-axis, perpendicular to the z-axis, or corners, the PML parameters are given by:

$$s_x = 1, \quad s_z = s \quad \text{regions} \perp x \text{ axis}; \qquad (4.159a)$$

$$s_x = s, \quad s_z = 1 \quad \text{regions} \perp z \text{ axis}; \qquad (4.159b)$$

$$s_x = 1, \quad s_z = 1 \quad \text{corners}. \qquad (4.159c)$$

Figure 4.40 shows the evolution of the normalized pulse power given in the previous example, for TE and TM pulses, when TBC or PML boundary conditions are implemented ($d_{PML} = 1$ μm, $R = 10^{-8}$). As it can be seen, both algorithms give very low values for the reflectivity at the boundary, close to $R = 10^{-7}$.

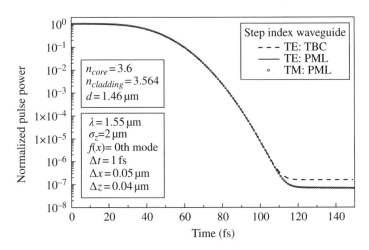

Figure 4.40   Time evolution of the normalized remaining power for pulse propagation in a slab waveguide

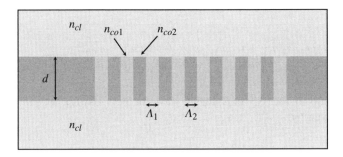

Figure 4.41   Optical waveguide grating with modulated refractive indices

The waveguide grating shown in Figure 4.41, which exhibits polarization dependence in a strong guiding structure, is now analysed, where PML boundary conditions are implemented. The refractive indices of the structure are $n_{co1} = 1.45$, $n_{co2} = 1.55$ and $n_{cl} = 1$ and the waveguide width is $d = 0.3$ μm. The grating periods are $\Lambda_1 = 0.328$ μm, $\Lambda_1 = 0.316$ μm and there are eight periods in the waveguide core. The original structure is surrounded by PML layers with $d_{PML} = 1$ μm and $R_{theo} = 10^{-8}$.

The reflected field $u(x,z,t)$ at a monitoring plane $z = z_1$ is recorded while the time marching proceeds. By performing Fourier transform and the overlap integral, the power reflectivity of the structure can be calculated, which is given by [48]:

$$R(\Delta\omega) = \frac{\left| \iint u(x,z_1,t) f_j(x) e^{i\Delta\omega t} dx dt \right|^2}{\left| \iint u_{inc}(x,z_1,t) f_j(x) e^{i\Delta\omega t} dx dt \right|}. \tag{4.160}$$

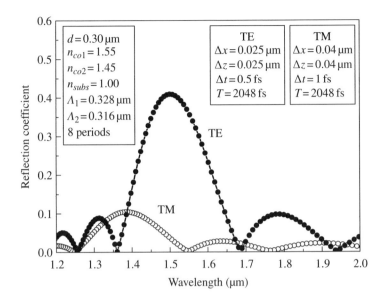

Figure 4.42 Spectral response of the reflection coefficient of the step-index waveguide grating for TE (continuous line) and TM (dashed line) propagation

where $f_j(x)$ is the jth mode profile of the waveguide at frequency ω and $u_{inc}(x,z,t)$ is the incident field.

Figure 4.42 shows the reflection coefficient for the waveguide grating for TE (full circles) and TM propagation (open circles). As can be seen, this waveguide grating is strongly polarization dependent because of the high index contrast of the structure and thus the maximum reflection coefficient are different for TE and TM propagation and also the position of the maxima depends on the polarization ($\lambda_{max,TE} \approx 1.50\,\mu m$, $\lambda_{max,TM} \approx 1.38\,\mu m$). These TD-BPM results are in good agreement with the results provided by using more accurate methods [38] around the central wavelength. Also, they exhibit slight spectral shifts from the 'exact' values away from the centre wavelength, due to the omission of the temporal second derivatives based on the SVEA.

4.8 Finite-Difference Time-Domain Method (FD-TD)

Today, the FDTD analysis is one of the most popular numerical methods for several classes of electromagnetic problems. Its many advantages have made it the method of choice in modelling and designing complicated RF and microwave circuits with dimensions comparable to radiation wavelength. In order to maintain high accuracy and numerical stability at optical frequencies, grid cell size and time step are of the order of tens of nanometres and tens of attoseconds, respectively. Thus, until recently, most optical devices had distance and/or time scales that made them inappropriate for FDTD analysis. However, for the emerging class of micrometre and nanometre-scale integrated optical devices, FDTD modelling has the potential to play a useful and practical design role [49, 50].

The FDTD method is an explicit grid-based technique for the direct solution of the fundamental Maxwell's time-dependent curl equations. It applies simple second-order accurate central difference approximations for the space and time derivatives of the electric and magnetic fields directly to the respective differential operators of the curl equations [51]. Electric and magnetic field components are interleaved in space to permit a natural satisfaction of tangential field continuity conditions at media interfaces. The FDTD is a marching-in-time procedure and at each time step the system of equations to update the field components is fully explicit, so there is no need to solve a system of linear simultaneous equations. As a consequence, the required computer storage and running time is proportional to N, where N is the number of electromagnetic field unknowns in the volume modelled.

Whereas BPM frequency-domain solutions are limited to sinusoidal steady state conditions, the FDTD method permits transient electromagnetic excitations and the computation of impulse responses. Furthermore, in contrast to conventional BPM numerical methods, FDTD makes no assumptions about the direction of wave propagation or the time-varying nature of the fields and in particular, there are no SVEAs. Since FDTD rigorously enforces boundary conditions at material interfaces, it is capable of modelling the geometrical complexities and material inhomogeneities of realistic micron-scale photonic and optoelectronic devices, including side-wall roughness, curved guiding structures with small bend radii, tapered waveguides, deep gratings and novel cavity shapes (rings, disks, cylindrical posts, defects in photonic crystals etc.). Consequently, the optical wave phenomena inherent in photonic structures that are comparable in size to the wavelength are fully accounted for. Also, FDTD can be used for optical waveguide modal analysis [52, 53] (see Appendix J).

In what follows, the derivation of the basic FDTD equations is presented. The standard algorithm is based on centred finite differences for space and time derivatives, which is second-order accurate in space and time. The leapfrog integration scheme marches the discretized electric and magnetic fields forward in time.

4.8.1 Finite-Difference Expressions for Maxwell's Equations in Three Dimensions

Maxwell's curl equations in linear, isotropic, nondispersive, lossy materials are given by [39]:

$$\frac{\partial \mathcal{H}}{\partial t} = -\frac{1}{\mu} \nabla \times \mathcal{E} - \frac{1}{\mu}(\mathcal{M}_{source} + \rho \mathcal{H}); \tag{4.161a}$$

$$\frac{\partial \mathcal{E}}{\partial t} = \frac{1}{\varepsilon} \nabla \times \mathcal{H} - \frac{1}{\varepsilon}(\mathcal{J}_{source} + \sigma \mathcal{E}), \tag{4.161b}$$

where σ denotes the electric conductivity (siemens/metre) and ρ is the magnetic conductivity (ohms/metre), which take account for electric and magnetic energy losses, respectively. The electric current density $\mathcal{J}$ (amperes/m^2) and the equivalent magnetic current density $\mathcal{M}$ (volts/m^2) have been expressed as:

$$\mathcal{J} = \mathcal{J}_{source} + \sigma \mathcal{E}, \tag{4.162a}$$

$$\mathcal{M} = \mathcal{M}_{source} + \rho \mathcal{H}. \tag{4.162b}$$

The two vectorial equations (4.161a) and (4.161ba) yield to the following system of six coupled scalar equations:

$$\frac{\partial E_x}{\partial t} = \frac{1}{\varepsilon}\left[\frac{\partial H_z}{\partial y} - \frac{\partial H_y}{\partial z} - \left(J_{source,x} + \sigma E_x\right)\right]; \tag{4.163a}$$

$$\frac{\partial E_y}{\partial t} = \frac{1}{\varepsilon}\left[\frac{\partial H_x}{\partial z} - \frac{\partial H_z}{\partial x} - \left(J_{source,y} + \sigma E_y\right)\right]; \tag{4.163b}$$

$$\frac{\partial E_z}{\partial t} = \frac{1}{\varepsilon}\left[\frac{\partial H_y}{\partial x} - \frac{\partial H_x}{\partial y} - \left(J_{source,z} + \sigma E_z\right)\right]; \tag{4.163c}$$

$$\frac{\partial H_x}{\partial t} = \frac{1}{\mu}\left[\frac{\partial E_y}{\partial z} - \frac{\partial E_z}{\partial y} - \left(M_{source,x} + \rho H_x\right)\right]; \tag{4.163d}$$

$$\frac{\partial H_y}{\partial t} = \frac{1}{\mu}\left[\frac{\partial E_z}{\partial x} - \frac{\partial E_x}{\partial z} - \left(M_{source,y} + \rho H_y\right)\right], \tag{4.163e}$$

$$\frac{\partial H_z}{\partial t} = \frac{1}{\mu}\left[\frac{\partial E_x}{\partial y} - \frac{\partial E_y}{\partial x} - \left(M_{source,z} + \rho H_z\right)\right]. \tag{4.163f}$$

Let us remember that in the derivation of Eqs. (4.163a–f) no assumption has been made *a priori* about the spatial or the temporal dependences of the fields. These equations are suitable for deriving finite difference schemes that allow to us obtain the temporal evolution of the electric and magnetic fields. For doing that, Yee proposed to discretize the computational space in elementary cells and appropriately situate the field components in it to achieve a high degree of accuracy [54]. Figure 4.43

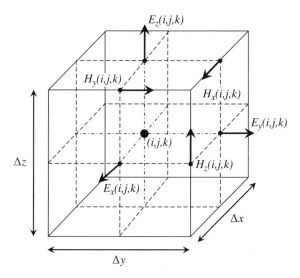

Figure 4.43 Yee's cell for 3D-light propagation. The full circle denotes the centre of the unit cell, where it is surrounded by the electric and magnetic field Cartesian components. Note the different shift with respect to the cell centre of the different field components

shows the Yee's cell used to discretize the fields in Eqs. (4.163a–f). Also, the time marching scheme is designed in such a manner that high accuracy is obtained.

The full evolution scheme of the fields is [39]:

$$
E_x\Big|_{i+1/2,j,k}^{n+1} = \left(\frac{1-\dfrac{\sigma_{i+1/2,j,k}\Delta t}{2\varepsilon_{i+1/2,j,k}}}{1+\dfrac{\sigma_{i+1/2,j,k}\Delta t}{2\varepsilon_{i+1/2,j,k}}}\right) E_x\Big|_{i+1/2,j,k}^{n} + \left(\frac{\dfrac{\Delta t}{\varepsilon_{i+1/2,j,k}}}{1+\dfrac{\sigma_{i+1/2,j,k}\Delta t}{2\varepsilon_{i+1/2,j,k}}}\right).
$$

$$
\left(\frac{H_z\Big|_{i+1/2,j+1/2,k}^{n+1/2}-H_z\Big|_{i+1/2,j-1/2,k}^{n+1/2}}{\Delta y} - \frac{H_y\Big|_{i+1/2,j,k+1/2}^{n+1/2}-H_y\Big|_{i+1/2,j,k-1/2}^{n+1/2}}{\Delta z} - J_{source,x}\Big|_{i+1/2,j,k}^{n+1/2}\right)
$$

$$(4.164a)$$

$$
E_y\Big|_{i,j+1/2,k}^{n+1} = \left(\frac{1-\dfrac{\sigma_{i,j+1/2,k}\Delta t}{2\varepsilon_{i,j+1/2,k}}}{1+\dfrac{\sigma_{i,j+1/2,k}\Delta t}{2\varepsilon_{i,j+1/2,k}}}\right) E_y\Big|_{i,j+1/2,k}^{n} + \left(\frac{\dfrac{\Delta t}{\varepsilon_{i,j+1/2,k}}}{1+\dfrac{\sigma_{i,j+1/2,k}\Delta t}{2\varepsilon_{i,j+1/2,k}}}\right).
$$

$$
\left(\frac{H_x\Big|_{i,j+1/2,k+1/2}^{n+1/2}-H_x\Big|_{i,j+1/2,k-1/2}^{n+1/2}}{\Delta z} - \frac{H_z\Big|_{i+1/2,j+1/2,k}^{n+1/2}-H_z\Big|_{i-1/2,j+1/2,k}^{n+1/2}}{\Delta x} - J_{source,y}\Big|_{i,j+1/2,k}^{n+1/2}\right)
$$

$$(4.164b)$$

$$
E_z\Big|_{i,j,k+1/2}^{n+1} = \left(\frac{1-\dfrac{\sigma_{i,j,k+1/2}\Delta t}{2\varepsilon_{i,j,k+1/2}}}{1+\dfrac{\sigma_{i,j,k+1/2}\Delta t}{2\varepsilon_{i,j,k+1/2}}}\right) E_z\Big|_{i,j,k+1/2}^{n} + \left(\frac{\dfrac{\Delta t}{\varepsilon_{i,j,k+1/2}}}{1+\dfrac{\sigma_{i,j,k+1/2}\Delta t}{2\varepsilon_{i,j,k+1/2}}}\right).
$$

$$
\left(\frac{H_y\Big|_{i+1/2,j,k+1/2}^{n+1/2}-H_y\Big|_{i-1/2,j,k+1/2}^{n+1/2}}{\Delta x} - \frac{H_x\Big|_{i,j+1/2,k+1/2}^{n+1/2}-H_x\Big|_{i,j-1/2,k+1/2}^{n+1/2}}{\Delta y} - J_{source,z}\Big|_{i,j,k+1/2}^{n+1/2}\right)
$$

$$(4.164c)$$

$$
H_x\Big|_{i,j+1/2,k+1/2}^{n+1/2} = \left(\frac{1-\dfrac{\rho_{i,j+1/2,k+1/2}\Delta t}{2\mu_{i,j+1/2,k+1/2}}}{1+\dfrac{\rho_{i,j+1/2,k+1/2}\Delta t}{2\mu_{i,j+1/2,k+1/2}}}\right) H_x\Big|_{i,j+1/2,k+1/2}^{n-1/2} + \left(\frac{\dfrac{\Delta t}{\mu_{i,j+1/2,k+1/2}}}{1+\dfrac{\rho_{i,j+1/2,k+1/2}\Delta t}{2\mu_{i,j+1/2,k+1/2}}}\right).
$$

$$
\left(\frac{E_y\Big|_{i,j+1/2,k+1}^{n}-E_y\Big|_{i,j+1/2,k}^{n}}{\Delta z} - \frac{E_z\Big|_{i,j+1,k+1/2}^{n}-E_z\Big|_{i,j,k+1/2}^{n}}{\Delta y} - M_{source,x}\Big|_{i,j+1/2,k+1/2}^{n}\right)
$$

$$(4.164d)$$

$$H_y\Big|_{i+1/2,j,k+1/2}^{n+1/2} = \left(\frac{1-\dfrac{\rho_{i+1/2,j,k+1/2}\Delta t}{2\mu_{i+1/2,j,k+1/2}}}{1+\dfrac{\rho_{i+1/2,j,k+1/2}\Delta t}{2\mu_{i+1/2,j,k+1/2}}}\right) H_y\Big|_{i+1/2,j,k+1/2}^{n-1/2} + \left(\frac{\dfrac{\Delta t}{\mu_{i+1/2,j,k+1/2}}}{1+\dfrac{\rho_{i+1/2,j,k+1/2}\Delta t}{2\mu_{i+1/2,j,k+1/2}}}\right).$$

$$\left(\frac{E_z\Big|_{i+1,j,k+1/2}^{n}-E_z\Big|_{i,j,k+1/2}^{n}}{\Delta x} - \frac{E_x\Big|_{i+1/2,j,k+1}^{n}-E_x\Big|_{i+1/2,j,k}^{n}}{\Delta z} - M_{source,y}\Big|_{i+1/2,j,k+1/2}^{n}\right)$$

$$(4.164e)$$

$$H_z\Big|_{i+1/2,j+1/2,k}^{n+1/2} = \left(\frac{1-\dfrac{\rho_{i+1/2,j+1/2,k}\Delta t}{2\mu_{i+1/2,j+1/2,k}}}{1+\dfrac{\rho_{i+1/2,j+1/2,k}\Delta t}{2\mu_{i+1/2,j+1/2,k}}}\right) H_z\Big|_{i+1/2,j+1/2,k}^{n-1/2} + \left(\frac{\dfrac{\Delta t}{\mu_{i+1/2,j+1/2,k}}}{1+\dfrac{\rho_{i+1/2,j+1/2,k}\Delta t}{2\mu_{i+1/2,j+1/2,k}}}\right).$$

$$\left(\frac{E_x\Big|_{i+1/2,j+1,k}^{n}-E_x\Big|_{i+1/2,j,k}^{n}}{\Delta y} - \frac{E_y\Big|_{i+1,j+1/2,k}^{n}-E_y\Big|_{i,j+1/2,k}^{n}}{\Delta x} - M_{source,z}\Big|_{i+1/2,j+1/2,k}^{n}\right)$$

$$(4.164f)$$

Δt being the time step and Δx, Δy and Δz the spatial discretization steps.

The magnetic field components at time $(n + 1/2)\Delta t$ are calculated using the previously calculated and stored magnetic field components at time $(n - 1/2)\Delta t$ and electric field components at time step $n\Delta t$. On the other hand, updating the electric field components at time $(n + 1)\Delta t$ requires the previously calculated and stored electric field components at time $n\Delta t$ and magnetic field components at time step $(n + 1/2)\Delta t$.

This time marching scheme is fully explicit and is stable for time steps that fulfil the stability criterion given by [55]:

$$\Delta t < \frac{1}{v_{max}\sqrt{\dfrac{1}{(\Delta x)^2} + \dfrac{1}{(\Delta y)^2} + \dfrac{1}{(\Delta z)^2}}}, \tag{4.165}$$

where v_{max} is the maximum wave phase velocity within the modelled structure. This condition is the so-called CFL criterion.

Regarding the spatial increments, as a rule of thumb, a good choice is to take Δx, Δy, Δz $< \lambda/20$. In addition, the discretization level should be fine enough to reproduce the optical structure to be modelled.

For computational purposes, the field components are allocated in memory as:

$$E_x\big|_{i+1/2,j,k} \Rightarrow E_x(i,j,k) \quad H_x\big|_{i,j+1/2,k+1/2} \Rightarrow H_x(i,j,k)$$

$$E_y\big|_{i,j+1/2,k} \Rightarrow E_y(i,j,k) \quad H_y\big|_{i+1/2,j,k+1/2} \Rightarrow H_y(i,j,k)$$

$$E_z\big|_{i,j,k+1/2} \Rightarrow E_z(i,j,k) \quad H_z\big|_{i+1/2,j+1/2,k} \Rightarrow H_z(i,j,k)$$

4.8.2   Truncation of the Computational Domain

At the limits of the computational region some special conditions must be provided. Setting the tangential components of the electric field to zero at the window boundaries implies the existence of a perfect electric conductor and its effect is to reflect the waves back to the computational domain, thus spoiling the correct wave propagation description for long-time simulations. Although several algorithms have been proposed to absorb waves from the boundaries [56, 57], here we only describe the PML implementation [58]. The PML is an artificial lossy material used for the truncation of numerical computation domains when using the FDTD method, being a very efficient algorithm for avoiding unwanted reflections from computational boundaries. With the PML technique, the 6 components of the electromagnetic field are split into 12 subcomponents and the 6 scalar Maxwell's equations are replaced by 12 scalar equations [39]. By properly associating different electric and magnetic conductivities to the different subcomponents, theoretically, the PML would not cause any reflection at an interface with the free space for incident fields of any frequencies and incident angles.

Suppose that in the x-y plane a semi-infinite PML medium is interfaced with a semi-infinite free space at the surface $x = 0$. The PML is in the region $x > 0$. Based on the criteria described by Berenger for determining the electric and magnetic losses in PML, the electric and magnetic conductivities, σ_y and ρ_y, should be zero. Additionally, the electric and magnetic conductivities, σ_x and ρ_x, should satisfy the condition:

$$\frac{\sigma_x}{\varepsilon_0} = \frac{\rho_x}{\mu_0}. \tag{4.166}$$

Under these circumstances, the free-space-PML interface has theoretically no reflection to incident fields of any frequencies and incident angles. In numerical computations, the thickness of PML is finite and the PML is usually terminated by an electric wall, actuating as a perfectly electric conductor. As the wave in PML reaches the terminating boundary and gets reflected, it is further attenuated as it propagates back toward the free-space-PML interface. By adjusting the value of conductivity one can control the amount of attenuation in PML, that is, the amount of reflection from PML. For any thickness of PML, ideally one can choose σ to be sufficiently large enough to make reflection from PML as small as wanted. However, this is not the case in actual numerical computations. In fact, any discontinuities in material properties, such as the conductivity σ, will cause numerical reflection. Therefore, for practical purposes, the conductivity σ is varied continuously with space, from zero at the free-space-PML interface to its maximum value σ_{max} at the outer side of the PML. If the thickness of PML is denoted by δ, the conductivity is usually written in the form:

$$\sigma(x) = \sigma_{max} \left(\frac{x}{\delta}\right)^p, \tag{4.167}$$

the power p being an integer, usually in the range 2–4. For PML terminated by an electric conducting wall, the theoretical reflection r (when the reflection due to the spatial variation of conductivity is ignored), can be calculated as:

$$r(\varphi) = \exp\left[-\frac{2}{p+1}\frac{\sigma_{max}\delta}{c\varepsilon_0}\cos\varphi\right], \tag{4.168}$$

where φ is the incident angle at the interface. For a given reflection coefficient at normal incidence and a given power p, the maximum value of the conductivity σ_{max} can be found to be:

$$\sigma_{max} = -\frac{(p+1)c\varepsilon_0}{2\delta}\log(r).$$

(4.169)

4.8.3 Two-Dimensional FDTD: TM Case

Let us assume that the structure being modelled extends to infinity in the y direction with no change in the shape or position of its transverse cross section (Figure 4.44). If the incident wave is also uniform in the y direction, then all partial derivatives of the fields with respect to y must equal zero. In addition, let us assume that the medium has no free electric or magnetic current sources. Under these conditions, the full set of Maxwell's curl equations reduces to:

$$\frac{\partial E_x}{\partial t} = \frac{1}{\varepsilon}\left[-\frac{\partial H_y}{\partial z} - \sigma E_x\right];$$

(4.170a)

$$\frac{\partial E_y}{\partial t} = \frac{1}{\varepsilon}\left[-\frac{\partial H_x}{\partial z} - \frac{\partial H_z}{\partial x} - \sigma E_y\right];$$

(4.170b)

$$\frac{\partial E_z}{\partial t} = \frac{1}{\varepsilon}\left[\frac{\partial H_y}{\partial x} - \sigma E_z\right];$$

(4.170c)

$$\frac{\partial H_x}{\partial t} = \frac{1}{\mu}\left[\frac{\partial E_y}{\partial z}\rho H_x\right];$$

(4.170d)

$$\frac{\partial H_y}{\partial t} = \frac{1}{\mu}\left[\frac{\partial E_z}{\partial x} - \frac{\partial E_x}{\partial z} - \rho H_y\right];$$

(4.170e)

$$\frac{\partial H_z}{\partial t} = \frac{1}{\mu}\left[-\frac{\partial E_y}{\partial x} - \rho H_z\right];$$

(4.170f)

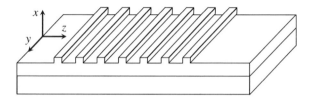

Figure 4.44 Example of a three-dimensional structure that is invariant along the y direction

TM-polarized modes have the E_x, E_z and H_y as non-null components and, therefore, the equations that govern the propagation of such fields are:

$$\frac{\partial E_x}{\partial t} = \frac{1}{\varepsilon}\left[-\frac{\partial H_y}{\partial z} - \sigma E_x\right];\qquad\qquad(4.171a)$$

$$\frac{\partial E_z}{\partial t} = \frac{1}{\varepsilon}\left[-\frac{\partial H_y}{\partial x} - \sigma E_z\right];\qquad\qquad(4.171b)$$

$$\frac{\partial H_y}{\partial t} = \frac{1}{\mu}\left[\frac{\partial E_z}{\partial x} - \frac{\partial E_x}{\partial z} - \rho H_y\right].\qquad\qquad(4.171c)$$

It is worth commenting that TM propagation (2D waveguides) is called TE_y propagation in FDTD formulation, as the electric field is transverse to the y-axis.

On the other hand, light propagation as TE-modes in two dimensions involves only the E_y, H_x and H_z components of the fields. In this case the equations to be considered are:

$$\frac{\partial E_y}{\partial t} = \frac{1}{\varepsilon}\left[\frac{\partial H_x}{\partial z} - \frac{\partial H_z}{\partial x} - \sigma E_y\right];\qquad\qquad(4.172a)$$

$$\frac{\partial H_x}{\partial t} = \frac{1}{\mu}\left[\frac{\partial E_y}{\partial z} - \rho H_x\right];\qquad\qquad(4.172b)$$

$$\frac{\partial H_z}{\partial t} = \frac{1}{\mu}\left[-\frac{\partial E_y}{\partial x} - \rho H_z\right].\qquad\qquad(4.172c)$$

The TE-modes are referred to as TM_y propagation in FDTD nomenclature, because this polarized light has the magnetic field transversal to the y-axis.

The TM and TE waves contain no common field vector components and thus they can exist simultaneously with no mutual interactions for structures composed of isotropic materials (or anisotropic materials having no off-diagonal components in the constitutive tensors).

4.8.3.1   Implementation of Finite Difference Scheme for TM Propagation

Let us now derive the finite difference scheme for TM propagation in two-dimensional structures, including the implementation of PML at the boundaries of the computational domain. One example of 2D-structures is given in Figure 4.45, which is the cross-section of the structure given in Figure 4.44. TM propagation requires the updating of the E_x, E_z and H_y components. For 2D-FDTD implementation, the fields are located at the grid positions shown in Figure 4.46. Let us observe that the computational domain is surrounded by a perfectly electric conductor, where the tangential electric fields are set to zero.

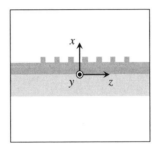

Figure 4.45 Cross-section of the two-dimensional structure shown in Figure 4.44. Light propagation in this structure can be separately treated as TE or TM fields

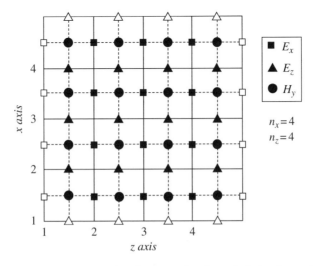

Figure 4.46 Yee's grid for TM propagation. Squares, triangles and circles denote the positions of the E_x, E_z and H_y components, respectively. Open symbols surrounding the computational region at $i = 1$, $i = nx + 1$, $k = 1$ and $k = nz + 1$ indicate zero fields, corresponding to the simulation of a perfect electric conductor

Direct finite differencing of Eqs. (4.171a–c) yields:

$$E_x\Big|_{i+1/2,k}^{n+1} = \left(\frac{1 - \frac{\sigma_{i+1/2,k}\Delta t}{2\varepsilon_{i+1/2,k}}}{1 + \frac{\sigma_{i+1/2,k}\Delta t}{2\varepsilon_{i+1/2,k}}}\right) \cdot E_x\Big|_{i+1/2,k}^{n} - \left(\frac{\frac{\Delta t}{\varepsilon_{i+1/2,k}}}{1 + \frac{\sigma_{i+1/2,k}\Delta t}{2\varepsilon_{i+1/2,k}}}\right) \cdot \left(\frac{H_y\Big|_{i+1/2,k+1/2}^{n+1/2} - H_y\Big|_{i+1/2,k-1/2}^{n+1/2}}{\Delta z}\right);$$

$$(4.173a)$$

$$E_z\Big|_{i,k+1/2}^{n+1} = \left(\frac{1 - \frac{\sigma_{i,k+1/2}\Delta t}{2\varepsilon_{i,k+1/2}}}{1 + \frac{\sigma_{i,k+1/2}\Delta t}{2\varepsilon_{i,k+1/2}}}\right) \cdot E_z\Big|_{i,k+1/2}^{n} - \left(\frac{\frac{\Delta t}{\varepsilon_{i,k+1/2}}}{1 + \frac{\sigma_{i,k+1/2}\Delta t}{2\varepsilon_{i,k+1/2}}}\right) \cdot \left(\frac{H_y\Big|_{i+1/2,k+1/2}^{n+1/2} - H_y\Big|_{i-1/2,k+1/2}^{n+1/2}}{\Delta x}\right);$$

$$(4.173b)$$

$$H_y\Big|_{i+1/2,k+1/2}^{n+1/2} = \left(\frac{1-\dfrac{\rho_{i+1/2,k+1/2}\Delta t}{2\mu_{i+1/2,k+1/2}}}{1+\dfrac{\rho_{i+1/2,k+1/2}\Delta t}{2\mu_{i+1/2,k+1/2}}}\right)\cdot H_y\Big|_{i+1/2,k+1/2}^{n-1/2}$$

$$(4.173c)$$

$$+\left(\frac{\dfrac{\Delta t}{\mu_{i+1/2,k+1/2}}}{1+\dfrac{\rho_{i+1/2,k+1/2}\Delta t}{2\mu_{i+1/2,k+1/2}}}\right)\cdot\left(\frac{E_z\Big|_{i+1,k+1/2}^{n}-E_z\Big|_{i,k+1/2}^{n}}{\Delta x}-\frac{E_x\Big|_{i+1/2,k+1}^{n}-E_z\Big|_{i+1/2,k}^{n}}{\Delta z}\right)$$

This field updating is of general use and these are stable providing the time step is chosen to fulfil the CFL criterion. Nevertheless, Maxwell's curl equations (4.171a–c) can be artificially modified to significantly reduce the reflections at the outer boundaries. In the case of TM propagation, this can be achieved by splitting the magnetic field component H_y in to two parts [39] and rewriting Maxwell's curl equation in the form:

$$\frac{\partial E_x}{\partial t} = \frac{1}{\varepsilon}\left[\frac{\partial H_y}{\partial z} - \sigma_z E_x\right];$$

$$(4.174a)$$

$$\frac{\partial E_x}{\partial t} = \frac{1}{\varepsilon}\left[\frac{\partial H_y}{\partial z} - \sigma_z E_x\right];$$

$$(4.174b)$$

$$\frac{\partial H_{yx}}{\partial t} = \frac{1}{\mu}\left[\frac{\partial E_z}{\partial x} - \rho_x H_{yx}\right];$$

$$(4.174c)$$

$$\frac{\partial H_{yz}}{\partial t} = \frac{1}{\mu}\left[-\frac{\partial E_x}{\partial z} - \rho_z H_{yz}\right];$$

$$(4.174d)$$

$$H_y = H_{yx} + H_{yz},$$

$$(4.174e)$$

where the anisotropic electric and magnetic conductivities should be appropriately set at the PMLs.

The finite difference of Eq. (4.173a) gives:

$$E_x\Big|_{i+1/2,k}^{n+1} = \left(\frac{1-\dfrac{\sigma_{i+1/2,k}\Delta t}{2\varepsilon_{i+1/2,k}}}{1+\dfrac{\sigma_{i+1/2,k}\Delta t}{2\varepsilon_{i+1/2,k}}}\right)E_x\Big|_{i+1/2,k}^{n}-\left(\frac{\dfrac{\Delta t}{\varepsilon_{i+1/2,k}}}{1+\dfrac{\sigma_{i+1/2,k}\Delta t}{2\varepsilon_{i+1/2,k}}}\right)\cdot\left(\frac{H_y\Big|_{i+1/2,k+1/2}^{n+1/2}-H_y\Big|_{i+1/2,k-1/2}^{n+1/2}}{\Delta z}\right).$$

$$(4.175)$$

For computational purposes, we use the following notation (see Figure 4.47):

$$E_x\Big|_{i+1/2,k} \Rightarrow E_x(i,k);$$

$$(4.176a)$$

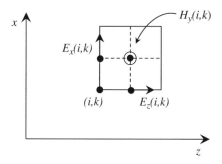

Figure 4.47 Location for the field components in the Yee's grid for performing the spatial finite differences for the TM case

$$\left(\frac{1 - \frac{\sigma_{i+1/2,k}\Delta t}{2\varepsilon_{i+1/2,k}}}{1 + \frac{\sigma_{i+1/2,k}\Delta t}{2\varepsilon_{i+1/2,k}}} \right) \Rightarrow C_{A,Ex}(i,k); \tag{4.176b}$$

$$\left(\frac{\frac{\Delta t}{\varepsilon_{i+1/2,k}\Delta z}}{1 + \frac{\sigma_{i+1/2,k}\Delta t}{2\varepsilon_{i+1/2,k}}} \right) \Rightarrow C_{B,Ex}(i,k), \tag{4.176c}$$

giving:

$$E_x^{n+1}(i:1,nx;k:2,nz) = C_{A,Ex}(i,k)E_x^n(i,k) - C_{B,Ex}(i,k)\left[H_y^{n+1/2}(i,k-1)\right]. \tag{4.177}$$

Similarly, the finite difference form of Eq. (4.174b) is:

$$E_z\Big|_{i,k+1/2}^{n+1} = \left(\frac{1 - \frac{\sigma_{i,k+1/2}\Delta t}{2\varepsilon_{i,k+1/2}}}{1 + \frac{\sigma_{i,k+1/2}\Delta t}{2\varepsilon_{i,k+1/2}}} \right) E_z\Big|_{i,k+1/2}^{n} - \left(\frac{\frac{\Delta t}{\varepsilon_{i,k+1/2}}}{1 + \frac{\sigma_{i,k+1/2}\Delta t}{2\varepsilon_{i,k+1/2}}} \right) \cdot \left(\frac{H_y\Big|_{i+1/2,k+1/2}^{n+1/2} - H_y\Big|_{i-1/2,k+1/2}^{n+1/2}}{\Delta x} \right), \tag{4.178}$$

with the notation:

$$E_z\Big|_{i,k+1/2} \Rightarrow E_z(i,k); \tag{4.179a}$$

$$\left(\frac{1 - \frac{\sigma_{i,k+1/2}\Delta t}{2\varepsilon_{i,k+1/2}}}{1 + \frac{\sigma_{i,k+1/2}\Delta t}{2\varepsilon_{i,k+1/2}}} \right) \Rightarrow C_{A,Ez}(i,k); \tag{4.179b}$$

$$\left(\frac{\frac{\Delta t}{\varepsilon_{i,k+1/2}\Delta x}}{1 + \frac{\sigma_{i,k+1/2}\Delta t}{2\varepsilon_{i,k+1/2}}} \right) \Rightarrow C_{B,Ez}(i,k); \tag{4.179c}$$

$$H_y\big|_{i+1/2,k+1/2} \Rightarrow H_y(i,k);$$
(4.179d)

Using this notation, Eq. (4.178) is written as:

$$E_z^{n+1}(i:2,nx;k:1,nz) = C_{A,Ex}(i,k)E_z^n(i,k) - C_{B,Ez}(i,k)\left[H_y^{n+1/2}(i,k) - H_y^{n+1/2}(i-1,k)\right],$$
(4.180)

nx and nz being the numbers of cells along the x and z directions.

The updating of the first sub-component of the magnetic field, according to Eq. (4.174c), is:

$$H_{yx}\big|_{i+1/2,k+1/2}^{n+1/2} = \left(\frac{1 - \frac{\rho_{i+1/2,k+1/2}\Delta t}{2\mu_{i+1/2,k+1/2}}}{1 + \frac{\rho_{i+1/2,k+1/2}\Delta t}{2\mu_{i+1/2,k+1/2}}}\right) \cdot H_{yx}\big|_{i+1/2,k+1/2}^{n-1/2}$$

$$+ \left(\frac{\frac{\Delta t}{\mu_{i+1/2,k+1/2}}}{1 + \frac{\rho_{i+1/2,k+1/2}\Delta t}{2\mu_{i+1/2,k+1/2}}}\right) \cdot \left(\frac{E_z\big|_{i+1,k+1/2}^n - E_z\big|_{i,k+1/2}^n}{\Delta x}\right)$$
(4.181)

where we have utilized the notation:

$$H_{yx}\big|_{i+1/2,k+1/2} \Rightarrow H_{yx}(i,k);$$
(4.182a)

$$\left(\frac{1 - \frac{\rho_{i+1/2,k+1/2}\Delta t}{2\mu_{i+1/2,k+1/2}}}{1 + \frac{\rho_{i+1/2,k+1/2}\Delta t}{2\mu_{i+1/2,k+1/2}}}\right) \Rightarrow D_{A,Hyx}(i,k);$$
(4.182b)

$$\left(\frac{\frac{\Delta t}{\mu_{i+1/2,k+1/2}\Delta x}}{1 + \frac{\rho_{i+1/2,k+1/2}\Delta t}{2\mu_{i+1/2,k+1/2}}}\right) \Rightarrow D_{B,Hyx}(i,k),$$
(4.182c)

yielding:

$$H_{yx}^{n+1/2}(i:1,nx;k:1,nz) = D_{A,Hyx}(i,k)H_{yx}^{n-1/2}(i,k) + D_{B,Hyx}(i,k)\left[E_z^n(i+1,k) - E_z^n(i,k)\right].$$
(4.183)

In a very similar way, Eq. (4.174d) is written as:

$$H_{yz}\big|_{i+1/2,k+1/2}^{n+1/2} = \left(\frac{1 - \frac{\rho_{i+1/2,k+1/2}\Delta t}{2\mu_{i+1/2,k+1/2}}}{1 + \frac{\rho_{i+1/2,k+1/2}\Delta t}{2\mu_{i+1/2,k+1/2}}}\right) \cdot H_{yz}\big|_{i+1/2,k+1/2}^{n-1/2}$$

$$- \left(\frac{\frac{\Delta t}{\mu_{i+1/2,k+1/2}}}{1 + \frac{\rho_{i+1/2,k+1/2}\Delta t}{2\mu_{i+1/2,k+1/2}}}\right) \cdot \left(\frac{E_x\big|_{i+1,k+1/2}^n - E_x\big|_{i+1/2,k}^n}{\Delta z}\right),$$
(4.184)

and the notation for the H_{yz} sub-component and field coefficients are:

$$H_{yz}\big|_{i+1/2,k+1/2} \Rightarrow H_{yz}(i,k);$$ (4.185a)

$$\left(\frac{1 - \frac{\rho_{i+1/2,k+1/2}\Delta t}{2\mu_{i+1/2,k+1/2}}}{1 + \frac{\rho_{i+1/2,k+1/2}\Delta t}{2\mu_{i+1/2,k+1/2}}} \right) \Rightarrow D_{A,Hyz}(i,k);$$ (4.185b)

$$\left(\frac{\frac{\Delta t}{\mu_{i+1/2,k+1/2}\Delta z}}{1 + \frac{\rho_{i+1/2,k+1/2}\Delta t}{2\mu_{i+1/2,k+1/2}}} \right) \Rightarrow D_{B,Hyz}(i,k),$$ (4.185c)

resulting in:

$$H_{yz}^{n+1/2}(i:1,nx;k:1,nz) = D_{A,Hyx}(i,k)H_{yz}^{n-1/2}(i,k) - D_{B,Hyz}(i,k)\left[E_x^n(i,k+1) - E_x^n(i,k)\right].$$ (4.186)

Finally, the updating of the magnetic field component is simply given by:

$$H_y\big|_{i+1/2,k+1/2}^{n+1/2} = H_{yx}\big|_{i+1/2,k+1/2}^{n+1/2} + H_{yz}\big|_{i+1/2,k+1/2}^{n+1/2},$$ (4.187)

that, according with this quoted notation, results in:

$$H_y^{n+1/2}(i:1,nx;k:1,nz) = H_{yx}^{n+1/2}(i,k) + H_{yz}^{n+1/2}(i,k).$$ (4.188)

Outside the PML region, and if the electric and magnetic conductivities of the media are zero, the coefficients of the fields simplify to:

$$C_{A,Ex}(i,k) = 1;$$ (4.189a)

$$C_{B,Ex}(i,k) = \frac{\Delta t}{\varepsilon_{i+1/2,k}\Delta z};$$ (4.189b)

$$C_{A,Ez}(i,k) = 1;$$ (4.189c)

$$C_{B,Ez}(i,k) = \frac{\Delta t}{\varepsilon_{i,k+1/2}\Delta x};$$ (4.189d)

$$D_{A,Hyx}(i,k) = 1;$$ (4.189e)

$$D_{B,Hyx}(i,k) = \frac{\Delta t}{\mu_{i+1/2,k+1/2}\Delta x};$$ (4.189f)

$$D_{A,Hyz}(i,k) = 1;$$ (4.189g)

$$D_{B,Hyz}(i,k) = \frac{\Delta t}{\mu_{i+1/2,k+1/2}\Delta z}.$$ (4.189h)

Now, we present the calculation of these field coefficients at the perfectly matched region (Berenger's layers). First, we must set the number of perfectly matched layers (n_{PML}) and the maximum reflection coefficient at the PML region (r_0). The thicknesses of the PML regions are then:

$$d_{PMLx} = n_{PML}\Delta x; \tag{4.190a}$$

$$d_{PMLz} = n_{PML}\Delta z. \tag{4.190b}$$

On the other hand, the maximum electric and magnetic anisotropic conductivities in the PML regions (for a power exponent of $p = 2$) are given by:

$$\sigma_{x\max} = -\frac{3\varepsilon_0 c \log(r_0)}{2d_{PMLx}}; \tag{4.191a}$$

$$\rho_{x\max} = -\frac{3\mu_0 c \log(r_0)}{2d_{PMLx}}; \tag{4.191b}$$

$$\sigma_{z\max} = -\frac{3\varepsilon_0 c \log(r_0)}{2d_{PMLz}}; \tag{4.191c}$$

$$\rho_{z\max} = -\frac{3\mu_0 c \log(r_0)}{2d_{PMLz}}. \tag{4.191d}$$

Now, the electric and magnetic conductivities at the PML regions depend only on the distance from the boundaries. According to the field locations within Lee's cell, these conductivities are expressed by:

$$\sigma_x(m:1,n_{PML}) = \sigma_{x\max}\left(\frac{n_{PML}-m+0.5}{n_{PML}+0.5}\right)^2; \tag{4.192a}$$

$$\rho_x(m:1,n_{PML}) = \rho_{x\max}\left(\frac{n_{PML}-m+1}{n_{PML}+0.5}\right)^2; \tag{4.192b}$$

$$\sigma_z(m:1,n_{PML}) = \sigma_{z\max}\left(\frac{n_{PML}-m+0.5}{n_{PML}+0.5}\right)^2; \tag{4.192c}$$

$$\rho_z(m:1,n_{PML}) = \rho_{z\max}\left(\frac{n_{PML}-m+1}{n_{PML}+0.5}\right)^2; \tag{4.192d}$$

where the integer m (from 1 to n_{PML}) denotes the distance from the boundary in cell units.

Finally, the coefficients of the fields inside the PML region, assuming a non-magnetic media ($\mu = \mu_0$), can be calculated according to:

Bottom PML region:

$$C_{A,Ez}(i;2,n_{PML}+1;k:1,nz) = \left(\frac{1-\frac{\sigma_x(i-1)\Delta t}{2\varepsilon_{i,k+1/2}}}{1+\frac{\sigma_x(i-1)\Delta t}{2\varepsilon_{i,k+1/2}}}\right); \tag{4.193a}$$

$$C_{B,Ez}(i:2,n_{PML}+1;k:1,nz) = \left(\frac{\frac{\Delta t}{2\varepsilon_{i,k+1/2}}}{1 + \frac{\sigma_x(i-1)\Delta t}{2\varepsilon_{i,k+1/2}}} \right); \tag{4.193b}$$

$$D_{A,Hyx}(i;1,n_{PML};k:1,nz) = \left(\frac{1 - \frac{\rho_x(i)\varepsilon_0\Delta t}{2\mu_0\varepsilon_{i+1/2,k+1/2}}}{1 + \frac{\rho_x(i)\varepsilon_0\Delta t}{2\mu_0\varepsilon_{i+1/2,k+1/2}}} \right); \tag{4.193c}$$

$$D_{B,Hyx}(i;1,n_{PML};k:1,nz) = \left(\frac{\frac{\Delta t}{\mu_0\Delta x}}{1 + \frac{\rho_x(i)\varepsilon_0\Delta t}{2\mu_0\varepsilon_{i+1/2,k+1/2}}} \right). \tag{4.193d}$$

Top PML region:

$$C_{A,Ez}(i;nx-n_{PML}+1,nx;k:1,nz) = \left(\frac{1 - \frac{\sigma_x(nx-i+1)\Delta t}{2\varepsilon_{i,k+1/2}}}{1 + \frac{\sigma_x(nx-i+1)\Delta t}{2\varepsilon_{i,k+1/2}}} \right); \tag{4.194a}$$

$$C_{B,Ez}(i;nx-n_{PML}+1;k:1,nz) = \left(\frac{\frac{\Delta t}{\varepsilon_{i,k+1/2}\Delta x}}{1 + \frac{\sigma_x(nx-i+1)\Delta t}{2\varepsilon_{i,k+1/2}}} \right); \tag{4.194b}$$

$$D_{A,Hyx}(i:nx-n_{PML}+1,nx;k:1,nz) = \left(\frac{1 - \frac{\rho_x(nx-i+1)\varepsilon_0\Delta t}{2\mu_0\varepsilon_{i+1/2,k+1/2}}}{1 + \frac{\rho_x(nx-i+1)\varepsilon_0\Delta t}{2\mu_0\varepsilon_{i+1/2,k+1/2}}} \right); \tag{4.194c}$$

$$D_{B,Hyx}(i:nx-n_{PML}+1,nx;k:1,nz) = \left(\frac{\frac{\Delta t}{\mu_0\Delta x}}{1 + \frac{\rho_x(nx-i+1)\varepsilon_0\Delta t}{2\mu_0\varepsilon_{i+1/2,k+1/2}}} \right). \tag{4.194d}$$

Left PML region:

$$C_{A,Ex}(i:1,nx;k:2,n_{PML}+1) = \left(\frac{1 - \frac{\sigma_z(k-1)\Delta t}{2\varepsilon_{i+1/2,k}}}{1 + \frac{\sigma_z(k-1)\Delta t}{2\varepsilon_{i+1/2,k}}} \right); \tag{4.195a}$$

$$C_{B,Ex}(i:1,nx;k:2,n_{PML}+1) = \left(\frac{\frac{\Delta t}{\varepsilon_{i+1/2,k}\Delta z}}{1 + \frac{\sigma_z(k-1)\Delta t}{2\varepsilon_{i+1/2,k}}} \right); \tag{4.195b}$$

$$D_{A,Hyz}(i:1,nx;k:1,n_{PML}) = \left(\frac{1 - \frac{\rho_z(k)\varepsilon_0 \Delta t}{2\mu_0 \varepsilon_{i+1/2,k+1/2}}}{1 + \frac{\rho_z(k)\varepsilon_0 \Delta t}{2\mu_0 \varepsilon_{i+1/2,k+1/2}}} \right) ; \qquad (4.195c)$$

$$D_{B,Hyz}(i:1,nx;k:1,n_{PML}) = \left(\frac{\frac{\Delta t}{\mu_0 \Delta z}}{1 + \frac{\rho_z(k)\varepsilon_0 \Delta t}{2\mu_0 \varepsilon_{i+1/2,k+1/2}}} \right) ; \qquad (4.195d)$$

Right PML region:

$$C_{A,Ex}(i:1,nx;k:nz-n_{PML}+1,nz) = \left(\frac{1 - \frac{\sigma_z(nz-k+1)\Delta t}{2\varepsilon_{i+1/2,k}}}{1 + \frac{\sigma_z(nz-k+1)\Delta t}{2\varepsilon_{i+1/2,k}}} \right) ; \qquad (4.196a)$$

$$C_{B,Ex}(i:1,nx;k:nz-n_{PML}+1,nz) = \left(\frac{\frac{\Delta t}{\varepsilon_{i+1/2,k}\Delta z}}{1 + \frac{\sigma_x(nz-k+1)\Delta t}{2\varepsilon_{i+1/2,k}}} \right) ; \qquad (4.196b)$$

$$D_{A,Hyz}(i:1,nx;k:nx-n_{PML}+1,nz) = \left(\frac{1 - \frac{\rho_z(nz-k+1)\varepsilon_0 \Delta t}{2\mu_0 \varepsilon_{i+1/2,k+1/2}}}{1 + \frac{\rho_z(nz-k+1)\varepsilon_0 \Delta t}{2\mu_0 \varepsilon_{i+1/2,k+1/2}}} \right) ; \qquad (4.196c)$$

$$D_{B,Hyz}(i:1,nx;k:nx-n_{PML}+1,nz) = \left(\frac{\frac{\Delta t}{\mu_0 \Delta z}}{1 + \frac{\rho_x(nz-k+1)\varepsilon_0 \Delta t}{2\mu_0 \varepsilon_{i+1/2,k+1/2}}} \right) . \qquad (4.196d)$$

4.8.4 Setting the Field Source

To start the propagation, we must choose the appropriate field source, either a pulsed or continuous wave source, having in mind that the source should have a smooth rise time to avoid unwanted behaviour. In waveguide propagation problems one usually wants to excite a waveguide mode (for instance, the fundamental mode) and in such cases the transversal dependence of the field source should be chosen as corresponding to that particular waveguide mode [59, 60].

The following expression simulates a pulsed source of temporal width τ and central frequency ω_0 propagating along the z direction:

$$H_y(x,z_0,t) = f(x) \cdot \exp\left[-\left(\frac{t-t_{start}}{\tau}\right)^2\right] \cdot \sin(\omega_0 t), \qquad (4.197)$$

where the input source H_y is supposed to be located at the position $z = z_0$. This source, which represents a Gaussian modulated sinusoidal wave, reaches its maximum at t_{start}, which should be ~ 20 times greater than the pulse width τ. For a plane wave source the function $f(x)$ is simply equal to 1. For the excitation of a waveguide mode, $f(x)$ is just the transversal field profile of the mode. Let us observe that this excitation source contains a wide range of frequencies and thus if the temporal pulse width is short (compared with $1/\omega_0$), the term 'waveguide mode' loses its validity as, strictly speaking, it can only be defined for monochromatic waves.

For a continuous wave input source, we can use the following expression for the field component H_y:

$$H_y(x,z_0,t) = f(x) \cdot \left(1-\exp\left[-(t/\tau)^2\right]\right) \cdot \sin(\omega_0 t). \qquad (4.198)$$

This excitation source, for sufficient long times, has a narrow range of frequencies and thus it can reproduce quite well a waveguide mode at the frequency ω_0.

4.8.5 Total-Field/Scattered-Field Formulation

If the structure to be modelled provokes significant reflection, the radiation can return to the position where the source is located, resulting in a degradation of the modelling. In addition, the hard sources shown in Eqs. (4.193) and (4.194) generate two identical waves that propagate both in the $+z$ and $-z$ directions. One way to overcome these two problems is to use the total-field, TF/ scattered-field, SF formulation. For doing that, the computational region is separated into two sub-regions: the total field region and the scattered (or reflected) field region [59]. The interface separating these two regions is the incident plane where the source is located. The structure of interest, such as a waveguide grating, is defined in the total field region. The interaction between the incident wave and the structure will take place in this region so that its field quantities must retain the information of both the incident and scattered waves. In the scattered (or reflected) field region the geometry is uniform and the field quantities in this region are the reflections from the structure located into the total field region. Since there are no discontinuities in this region, these signals will not be reflected back to the total field region.

When the required spatial difference for the field updates is taken across the interface plane, separating the total-field region and the scattered-field region, a problem of consistency arises. That is, on the total-field region side of the interface the field to be used in the difference expression is assumed to be a total field, whereas on the scattered-field region side of the interface the field to be used in the difference expression is assumed to be a scattered field. Therefore, it would be inconsistent to perform an arithmetic difference between scattered- and total-field values. To recover consistency it is enough to add another term in the field update equations, that is, a function of the assumed-known incident wave. This correction must be implemented only at grid points located at the interface plane separating the total-field and the scattered-field.

In particular, for an excitation corresponding to a $+z$ directed wave, located at the plane defined by the grid point nm in the z-axis, Eq. (4.186) should be corrected by:

$$H_{yz}^{n+1/2}(i:1,nx;nm-1) = D_{A,Hyz}(i,nm-1)H_{yz}^{n-1/2}(i,nm-1) - D_{B,Hyz}(i,nm-1)\begin{bmatrix} E_x^n(i,nm) - \\ E_x^n(i,nm-1) - \\ Source\_E_x^n(i) \end{bmatrix},$$

(4.199)

where the additional term is defined by the source type. If the excitation consists of a waveguide mode with propagation constant β and a transversal H_z field distribution given by $\phi(i)$, the source term is given by:

$$Source\_E_x^n(i) = \frac{\beta}{\omega_0 \varepsilon(i,nm)} \phi(i) sin[\omega_0 n\Delta t - \beta\Delta z/2]\left(1 - e^{-n\Delta t/\tau}\right).$$

(4.200)

The update of the field component, E_x, given in Eq. (4.177) must be also corrected in the following the way:

$$E_x^{n+1}(i:1,nx;nm) = C_{A,Ex}(i,nm)E_x^n(i,nm) - C_{B,Ex}(i,nm)\begin{bmatrix} H_y^{n+1/2}(i,nm) - H_y^{n+1/2}(i,nm-1) \\ + Source\_H_y^{n+1/2}(i) \end{bmatrix}.$$

(4.201)

The additional term corresponding to the input source is given by:

$$Source\_H_y^{n+1/2}(i) = \phi(i)sin[\omega_0(n+1/2)\Delta t]\left(1 - e^{-(n+1/2)\Delta t/\tau}\right).$$

(4.202)

For simulating a plane wave CW source in a homogeneous medium, one can use Eqs. (4.200) and (4.202) with $\phi(i) = 1$ and β being the wavenumber in the medium.

The implementation of these formulae for the fields updates generates a $+z$ propagating wave at the plane position $k = nm$, with very little spurious backward field for $k < nm$. On the other hand, any reflected field induced by the structure at $z > nm$ can pass through the incident plane, allowing for the correct analysis of the reflected waves, without spurious interference of the incident field source.

Figure 4.48 shows the H_y amplitude for a plane wave, propagating in vacuum, as a function of the propagation distance at a fixed time. The input plane source is located at $z = -5$ μm and consists of a sinusoidal wave with $\omega_0 = 1.26 \times 10^{15}$ s^{-1} (expression 4.197), with $f(x) = 1$, $t_{start} = 20$ fs and $\tau = 4$ fs. The application of the TF/SF formulation generates a $+z$ propagating wave, with a negligible field in the left of the plane input source.

If the plane wave in the former example impinges a planar interface with a medium with refractive index $n = 3.6$, located at $z = +3$ μm, the wave will be partly transmitted and a fraction is reflected back to the vacuum region (Figure 4.49). In the region for $z > +3$ μm (situated in the total-field region) only the transmitted field is present. In the region -5 μm $< z < +3$ μm (located also in the total field region) the field is the superposition of the incident field and the reflected wave from the interface. At the left of the input plane (scattered field region) only the reflected field from

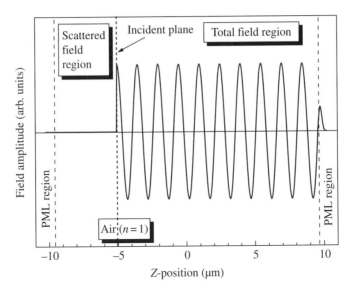

Figure 4.48 H_y amplitude of a plane wave propagating in vacuum, computed by FDTD using the scattered field/total field formulation. The source at $z = -5\,\mu m$ consists of a sinusoidal wave with $\omega_0 = 1.26 \times 10^{15}\,s^{-1}$, generating a $+z$ propagating wave. At grid locations for $z < -5\,\mu m$ the field is negligible. Simulation parameters: $\Delta x = \Delta z = 40\,nm$, $\Delta t = 0.08\,fs$

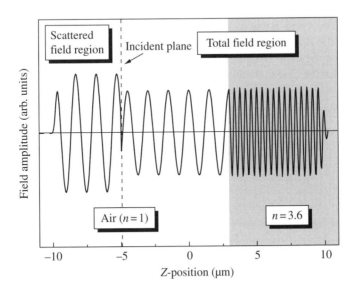

Figure 4.49 H_y amplitude of a plane wave computed by FDTD using the scattered-field/total-field formulation. The source, located at $z = -5\,\mu m$, consists of a $+z$ propagating sinusoidal wave with $\omega_0 = 1.26 \times 10^{15}\,s^{-1}$. It propagates in vacuum until at $z = +3\,\mu m$ it reaches a medium with $n = 3.6$, generating a reflected and a transmitted wave. At the region $z < -5\,\mu m$ only the reflected wave is present

the interface exists. Thus the TF/SF formulation allows to the scattered radiation (reflected field from the structure in the total-field region) to pass through the incident plane, which permits us to analyse the backscattered field without spurious effect from the input source.

The next example shows the applicability of the TF/SF formulation to a waveguide problem. The structure consists of a 300 nm width symmetric step-index planar waveguide, with core refractive index of 3.60 and cladding index of 3.24 (Figure 4.50a) [61]. Using TF/SF formulation, the fundamental TM waveguide mode at $\lambda = 800$ nm is launched using expressions (4.200) and (4.202), where the field distribution $\phi(i)$ and the propagation constant β of the mode have been previously calculated using standard mode solvers. As can be seen in Figure 4.50b, the field distribution is invariant along the propagation in the $+z$ direction, while a negligible field appears on the left of the incident plane. This excitation scheme is also now used to analyse the reflection of the waveguide mode from a waveguide facet with air (Figure 4.50c). The input plane is situated at 1 μm on the left of the waveguide–air interface. When radiation reaches the waveguide facet, part of it is transmitted into the air region (situated in the total-field region), where it suffers diffraction (Figure 4.50d). The waveguide–air interface provokes reflection and part of it returns as the waveguide mode, where it interferes with the input mode giving rise to an interference pattern, also in the total-field region. On the other hand, only radiation scattered from the waveguide facet exists on the left of the incident plane, which corresponds to the scattered-field region.

4.8.6   Two-Dimensional FDTD: TE Case

The equations in finite differences are now deduced for TE propagation (TM$_y$ waves) in two-dimensional structures. In this case, the non-null components of the fields are E_y, H_x and H_z (referred to Figure 4.44). Yee's grid in this case is depicted in Figure 4.51, where a grid of 4×4 cells is presented. Note the zero electric field components surrounding the computational domain.

Now, following the same procedure developed for TM propagation, we derive the finite difference scheme for the time marching of field components for TE waves. The artificially modified Maxwell's curl equations for reducing significantly the reflections at boundaries, in the case of TE propagation, are rewritten in the form:

$$\frac{\partial H_x}{\partial t} = \frac{1}{\mu}\left[\frac{\partial E_y}{\partial z} - \rho_z H_x\right]; \tag{4.203a}$$

$$\frac{\partial H_z}{\partial t} = \frac{1}{\mu}\left[\frac{\partial E_y}{\partial z} - \rho_x H_z\right]; \tag{4.203b}$$

$$\frac{\partial E_{yx}}{\partial t} = \frac{1}{\varepsilon}\left[-\frac{\partial H_z}{\partial x} - \sigma_x E_{yx}\right]; \tag{4.203c}$$

$$\frac{\partial E_{yz}}{\partial t} = \frac{1}{\varepsilon}\left[\frac{\partial H_x}{\partial z} - \sigma_z E_{yz}\right]; \tag{4.203d}$$

$$E_y = E_{yx} + E_{yz}. \tag{4.203e}$$

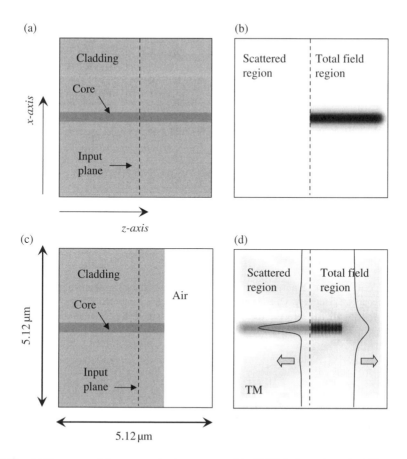

Figure 4.50 (a) Geometry of the symmetric planar waveguide. (b) Light intensity at $\lambda = 800$ nm computed by FDTD using the TF/SF formulation, using a waveguide mode as input source. (c) Waveguide facet situated at 1 μm on the right with respect to the incident plane. (d) Propagation of the fundamental mode along the waveguide until it reaches the facet (air), generating reflected and transmitted waves. Simulation parameters: $\Delta x = \Delta z = 20$ nm, $\Delta t = 0.04$ fs

Here, we observe the splitting of the electric field component E_y into two parts (E_{yx} and E_{yz}) and the incorporation of anisotropic electric and magnetic conductivities.

The finite differences of equations in (4.203) yield:

$$H_x\Big|_{i,k+1/2}^{n+1/2} = \left(\frac{1-\frac{\rho_{i,k+1/2}\Delta t}{2\mu_{ik+1/2}}}{1+\frac{\rho_{i,k+1/2}\Delta t}{2\mu_{i,k+1/2}}}\right) \cdot H_x\Big|_{i,k+1/2}^{n-1/2} + \left(\frac{\frac{\Delta t}{\mu_{i,k+1/2}}}{1+\frac{\rho_{i,k+1/2}\Delta t}{2\mu_{i,k+1/2}}}\right) \cdot \left(\frac{E_y\Big|_{i,k+1/2}^{n} - E_y\Big|_{i,k}^{n}}{\Delta z}\right); \quad (4.204a)$$

$$H_z\Big|_{i+1/2,k}^{n+1/2} = \left(\frac{1-\frac{\rho_{i+1/2,k}\Delta t}{2\mu_{i+1/2,k}}}{1+\frac{\rho_{i+1/2,k}\Delta t}{2\mu_{i+1/2,k}}}\right) H_z\Big|_{i+1/2,k}^{n-1/2} + \left(\frac{\frac{\Delta t}{\mu_{i+1/2,k}}}{1+\frac{\rho_{i+1/2,k}\Delta t}{2\mu_{i+1/2,k}}}\right) \cdot \left(\frac{E_y\Big|_{i+1,k}^{n} - E_y\Big|_{i,k}^{n}}{\Delta x}\right); \quad (4.204b)$$

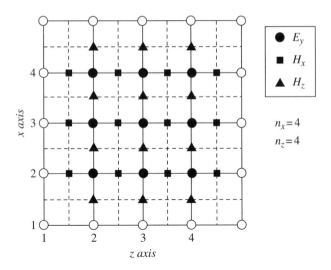

Figure 4.51 Yee's grid for TE propagation. Circle, squares and triangles denote the positions of the E_y, H_x and H_z components, respectively. Open symbols surrounding the computational region at $i = 1$, $i = nx + 1$, $k = 1$ and $k = nz + 1$ indicate zero electric fields, corresponding to the simulation of a perfect electric conductor

$$E_{yx}\Big|_{i,k}^{n+1} = \left(\frac{1 - \frac{\sigma_{i,k}\Delta t}{2\varepsilon_{i,k}}}{1 + \frac{\sigma_{i,k}\Delta t}{2\varepsilon_{i,k}}}\right) E_{yx}\Big|_{i,k}^{n} + \left(\frac{\frac{\Delta t}{\varepsilon_{i,k}}}{1 + \frac{\sigma_{i,k}\Delta t}{2\varepsilon_{i,k}}}\right) \cdot \left(\frac{H_z\Big|_{i+1/2,k}^{n+1/2} - H_z\Big|_{i-1/2,k}^{n+1/2}}{\Delta x}\right); \qquad (4.204c)$$

$$E_{yz}\Big|_{i,k}^{n+1} = \left(\frac{1 - \frac{\sigma_{i,k}\Delta t}{2\varepsilon_{i,k}}}{1 + \frac{\sigma_{i,k}\Delta t}{2\varepsilon_{i,k}}}\right) E_{yz}\Big|_{i,k}^{n} + \left(\frac{\frac{\Delta t}{\varepsilon_{i,k}}}{1 + \frac{\sigma_{i,k}\Delta t}{2\varepsilon_{i,k}}}\right) \cdot \left(\frac{H_x\Big|_{i,k+1/2}^{n+1/2} - H_x\Big|_{i,k-1/2}^{n+1/2}}{\Delta z}\right); \qquad (4.204d)$$

$$E_y\Big|_{i,k}^{n+1} = E_{yz}\Big|_{i,k}^{n+1} + E_{yx}\Big|_{i,k}^{n+1}. \qquad (4.204e)$$

The field components in the Yee's grid (see Figure 4.52) are stored in the computer memory as:

$$H_x|_{i,k+1/2} \Rightarrow H_x(i,k) \qquad (4.205a)$$

$$H_Z|_{i+1/2,k} \Rightarrow H_z(i,k) \qquad (4.205b)$$

$$E_{yx}|_{i,k} \Rightarrow E_{yx}(i,k) \qquad (4.205c)$$

$$E_{yz}|_{i,k} \Rightarrow E_{yz}(i,k) \qquad (4.205d)$$

$$E_y|_{i,k} \Rightarrow E_y(i,k) \qquad (4.205e)$$

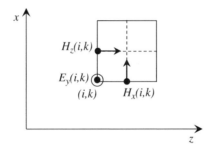

Figure 4.52 Notation used for the field components in the Yee's grid for performing the spatial finite differences for TE propagation

Thus, Eqs. (4.204a) to (4.204e) are implemented in the following way:

$$H_x^{n+1/2}(i:2,nx;k:1,nz) = D_{A,Hx}(i,k)H_x^{n-1/2}(i,k) - D_{B,Hx}(i,k)\left[E_y^n(i,k+1) - E_y^n(i,k)\right];$$

$$(4.206a)$$

$$H_z^{n+1/2}(i:1,nx;k:2,nz) = D_{A,Hz}(i,k)H_z^{n-1/2}(i,k) - D_{B,Hz}(i,k)\left[E_y^n(i+1,k) - E_y^n(i,k)\right];$$

$$(4.206b)$$

$$E_{yx}^{n+1}(i:2,nx;k:2,nz) = C_{A,Eyx}(i,k)E_{yx}^n(i,k) - C_{B,Eyx}(i,k)\left[H_z^{n+1/2}(i,k) - H_z^{n+1/2}(i-1,k)\right];$$

$$(4.206c)$$

$$E_{yz}^{n+1}(i:2,nx;k:2,nz) = C_{A,Eyz}(i,k)E_{yz}^n(i,k) - C_{B,Eyz}(i,k)\left[H_x^{n+1/2}(i,k) - H_x^{n+1/2}(i,k-1)\right];$$

$$(4.206d)$$

$$E_y^{n+1}(i:2,nx;k:2,nz) = E_{yx}^{n+1}(i,k) + E_{yz}^{n+1}(i,k).$$
$$(4.206e)$$

Where the field coefficients are given by:

$$D_{A,Hx}(i,k) = \left(1 - \frac{\rho_{i,k+1/2}\Delta t}{2\mu_{i,k+1/2}}\right) \cdot \left(1 - \frac{\rho_{i,k+1/2}\Delta t}{2\mu_{i,k+1/2}}\right)^{-1}; \qquad (4.207a)$$

$$D_{B,Hx}(i,k) = \left(\frac{\Delta t}{\mu_{i,k+1/2}\Delta z}\right) \cdot \left(1 + \frac{\rho_{i,k+1/2}\Delta t}{2\mu_{i,k+1/2}}\right)^{-1}; \qquad (4.207b)$$

$$D_{A,Hz}(i,k) = \left(1 - \frac{\rho_{i+1/2,k}\Delta t}{\mu_{i+1/2,k}}\right) \cdot \left(1 + \frac{\rho_{i+1/2,k}\Delta t}{2\mu_{i+1/2,k}}\right)^{-1}; \qquad (4.207c)$$

$$D_{B,Hz}(i,k) = \left(\frac{\Delta t}{\mu_{i+1/2,k}\Delta x}\right) \cdot \left(1 + \frac{\rho_{i+1/2,k}\Delta t}{2\mu_{i+1/2,k}}\right)^{-1}; \qquad (4.207d)$$

$$C_{B,Eyx}(i,k) = \left(1 - \frac{\sigma_{i,k}\Delta t}{2\varepsilon_{i,k}}\right) \cdot \left(1 + \frac{\sigma_{i,k}\Delta t}{2\varepsilon_{i,k}}\right)^{-1}; \tag{4.207e}$$

$$C_{B,Eyx}(i,k) = \left(\frac{\Delta t}{\varepsilon_{i,k}\Delta x}\right) \cdot \left(1 + \frac{\sigma_{i,k}\Delta t}{2\varepsilon_{i,k}}\right)^{-1}; \tag{4.207f}$$

$$C_{A,Eyz}(i,k) = \left(1 - \frac{\sigma_{i,k}\Delta t}{2\varepsilon_{i,k}}\right) \cdot \left(1 + \frac{\sigma_{i,k}\Delta t}{2\varepsilon_{i,k}}\right)^{-1}; \tag{4.207g}$$

$$C_{B,Eyz}(i,k) = \left(\frac{\Delta t}{\varepsilon_{i,k}\Delta z}\right) \cdot \left(1 + \frac{\sigma_{i,k}\Delta t}{2\varepsilon_{i,k}}\right)^{-1}; \tag{4.207h}$$

Using the anisotropic electric and magnetic conductivities defined in Eq. (4.192), the coefficients of the fields inside the PML layers, introduced by avoiding reflections at the boundaries of the computational domain and assuming a non-magnetic medium, must be modified according to:

Bottom PML region:

$$D_{A,Hz}(i:1,n_{PML};k:2,nz) = \left(\frac{1 - \frac{\rho_x(i)\varepsilon_0\Delta t}{2\mu_0\varepsilon_{i+1/2,k}}}{1 - \frac{\rho_x(i)\varepsilon_0\Delta t}{2\mu_0\varepsilon_{i+1/2,k}}}\right); \tag{4.208a}$$

$$D_{B,Hz}(i:1,n_{PML};k:2,nz) = \left(\frac{\frac{\Delta t}{\mu_0\Delta x}}{1 + \frac{\rho_x(i)\varepsilon_0\Delta t}{2\mu_0\varepsilon_{i+1/2,k}}}\right); \tag{4.208b}$$

$$C_{A,Eyx}(i:2,n_{PML}+1;k:2,nz) = \left(\frac{1 - \frac{\sigma_x(i-1)\Delta t}{2\varepsilon_{i,k}}}{1 - \frac{\sigma_x(i-1)\Delta t}{2\varepsilon_{i,k}}}\right); \tag{4.208c}$$

$$C_{A,Eyx}(i:2,n_{PML}+1;k:2,nz) = \left(\frac{\frac{\Delta t}{\varepsilon_{i,k}\Delta x}}{1 + \frac{\sigma_z(i-1)\Delta t}{2\varepsilon_{i,k}}}\right). \tag{4.208d}$$

Top PML region:

$$D_{A,Hz}(i:nx-n_{PML}+1,nx;k:2,nz) = \left(\frac{1 - \frac{\rho_x(nx-i+1)\varepsilon_0\Delta t}{2\mu_0\varepsilon_{i+1/2,k}}}{1 - \frac{\rho_x(nx-i+1)\varepsilon_0\Delta t}{2\mu_0\varepsilon_{i+1/2,k}}}\right); \tag{4.209a}$$

$$D_{B,Hz}\left(i:nx-n_{PML}+1,nx;k:2,nz\right)=\left(\dfrac{\dfrac{\Delta t}{\mu_0\Delta x}}{1-\dfrac{\rho_x(nx-i+1)\varepsilon_0\Delta t}{2\mu_0\varepsilon_{i+1/2,k}}}\right);\qquad(4.209b)$$

$$C_{A,Eyx}\left(i:nx-n_{PML}+1,nx;k:2,nz\right)=\left(\dfrac{1-\dfrac{\sigma_x(nx-i+1)\Delta t}{2\varepsilon_{i,k}}}{1+\dfrac{\sigma_x(nx-i+1)\Delta t}{2\varepsilon_{i,k}}}\right);\qquad(4.209c)$$

$$C_{B,Eyx}\left(i:nx-n_{PML}+1,nx;k:2,nz\right)=\left(\dfrac{\dfrac{\Delta t}{\varepsilon_{i,k}\Delta x}}{1+\dfrac{\sigma_x(nx-i+1)\Delta t}{2\varepsilon_{i,k}}}\right).\qquad(4.209d)$$

Left PML region:

$$D_{A,Hx}\left(i:2,nx;k:1,n_{PML}\right)=\left(\dfrac{1-\dfrac{\rho_z(k)\varepsilon_0\Delta t}{2\mu_0\varepsilon_{i,k+1/2}}}{1-\dfrac{\rho_z(k)\varepsilon_0\Delta t}{2\mu_0\varepsilon_{i,k+1/2}}}\right);\qquad(4.210a)$$

$$D_{B,Hx}\left(i:2,nx;k:1,n_{PML}\right)=\left(\dfrac{\dfrac{\Delta t}{\mu_0\Delta z}}{1+\dfrac{\rho_z(k)\varepsilon_0\Delta t}{2\mu_0\varepsilon_{i,k+1/2}}}\right);\qquad(4.210b)$$

$$C_{A,Eyz}\left(i:2,nx;k:2,n_{PML}+1\right)=\left(\dfrac{1-\dfrac{\sigma_z(k-1)\Delta t}{2\varepsilon_{i,k}}}{1+\dfrac{\sigma_z(k-1)\Delta t}{2\varepsilon_{i,k}}}\right);\qquad(4.210c)$$

$$C_{B,Eyz}\left(i:2,nx;k:2,n_{PML}+1\right)=\left(\dfrac{\dfrac{\Delta t}{\varepsilon_{i,k}\Delta z}}{1+\dfrac{\sigma_z(k-1)\Delta t}{2\varepsilon_{i,k}}}\right).\qquad(4.210d)$$

Right PML region:

$$D_{A,Hx}\left(i:2,nx;k:nz-n_{PML}+1,nz\right)=\left(\dfrac{1-\dfrac{\rho_z(nz-k+1)\varepsilon_0\Delta t}{2\mu_0\varepsilon_{i,k+1/2}}}{1+\dfrac{\rho_z(nz-k+1)\varepsilon_0\Delta t}{2\mu_0\varepsilon_{i,k+1/2}}}\right);\qquad(4.211a)$$

$$D_{B,Hx}(i:2,nx;k:nz-n_{PML}+1,nz) = \left(\frac{1 - \frac{\rho_z(nz-k+1)\varepsilon_0 \Delta t}{2\mu_0 \varepsilon_{i,k+1/2}}}{1 + \frac{\rho_z(nz-k+1)\varepsilon_0 \Delta t}{2\mu_0 \varepsilon_{i,k+1/2}}} \right); \qquad (4.211b)$$

$$C_{A,Eyz}(i:2,nx;k:nz-n_{PML}+1,nz) = \left(\frac{1 - \frac{\sigma_z(nz-k+1)\Delta t}{2\varepsilon_{i,k}}}{1 + \frac{\sigma_z(nz-k+1)\Delta t}{2\varepsilon_{i,k}}} \right); \qquad (4.211c)$$

$$C_{B,Eyz}(i:2,nx;k:nz-n_{PML}+1,nz) = \left(\frac{\frac{\Delta t}{\varepsilon_{i,k}\Delta z}}{1 + \frac{\sigma_z(nz-k+1)\Delta t}{2\varepsilon_{i,k}}} \right). \qquad (4.211d)$$

To apply the TF/SF formulation to TE-propagation, some of these equations must be properly modified. In particular, for an input that simulates a $+z$ directed wave, located at the plane defined by the grid point nm in the z-axis, Eq. (4.206a) should be corrected by:

$$H_x^{n+1/2}(i:2,nx;nm) = D_{A,Hx}(i,nm)H_x^{n-1/2}(i,nm) - D_{B,Hx}(i,nm) \begin{bmatrix} E_y^n(i,nm+1) - E_y^n(i,nm) \\ +Source\_E_y^n(i) \end{bmatrix},$$
$$(4.212)$$

where the additional term is defined by the source type. To excite a waveguide mode with a transversal E_y field distribution given by $\phi(i)$, the source term is given by:

$$Source\_E_y^n(i) = \phi(i)\sin[\omega_0 n\Delta t]\left(1 - e^{-n\Delta t/\tau}\right). \qquad (4.213)$$

The update of the field component, E_{yz}, given in Eq. (4.206d) must be also corrected in the following way:

$$E_{yz}^{n+1}(i:2,nx;nm) = C_{A,Eyz}(i,nm)E_{yz}^n(i,nm) - C_{B,Eyz}(i,nm) \begin{bmatrix} H_x^{n+1/2}(i,nm) - H_x^{n+1/2}(i,nm-1) \\ +Source\_H_x^{n+1/2}(i) \end{bmatrix}$$
$$(4.214)$$

The additional term corresponding to the input source is given by:

$$Source\_H_x^{n+1/2}(i) = \frac{\varepsilon_0 c^2 \beta}{\omega_0}\phi(i)\sin[\omega_0(n+1/2)\Delta t - \beta\Delta z/2]\left(1 - e^{-(n+1/2)\Delta t/\tau}\right), \qquad (4.215)$$

where β is the propagation constant of the waveguide mode. For simulating a plane wave source in a homogeneous medium, one can use Eqs. (4.213) and (4.215) with $\phi(i) = 1$ and β being the wavenumber in the medium.

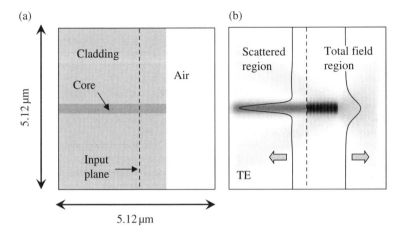

Figure 4.53 (a) Waveguide facet situated at 1 μm on the right with respect to the incident plane. (b) Light intensity pattern generated by an input TE-waveguide mode at $\lambda = 800$ nm computed by FDTD using the TF/SF formulation. Simulation parameters: $\Delta x = \Delta z = 20$ nm, $\Delta t = 0.04$ fs

To see the applicability of the 2D-FDTD for the propagation of TE-waves, including the implementation of PML and the TS/SF formalism, the scattered field of a TE-waveguide mode at the planar waveguide/air interface is examined (Figure 4.53), using the parameter structure given in the previous example. The region of interest is surrounded by 20 cells of PML and the input plane is located 1 μm from the interface. Only waves coming from the waveguide–air interface are allowed to enter into the scattered region, indicating the good performance of the TS/SF formalism. Most of the reflected radiation are in the form of backward waveguide mode, giving a value of $R = 0.42$, in accordance with calculations previously obtained by FD-BPM using the iterative bidirectional-BPM presented in Section 4.2.

References

[1] Hadley, G.R. (1992) Wide-angle beam propagation using Padé approximant operators. *Optics Letters* **17**, 1426–1428.

[2] Press, W.H., Flannery, B.P., Teukolsky, S.A. and Vettering, W.T. (1986) Evaluation of functions, in *Numerical Recipes: The Art of Scientific Computing*, Chapter 5. Cambridge University Press, New York.

[3] Hadley, G.R. (1992) Multistep method for wide-angle beam propagation. *Optics Letters* **17**, 1743–1745.

[4] Huang, W.P. and Xu, C.L. (1992) A wide-angle vector beam propagation method. *IEEE Photonics Technology Letters* **4**, 1118–1120.

[5] Shibayama, J., Takahashi, T., Yamauchi, J. and Nakano, H. (2006) A three-dimensional horizontally wide-angle noniterative beam-propagation method based on the alternating-direction implicit scheme. *IEEE Photonics Technology Letters* **18**, 661–663.

[6] Bhattacharya, D.B and Sharma, A. (2008) Finite difference split step method for non-paraxial semivectorial beam propagation in 3D. *Optical and Quantum Electronics* **40**, 933–942.

[7] Liu, Q.H. and Chew, W.C. (1991) Analysis of discontinuities in planar dielectric waveguides: an eigenmode propagation method. *IEEE Transaction on Microwave Theory and Techniques* **39**, 422–430.

[8] Chiou, Y. and Chang, H. (1997) Analysis of optical waveguide discontinuities using the Padé approximants. *IEEE Photonics Technology Letters* **9**, 964–966.

[9] El-Refaie, H., Yevick, D. and Betty, I. (1999) Padé approximation analysis of reflection at optical waveguide facets. Proceedings of 1999 Technology Digests Integrated Photon Research, pp. 104–106.

[10] Milinazzo, F.A., Zala, C.A. and Brooke, G.H. (1997) Rational square-root approximations for parabolic equation algorithms. *Journal of Acoustic Society of America* **101**, 760–766.

[11] Rao,H., Steel, M.J., Scarmozzino, R. and Osgood Jr., R.M. (2000) Complex propagators for evanescent waves in bidirectional beam propagation method. *Journal of Lightwave Technology* **18**, 1155–1160.

[12] Rao, H., Scarmozzino, R. and Osgood Jr., R.M. (1999) A bidirectional beam propagation method for multiple dielectric interfaces. *IEEE Photonics Technology Letters* **11**, 830–832.

[13] Ikegami, T. (1972) Reflectivity of mode at facet and oscillation mode in double-heterostructure injection lasers. *IEEE Journal of Quantum Electronics* **QE-8**, 470–476.

[14] Handelman, D., Hardy, A. and Katzir, A. (1986) Reflectivity of TE modes at the facets of buried heterostructure injection lasers. *IEEE Journal of Quantum Electronics* **22**, 498–500.

[15] Hong, J., Huang, W. and Makino, T. (1992) On the transfer matrix method for distributed-feedback waveguide devices. *Journal of Lightwave Technology* **10**, 1860–1868.

[16] Vallés, J.A., Lázaro, J.A. and Rebolledo, M.A. (1996) Modeling of integrated erbium-doped waveguide amplifiers with overlapping factors methods. *IEEE Journal of Quantum Electronics* **32**, 1685–1694.

[17] Cantelar, E., Nevado, R., Lifante, G. and Cussó, F. (2000) Modelling of optical amplification in Er/Yb co-doped LiNbO₃ waveguides. *Optical and Quantum Electronics* **32**, 819–827.

[18] Chryssou, C.E., Federighi, M.F. and Pitt, C.W. (1996) Lossless stripe waveguide optical beam splitter: modeling of the Y-structure. *Journal of Lightwave Technology* **14**, 1699–1703.

[19] Fedrighi, M., Massarek, I. and Trwoga, P.F. (1993) Optical amplification in thin optical waveguides with high Er concentration. *IEEE Photonics Technology Letters* **5**, 227–229.

[20] Caccavale, F., Segato, F. and Mansour, I. (1997) A numerical study of Erbium doped active LiNbO₃ waveguides by the beam propagation method. *Journal of Lightwave Technology* **15**, 2294–2300.

[21] Jose, G., Sorbello, G., Taccheo, S., Cianci, E., Foglietti, V. and Laporta, P. (2003) Active waveguide devices by Ag-Na ion exchange on erbium-ytterbium doped phosphate glasses. *Journal of Non-Crystalline Solids* **322**, 256–261.

[22] Liu, K. and Pun, E.Y.B. (2007) Modeling and experiments of packaged Er³⁺-Yb³⁺ co-doped glass waveguide amplifiers. *Optics Communications* **273**, 413–420.

[23] Vallés, J.A., Rebolledo, M.A. and Cortés, J. (2006) Full characterization of packaged Er-Yb-codoped phosphate glass waveguides. *IEEE Journal of Quantum Electronics* **42**, 152–159.

[24] Hoekstra, H.J.W.M., Noordman, O., Krijnen, G.J.M., Varshney, R.K. and Henselmans, E. (1997) Beam-propagation method for second-harmonic generation in waveguides with birefringent materials. *Journal of the Optical Society of America B* **14**, 1823–1830.

[25] Maes,B., Bienstman,P., Baets,R., Hu,B., Sewell, P. and Benson,T. (2008) Modeling comparison of second harmonic generation in high-index-contrast devices. *Optical and Quantum Electronics* **40**, 13–22.

[26] Chou,H.F., Lin, C.F. and Wang,G.C. (1998) An iterative finite difference beam propagation method for modeling second-order nonlinear effects in optical waveguides. *Journal of Lightwave Technology* **16**, 1686–1693.

[27] Chou, H.F., Lin, C.F. and Mou, S. (1999) Comparison of finite difference beam propagation methods for modeling second-order non-linear effects. *Journal of Lightwave Technology* **17**, 1481–1486.

[28] Masoudi, H.M. and Arnold, J.M. (1995) Parallel beam propagation method for the analysis of second harmonic generation. *IEEE Photonics Technology Letters* **7**, 400–402.

[29] Alsunaidi, M.A., Masoudi, H.M. and Arnold, J.M. (2000) A time domain algorithm for the analysis of second-harmonic generation in nonlinear optical structures. *IEEE Photonics Technology Letters* **12**, 395–397.

[30] Yariv, A. and Yeh, P. (1984) *Optical Waves in Crystals*. John Wiley & Sons, Inc., New York.

[31] Alferness, R.C. (1980) Efficient waveguide electro-optic TE-TM converter/wavelength filter. *Applied Physics Letters* **36**, 513–515.

[32] Fallahkhair, A.B., Li, K.S. and Murphy, T.E. (2008) Vector finite difference modesolver for anisotropic dielectric waveguides. *Journal of Lightwave Technology* **26**, 1423–1431.

[33] Xu, C.L., Huang, W.P., Chrostowski, J. and Chaudhuri, S.K. (1994) A full-vectorial beam propagation method for anisotropic waveguides. *Journal of Lightwave Technology* **12**, 1926–1931.

[34] Hempelmann, U. and Bersiner, L. (1993) Wave propagation in integrated acoustooptical anisotropic waveguides. *IEE Proceedings-J* **140**, 193–200.

[35] McKenna, J. and Reinhart, F.K. (1976) Double-heterostructure GaAs-Al$_x$Ga$_{1-x}$As [110] p-n-junction-diode modulator. *Journal of Applied Physics* **47**, 2069–2079.

[36] Davis, C.C. (1996) The electro-optic and acousto-optic effects and modulation of light beams, in *Lasers and Electro-Optics: Fundamentals and Engineering*. Cambridge University Press, p. 475.

[37] Shibayama, J., Takahashi, T., Yamauchi, J. and Nakano, H. (1999) Finite-difference time-domain beam propagation method for analysis of three-dimensional optical waveguides. *Electronics Letters* **35**, 1548–1549.

[38] Shibayama, J., Muraki, M., Yamauchi, J. and Nakano, H. (2005) Comparative study of several time-domain methods for optical waveguide analyses. *Journal of Lightwave Technology* **23**, 2285–2293.

[39] Taflove, A. and Hagness, S.C. (2005) in *Computational Electrodynamics: The Finite-Difference Time-Domain Method*, 3rd edn, Artech House Inc., Boston, MA/London.

[40] Masoudi, H.M., Al-Sunaidi, M.A. and Arnold, J.M. (2001) Efficient time-domain beam-propagation method for modelling integrated optical devices. *Journal of Lightwave Technology* **19**, 759–771.

[41] Liu, P.L., Zhao, Q. and Choa, F.S. (1995) Slow-wave finite-difference beam propagation method. *IEEE Photonics Technology Letters* **7**, 890–892.

[42] Feng, N.N., Zhou, G.R. and Huang, W.P. (2005) An efficient split-step time-domain beam-propagation method for modelling of optical waveguide devices. *Journal of Lightwave Technology* **23**, 2186–2191.

[43] Ma, F. (1997) Slowly varying envelope simulation of optical waves in time domain with transparent and absorbing boundary conditions. *Journal of Lightwave Technology* **15**, 1974–1985.

[44] Lim, J.J., Benson, T.M., Larkins, E.C. and Sewell, P. (2002) Wideband finite-difference-time-domain beam propagation method. *Microwave and Optical Technology Letters* **34**, 243–247.

[45] Shibayama, J., Takahashi, T., Yamauchi, J. and Nakano, H. (2000) Efficient time-domain finite-difference beam propagation methods for the analysis of slab and circularly symmetric waveguides. *Journal of Lightwave Technology* **18**, 437–442.

[46] Jin, G.H., Harari, J., Vilcot, J.P. and Decoster, D. (1997) An improved time-domain beam propagation method for integrated optics components. *IEEE Photonics Technology Letters* **9**, 348–350.

[47] Koshiba, M., Tsuji, Y. and Hikari, M. (2000) Time-domain beam propagation method and its application to photonic crystal circuits. *Journal of Lightwave Technology* **18**, 102–110.

[48] Feng, N.N. and Huang, W.P. (2004) Time-domain reflective beam propagation method. *IEEE Journal of Quantum Electronics* **40**, 778–783.

[49] Manolatou, C., Johnson, S.G., Fan, S., Villeneuve, P.R., Haus, H.A. and Joannopoulos, J.D. (1999) High-density integrated optics. *Journal of Lightwave Technology* **17**, 1682–1692.

[50] Stoffer, R., Hoekstra, H.J.W.M., de Ridder, R.M., van Groesen, E. and van Beckum, F.P.H. (2000) Numerical studies of 2D photonic crystals: waveguides, coupling between waveguides and filters. *Optical and Quantum Electronics* **32**, 947–961.

[51] Taflove, A. and Umashankar, K.R. (1989) Review of FD-TD numerical modeling of electromagnetic wave scattering and radar cross section. *Proceedings of the IEEE* **77**, 682–697.

[52] Chu, T.S. and Chaudhuri, S.K. (1990) Combining modal analysis and the finite-difference time domain method in the study of dielectric waveguide problems. *IEEE Transaction on Microwave Theory and Techonoly* **38**, 1755–1760.

[53] Zhou, G.R. and Li, X. (2004) Wave equation-based semivectorial compact 2D-FDTD method for optical waveguide modal analysis. *Journal of Lightwave Technology* **22**, 677–683.

[54] Yee, K.S. (1996) Numerical solution of initial boundary value problems involving Maxwell's equations in isotropic media. *IEEE Transaction on Antennas and Propagation* **14**, 302–307.

[55] Taflove, A. and Brodwin, M.E. (1975) Numerical solution of steady-state electromagnetics scattering problems using the time-dependent Maxwell's equations. *IEEE Transaction on Microwave Theory and Technology* **MTT-23**, 623–630.

[56] Mur, G. (1981) Absorbing boundary conditions for the finite-difference approximation of the time-domain electromagnetic field equation. *IEEE Transactions on Electromagnetic Compatibility* **23**, 377.

[57] Fang, J. and Mei, K.K. (1988) A super-absorbing boundary algorithm for solving electromagnetic problems by the time-domain finite-difference method. *IEEE Antennas and Propagation Symposium* **2**, 472–475.

[58] Berenger, J. (1994) A perfectly matched layer for the absorption of electromagnetic waves. *Journal of Computational Physics* **114**, 185–200.

[59] Chu, S.T., Huang, W.P. and Chaudhuri, S.K. (1991) Simulation and analysis of waveguide based optical integrated circuits. *Computer Physics Communications* **68**, 451–484.

[60] Brovko, A.V., Manenkov, A.B. and Manenkov, S.A. (2004) Diffraction of a guided mode of a dielectric waveguide. *Radiophysics and Quantum Electronics* **47**, 48–62.

[61] Brovko, A.V., Manenkov, A.B. and Rozhnev, A.G. (2003) FDTD-analysis of the wave diffraction from dielectric waveguide discontinuities. *Optical and Quantum Electronics* **35**, 395–406.

5

BPM Analysis of Integrated Photonic Devices

Introduction

The beam-propagation method (BPM) described in the preceding chapters supplies a universal numerical tool for describing the performance of a great variety of integrated optical devices. Although particular devices have specific routes to be modelled with their own constraints, the great advantage of the BPM lies in the fact that, as few approximations have been made for its derivation, its applicability is quite wide and almost any integrated photonic device can be modelled using it. For example, when modelling the performance of MMI (multimode interference) splitters by its specific theory, one must made the important assumption that many modes are involved in the waveguide transition. If it is not the case, the MMI theory gives inexact results. On the contrary, modelling MMI devices using BPM provides excellent results, as BPM can deal with arbitrary number of propagating modes, including radiation modes. Nevertheless, in some cases extensions of the BPM should be used, for example, to take into account backward propagating waves generated in the MMI device, which can be included in the modelling by using Bi-BPM or TD-BPM (time-domain beam-propagation method).

Here, we present some examples of integrated optical elements commonly used in practical integrated photonic devices and their performance and relevant characteristics are analysed by BPM numerical tools developed in previous chapters.

5.1 Curved Waveguides

To interconnect two points in an integrated optical chip, one should use a straight waveguide. If the waveguide is monomode and at the input the fundamental mode is launched, the exit should have obviously the same field profile distribution and intensity (assuming no absorption). In many cases, the two points to be connected by waveguides are separated with respect to the main axis of the chip and this offset has to be overcome by using curved waveguides,

Beam Propagation Method for Design of Optical Waveguide Devices, First Edition. Ginés Lifante Pedrola.

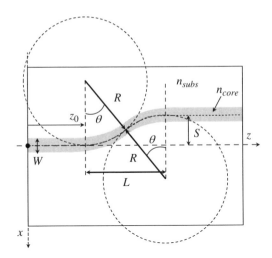

Figure 5.1   Geometry of a curved waveguide defined by two arcs of tangential circles

introducing some propagation losses. The correct design of the curved sections of the waveguide must be the carefully tailored in order to minimize the radiation losses.

Figure 5.1 shows the geometry of a curved waveguide starting at the position z_0, which is formed by two circle arcs of angle θ and radius R. The centres of the circles are at the positions (cx_1, cz_1) and (cx_2, cz_2), where these coordinates are given by:

$$cx_1 = -R \tag{5.1a}$$

$$cz_1 = z_0 \tag{5.1b}$$

$$cx_2 = R(2\cos\theta - 1) \tag{5.1c}$$

$$cz_2 = 2R\sin\theta + z_0 \tag{5.1d}$$

The transition from circle 1 to circle 2 occurs at the position $x = R(1 - \cos\theta)$, $z = R\sin\theta + z_0$, the total length of the curved region is $L = 2R\sin\theta$ and the offset of the waveguide along the x-direction is $S = 2R(1 - \cos\theta)$.

Often the wave equation of a bend waveguide is transformed from standard cylindrical coordinates r, φ and z, to the local coordinates x, y and z, with $r \rightarrow R + x$, $\varphi \rightarrow z/R$ and $z \rightarrow y$, where R is the bending radius. Due to the formal analogy to the wave equations of straight waveguides, this approach is called the equivalent straight-waveguide (ESW) approach [1, 2].

Using the conformal mapping technique [3], a curved waveguide with a symmetric index profile can be transformed into an ESW with a modified asymmetric index profile given by:

$$n_{eq}(x,y) = n(x,y)e^{x/R} \approx n(x,y)\left(1 + \frac{x}{R}\right) \tag{5.2}$$

where $n(x,y)$ is the cross-sectional refractive index distribution of the original straight waveguide, $n_{eq}(x,y)$ is the equivalent refractive index distribution of the curved waveguide (in

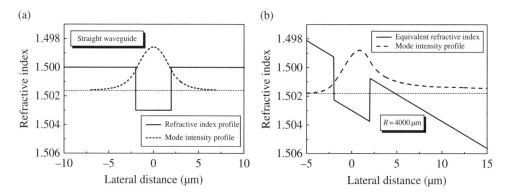

Figure 5.2   (a) Index profile and mode profile of a monomode step-index planar waveguide. (b) Equivalent graded index waveguide of the curved waveguide and its associated mode profile, where it can be observed that some losses will be present due to the non-negligible tunnelling to the region on the right

the x–z plane) transformed in a straight waveguide and R is the bend radius. Figure 5.2a shows the index profile of a step-index symmetric planar waveguide, with $n_{core} = 1.503$, $n_{cladding} = 1.500$ and a width of $W = 4\,\mu m$. Dashed line represent the intensity profile of the guided mode for the TE (transverse electric) field polarization at $\lambda = 1\,\mu m$ ($N_{eff} = 1.501594$). In the right part of Figure 5.2 the index profile of the ESW corresponding to a curved waveguide ($R = 4000\,\mu m$) is plotted, calculated by means of the ESW approach. Its associated intensity mode profile (dashed line) leaks to the right part of the structure, giving rise to propagation losses.

The propagation loss of the ESW structure is now analysed by means of BPM, which permits calculating the power loss as a function of the curvature radius R. To carry out this calculation, we proceed as follows:

1. Launch the fundamental eigenmode of the original structure, previously obtained by imaginary-distance BPM ($u(x,z = 0)$).
2. Propagate the field by standard BPM along the modified ESW structure and evaluate the remaining power in the fundamental mode as a function of the distance z by using the overlap integral $\Gamma(z)$ defined as:

$$\Gamma(z) = \frac{\left| \int_{-\infty}^{\infty} u(x,z)u^*(x,0)dx \right|^2}{\int_{-\infty}^{\infty} |u(x,0)|^2\, dx} \tag{5.3}$$

3. Plot $\log(\Gamma(z))$ as a function of the propagation distance z and evaluate the losses from the slope of the graph.

As one example, the intensity map of the ESW corresponding to a curvature radius of $R = 5\,mm$ generated by BPM is presented in Figure 5.3. It is observed the loss of power

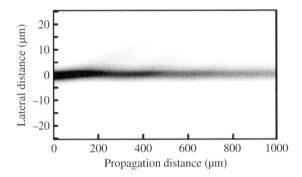

Figure 5.3 Propagation of light in a ESW obtained from a curved step-index waveguide with curvature radius of $R = 5$ mm. Parameters of the simulation: TE propagation, $\Delta x = 0.05$ µm, $\Delta z = 0.5$ µm, working wavelength $\lambda = 1$ µm

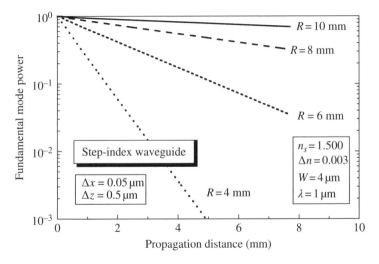

Figure 5.4 Fundamental mode power as a function of the propagation distance for an step index curved waveguide, calculated using the ESW approach and BPM, for various curvature radii. Parameters of the waveguide and the simulation are given in the figure

due to a continuous leakage induced by the tunnelling of the propagating mode presented in the Figure 5.2b.

Using the procedure described, Figure 5.4 shows the fundamental mode power remaining in the waveguide $\Gamma(z)$, in logarithmic scale as a function of the propagation distance z, for several curvature radii R. In this graph two different regions can be observed: a first region with transient behaviour, superimposed to some reduction in power and soon a second region of linear behaviour, indicating an exponential power loss as a function of the distance. From the slope of this second region the power loss rate is calculated.

The power-loss rate in the curved waveguide is now calculated as a function of the bend radius (Figure 5.5), indicating that, for the waveguide structure presented in Figure 5.2,

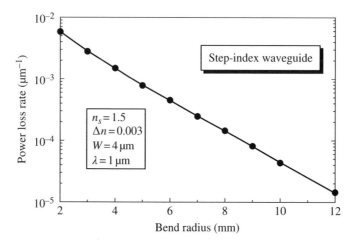

Figure 5.5   Calculated power-loss rate of a low-contrast step-index curved waveguide for different bend radii. Parameters are given in the figure

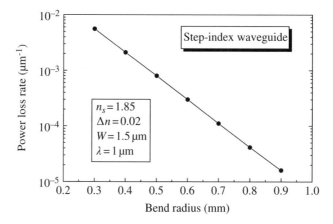

Figure 5.6   Calculated radiation losses of a medium-index contrast curved waveguide as a function of the bend radius. Parameters of the waveguide structure are indicated in the inset

assumable losses ($A < 1$ dB/cm) require curvature radius greater than $\sim$11 mm. If the waveguide presents higher index contrast, the curvature radius can be relaxed to lower values to maintain similar levels of power-loss. Figure 5.6 shows one example of medium-index contrast waveguide ($\Delta n = 0.02$), where a radius of curvature of $\sim$0.9 mm is enough in this case to avoid excessive attenuation induced by the curvature losses. Thus, for compact integrated photonic devices, high index contrast waveguide structures are desirable.

The losses induced by curvature of channel waveguides can be also computed by BPM after applying the ESW approach. One example of losses in a curved ridge waveguide with high index contrast is now presented [4]. The cross section of the ridge structure is sketched in Figure 5.7,

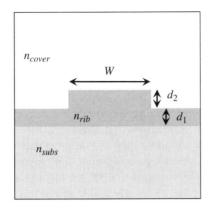

Figure 5.7   Cross-section of the ridge waveguide for curvature loss analysis

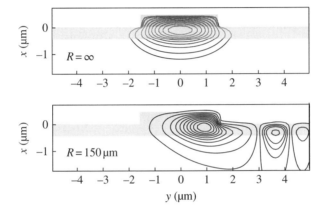

Figure 5.8   Contour lines of the distribution of the electric field absolute value over the cross section of a high index contrast ridge waveguide, corresponding to the fundamental quasi-TM mode. Upper figure: straight waveguide. Bottom figure: curved waveguide with $R = 150\,\mu m$

with the following parameters: $n_{cover} = 1.00$, $n_{rib} = 3.44$, $n_{subs} = 3.35$, $d_1 = 0.4\,\mu m$, $d_2 = 0.4\,\mu m$ and $W = 3\,\mu m$.

Figure 5.8 shows the field profile of the fundamental quasi-TM (TM, transverse magnetic) field mode of the ridge waveguide at $\lambda = 1.15\,\mu m$ for the unperturbed structure (straight wave-guide) and the TM-mode of the ESW corresponding to a curvature radius of $R = 150\,\mu m$, obtained by conventional semi-vectorial-BPM. Let us note that imaginary-distance BPM can-not be used to obtain the fundamental mode in the curved waveguide, due to the strong leakage of the field through the right side of the structure. This simulation uses a discretization window of $\Delta x = 0.025\,\mu m$ and $\Delta y = 0.1\,\mu m$ and a longitudinal step of $\Delta z = 0.1\,\mu m$.

By computing the power loss of the mode as a function of the propagation distance, as it was described previously for planar waveguides, the power loss rate can be calculated. A value of $3.4 \times 10^{-3}\,\mu m^{-1}$ is obtained for TM-propagation, while similar power loss rate ($3.6 \times 10^{-3}\,\mu m^{-1}$) is found for the TE-polarized fundamental mode.

5.2 Tapers: Y-Junctions

5.2.1 Taper as Mode-Size Converter

A tapered waveguide consists of a smooth transition from a wide to a narrow waveguide. One of the applications of the taper is as mode size converter [5]. If the input light comes from a wide waveguide, for instance, coming from an optical fibre and one wants to connect it with an integrated optical circuits with smaller waveguide transversal dimensions, there is need to adapt the input mode size to dimensions that match the waveguide circuit. The linear (or curved) taper is a good choice for performing this task, but careful design must be considered to avoid high losses. If the tapered region is abrupt, high losses will be induced at the transition. On the contrary, if the transition is smooth, the input mode can accommodate itself to the continuously diminishing the waveguide width and in this way very low losses can be achieved. In this case, the taper is said to be adiabatic. Therefore, depending on the waveguide parameters, a maximum taper angle is recommended to maintain the transition losses at a reasonable value.

A taper which presents a linear transition is characterized by the angle θ that forms with respect to the propagation axis. In a linear taper formed by an input step-index waveguide with a width W_{input} and an exit width W_{output} (Figure 5.9), the length of the transition L_{taper} reads as:

$$L_{taper} = \frac{W_{input} - W_{output}}{2 \tan \theta} \tag{5.4}$$

BPM can be used to track the light propagation along the taper structure [6]; if the taper presents abrupt transitions (large taper angles, $\theta > 30°$), wide-angle BPM algorithm should be used. Figure 5.10 shows the simulation of the light propagation along the transition between a wide-multimode-waveguide and a narrow monomode-waveguide, for two different taper

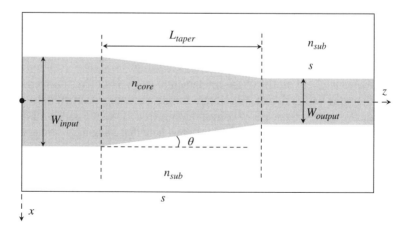

Figure 5.9   Schematic diagram of a linear taper based on step-index planar waveguides

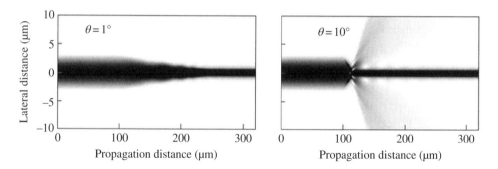

Figure 5.10 Light propagation in tapered step-index waveguides using BPM. Parameters of the taper structure: $W_{input} = 6$ μm, $W_{output} = 1$ μm, $n_{substrate} = 3.20$ and $n_{core} = 3.29$. The two simulations correspond to different taper angles, at a wavelength of $\lambda = 1.55$ μm

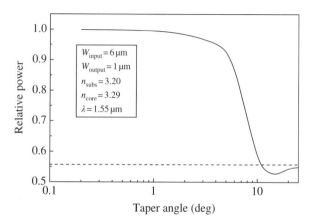

Figure 5.11 Relative power remaining in the output monomode waveguide in a tapered waveguide, as a function of the taper angle. Parameters of the taper are given in the inset

angles, using the TE-fundamental mode of the wide waveguide as the input field. The input and output waveguides are step index planar waveguides with widths of 6 and 1 μm, respectively. The core layer has a refractive index of 3.29 and the substrate has a value of 3.20, at a working wavelength of $\lambda = 1.55$ μm.

While for the taper with $\theta = 1°$ the light distribution accommodates itself along the smooth transition to the narrow waveguide, the structure with a taper with $\theta = 10°$ shows strong coupling to radiation modes, giving rise to large losses. Because of its behaviour, the structure taper with $\theta = 1°$ is said to be an adiabatic transition. Using these BPM numerical simulations, the guide power in the exit waveguide can be computed as a function of the angle θ of the taper (Figure 5.11). For this particular structure and working wavelength, the taper angle should be lower than ~5° to maintain the power loss at values lower than 0.5 dB.

The lower limit of the power transfer in a tapered structure can be calculated by the overlap integral between the fundamental mode of the input waveguide and the guided mode of the exit narrow-waveguide, given by (see Appendix C):

$$\Gamma = \frac{\left| \int_{-\infty}^{\infty} u_{input}(x) u_{output}^{*}(x) dx \right|^2}{\int_{-\infty}^{\infty} \left| u_{input}(x) \right|^2 dx \int_{-\infty}^{\infty} \left| u_{output}(x) \right|^2 dx} \tag{5.5}$$

The calculated limit using this formula gives a value of $\Gamma = 0.56$, which is just the asymptotic limit of the data in Figure 5.11 (dashed line).

5.2.2   Y-Junction as 1 × 2 Power Splitter

The easiest way to fabricate a beam splitter in waveguide circuits consists of designing a Y-junction (Figure 5.12), where an input waveguide (usually monomode at the working wavelength) splits through a tapered curved transition into two (or more) guides. If the Y-junction is symmetric with two exit waveguides, this element can work as a 3 dB power splitter (50% of input power in each output branch) [7]. To avoid high losses, the tapered region should be adiabatic, which implies the use of the correct curvature radius in the curved regions.

Figure 5.12 presents the layout of a 1 × 2 power splitter based on a symmetric Y-branch with two-exit ports. The transition between the input waveguide and one exit waveguide is designed following the path of two opposite arc circles, similar to the design of the curved waveguides shown in Section 5.1. Two main issues must be taken into account for a correct design of the power splitter: first, the curvature radius R should be large enough to avoid losses at the transition and second, the separation S between the output waveguides should be high enough to avoid coupling between them.

Figure 5.13 presents the BPM simulation of a Y-splitter made of step-index monomode planar waveguides, with the following parameters: $n_{core} = 1.87$, $n_{subs} = 1.85$ and a waveguide

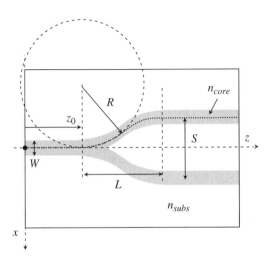

Figure 5.12   Geometrical design of a Y-junction based on step-index planar waveguides

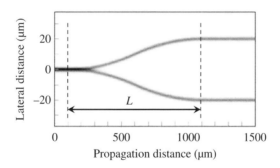

Figure 5.13 Light propagation along an adiabatic Y-junction. The transition starts at $z = 100$ μm and has a total length of 1000 μm. Parameters of the simulation: $\Delta x = 0.15$ μm, $\Delta z = 1$ μm; working wavelength $\lambda = 1.55$ μm

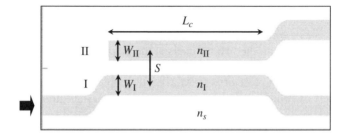

Figure 5.14 Layout of a directional coupler, based on the light coupling between two parallel waveguides. The length of the coupler is L_c and the parallel waveguides are separated a distance S, measured from their centres. At the exit, the waveguides are well separated using two opposite S-bends to avoid unwanted further coupling

width of $W = 2$ μm. The guided TE-mode at $\lambda = 1.55$ μm is used as the input beam. The parameters of the Y-branch are $L = 1000$ μm and $S = 40$ μm, which corresponds to a curvature radius of the circle of $R = 12.5$ mm ($R \approx L^2/2S$). This design provides a smooth structure, where the loss induced by the splitter, obtained by BPM simulation, is as low as 0.04 dB, indicating that this design is an adiabatic transition.

5.3 Directional Couplers

A directional coupler consists of two dielectric waveguides placed in close proximity (Figure 5.14). The interaction of evanescent fields of the guided modes in the individual waveguides causes power exchange between the coupled waveguides. This power exchange can be controlled by adjusting the synchronization and the coupling coefficient between the two guides. Several passive and functional devices can be envisaged based on the directional coupler configuration utilizing this coupling mechanism where BPM can be used to simulate the performance and characteristics of the directional coupled based devices. Most of the directional coupler devices are polarization sensitive. The polarization properties of the waveguide structure are especially important if the structures are strongly guided. Even for weakly guided structures, the polarization effect may still play an important role if the optical length of the device is long. Thus, for the correct simulation of directional coupler based devices, vectorial BPM should be used to obtain reliable results.

5.3.1 Polarization Beam-Splitter

A polarization beam-splitter can be realized based on a directional coupler made of two identical (and thus synchronous) waveguides (Figure 5.14). Under the appropriate design, the mixed TE/TM polarized modes at the input of waveguide I can be spatially separated into the orthogonal polarizations at the waveguides I and II outputs. This device is based on the fact that the TE and the TM modes usually have different coupling lengths due to the polarization effect. If the length of the directional coupler L_c is set to:

$$L_c = \frac{L_{TE}L_{TM}}{L_{TE}-L_{TM}} \tag{5.6}$$

where L_{TE} and L_{TM} are the coupling lengths for the TE and the TM modes, then the two polarizations will be separated into the two channels at the output of the coupler.

Figure 5.15 plots the field intensity of the TE and the TM beams as functions of the propagation distance, simulated by vectorial BPM, for a directional coupler used as a polarization beam-splitter, made of two parallel step-index symmetric slab waveguides [8]. The refractive

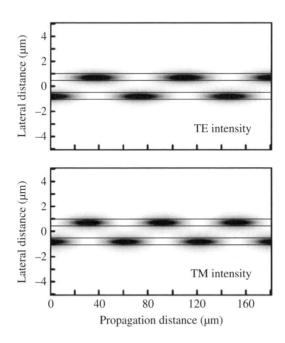

Figure 5.15 Upper figure: beam propagation simulation of TE light. Lower figure: propagation of TM light. At the input, the corresponding fundamental mode of one isolated waveguide is launched in the bottom waveguide. Observe the different coupling length between orthogonal polarizations

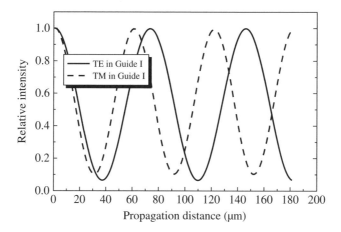

Figure 5.16   Relative intensity in waveguide I for TE light (continuous line) and for TM light (dashed line) as a function of the propagation distance

indices of the waveguide cores are $n_I = n_{II} = 1.5$ and of the substrate region is $n_s = 1.3$. The width of the guiding layer is $W_I = W_{II} = 0.5\,\mu\mathrm{m}$ and their separation is $S = 1.5\,\mu\mathrm{m}$. This directional coupler is designed to work at the wavelength of $\lambda = 1.5\,\mu\mathrm{m}$. The input beam is the fundamental mode (TE or TM) of one isolated waveguide, launched in the bottom waveguide.

As can be seen in the figure, the BPM numerical simulation indicates that at the end of the propagation ($\approx 181\,\mu\mathrm{m}$) the light is in the upper waveguide when TE light is launched in the bottom guide. By the contrary, when the TM mode is injected, the output light is in the bottom guide. Thus, the two polarized modes are separated into two different waveguides at the output. Figure 5.16 shows the guided power that remains in the input waveguide (guide I). At the input, the TE and the TM are equally excited. Just after propagating a distance $181\,\mu\mathrm{m}$, the two polarizations are split into two different waveguides. Figure 5.17 plots the field intensity profiles of the TE and TM beams at $z = 181\,\mu\mathrm{m}$. Let us note that considerable crosstalk occurs due to the close coupling between the two waveguides.

Let us now calculate theoretically the coupling length using the parameter structure of the polarization beam-splitter. If the guide I and II are well separated (independent waveguides), their effective indices at $\lambda = 1.5\,\mu\mathrm{m}$ for TE and TM polarization are $N_{TE} = 1.373151$ and $N_{TM} = 1.355569$, respectively. When the two identical waveguides are in close proximity, at a distance $S = 1.5\,\mu\mathrm{m}$ between their centres, the effective indices for the symmetric and anti-symmetric TE modes of the structure are $N_S = 1.381939$ and $N_A = 1.361285$, respectively. The corresponding effectives indices for TM polarized modes are $N_S = 1.365889$ and $N_A = 1.341077$. According with the formula of the coupling length:

$$L_c = \frac{\pi}{\Delta\beta} = \frac{\lambda}{2\Delta N} \tag{5.7}$$

where $\Delta N \equiv N_S - N_A$, the calculated coupling length for TE and TM polarized modes are 36.31 and 30.22 $\mu\mathrm{m}$, respectively. Using now the formula (5.6), the length of the directional coupler

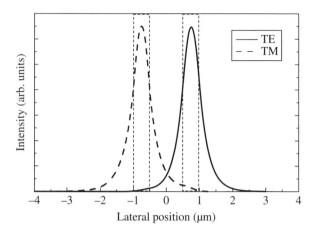

Figure 5.17   Intensity profiles at the output coupler for TE (continuous line) and TM (dashed line) polarization

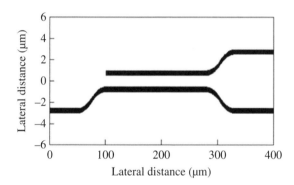

Figure 5.18   Layout of the polarization beam-splitter, including the two curved waveguides at the end of the parallel coupler

is calculated to be $L_c = 180.4\,\mu\text{m}$, in accordance with the results provided by vectorial BPM simulations.

To minimize coupling between the TE and TM beams at the end of the parallel waveguides, two curved sections should be included at the exit guides, as indicated in the layout sketched in Figure 5.14. The precise design for this particular coupler is drawn in Figure 5.18.

Following the design of the polarization beam-splitter presented in Figure 5.18, it is clear that at the beginning of the curved sections, where the two waveguides are still in close proximity, some coupling will still exist. Therefore, the length of the parallel section of the structure should be reduced accordingly to attain a correct spatial polarization splitting. Using a lateral shift of $2\,\mu\text{m}$ in the curved section for each waveguide and for a curve length of $50\,\mu\text{m}$, the length of the parallel waveguides is calculated be $170\,\mu\text{m}$, according with the numerical simulations supplied by vectorial BPM. These results are shown in Figure 5.19.

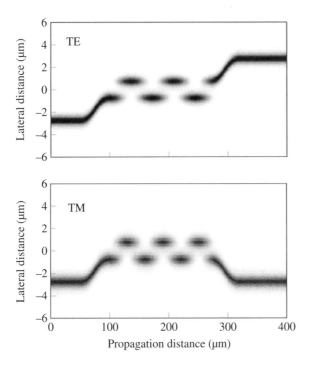

Figure 5.19 Intensity map of the light propagation for TE (upper) and TM (bottom) beams along the structure drawn in Figure 5.18, computed by vectorial BPM

5.3.2 Wavelength Filter

The directional coupler schematically illustrated in Figure 5.14 can be also used to separate two different wavelengths (λ_1 and λ_2) and thus this device can act as a wavelength filter. The idea that underlies the filter operation is to build the directional coupler by using two dissimilar waveguides in such a way that they are properly designed to be synchronized at a particular wavelength (for instance, λ_1) [8]. Thus, the beam at the wavelength λ_1 is completely transferred to the parallel waveguide. On the other hand, the coupler allows the guided beam with the other wavelength (λ_2) to remain in the input guide, because the two dissimilar waveguides are not synchronized at the wavelength λ_2.

The two slab waveguides with different widths and refractive indices have distinct dispersion curves which intersect with each other at the selected wavelength λ_1 (Figure 5.20, where material dispersion has been ignored). The widths of the two slabs are $W_1 = 1\,\mu\text{m}$ and $W_2 = 1.6\,\mu\text{m}$, and the refractive indices are $n_I = 1.5005$, $n_{II} = 1.4900$ and $n_s = 1.4500$. The phase-matching condition at this wavelength is satisfied if the propagation constants of the guided modes of the two individual (isolated) waveguides β_I and β_{II} are equal:

$$\beta_I(\lambda_1) = \beta_{II}(\lambda_1) \tag{5.8}$$

Thus maximum power exchange occurs at the wavelength λ_1 [9] and will cease to happen when the wavelength is detuned from λ_1.

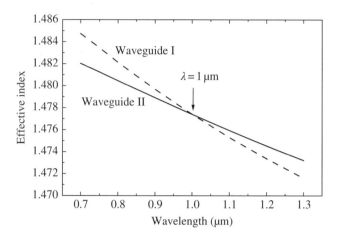

Figure 5.20 Dispersion curves for the two dissimilar step-index waveguides forming the asymmetric directional coupler based wavelength filter

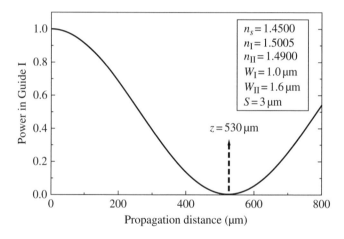

Figure 5.21 Power in waveguide I as a function of the coupler length for an asymmetric directional coupler using BPM simulations. At a propagation distance of 530 μm all the power is in waveguide II

The coupling length is determined at the wavelength λ_1 by:

$$L_c = \frac{\pi}{\beta_S(\lambda_1) - \beta_A(\lambda_1)} \tag{5.9}$$

At $\lambda_1 = 1$ μm and for a waveguide separation of $S = 3$ μm, the effective refractive indices for the symmetric and anti-symmetric TE modes are 1.477865 and 1.476921, respectively, calculated by the multilayer approximation method [9]. Using the formula (5.9), the calculated coupling length is $L_c = 529.7$ μm. Figure 5.21 shows the BPM results of the power remaining in

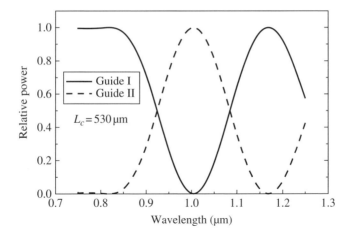

Figure 5.22 Relative power in waveguide I (continuous line) and waveguide II (dashed line) as a function of the wavelength, for a coupler length of 530 μm

waveguide I as a function of the propagation distance when the TE-fundamental mode is launched in guide I. The BPM simulation uses a grid spacing of $\Delta x = 0.0125$ μm and a propagation step of $\Delta z = 1$ μm. As can be observed, the propagation distance for the minimum power in guide I is obtained after a propagation distance of 530 μm, in agreement with the value obtained using Eq. (5.9).

The spectral response of the filter for the TE polarization is shown in Figure 5.22, were material dispersion was ignored. Continuous line represents the power in waveguide I at a position $z = 530$ μm, when the fundamental TE mode is launched in guide I, while the dashed line indicates the power that remains in guide II. A total transfer from guide I to guide II is observed for a wavelength at 1 μm. On the contrary, at wavelength around 0.81 μm the power is in guide I, indicating that this device can separate efficiently these two wavelengths.

For practical designs, the waveguides' ends must be separated to avoid further coupling. This can be realized by using two opposite curves waveguides. As the curved section also supplies some amount of power coupling, the length section of the parallel waveguides should reduce. This design can be easily modelled by means of BPM, in spite of that the coupling coefficient is now z-dependent. Figure 5.23 presents a practical design of the coupler, where the parameters of the structure are indicated on the figure. The separation of the waveguides at the end of the structure (7 μm) is large enough to avoid further coupling. The BPM simulations corresponding to this particular design are shown in Figure 5.24, where the filtering characteristic of the directional coupler based wavelength filter is demonstrated.

5.4 Multimode Interference Devices

5.4.1 Multimode Interference Couplers

MMI effect is based in the self-imaging effect. This effect is a property of multimode waveguides by which an input field profile is reproduced in single or multiple images at periodic intervals along the propagation direction of the guide. Integrated optical elements based on

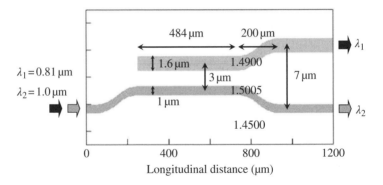

Figure 5.23 Geometric design of an asymmetric directional coupler working as wavelength filter

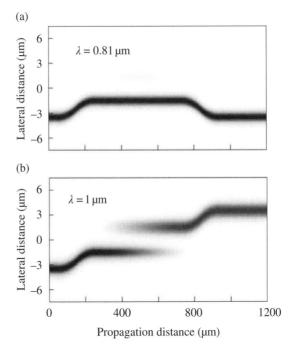

Figure 5.24 BPM simulation of the asymmetric directional coupler described in Figure 5.23 working as wavelength filter at $\lambda = 0.81$ μm (a) and $\lambda = 1.0$ μm (b)

the MMI effect (MMI devices) are small devices that allow a wide range of applications, from couplers, (de-) multiplexers, routers, switches and so on.

The central structure of a MMI device is a waveguide designed to support a large number of modes (typically greater than 3). In order to launch light into and recover light from that multimode waveguide, a number of access (usually single moded) waveguides are placed at its beginning and at its end. Such devices are generally referred to as $M \times N$ MMI couplers, where M and N are the number of input and output waveguides, respectively. Figure 5.25 shows an

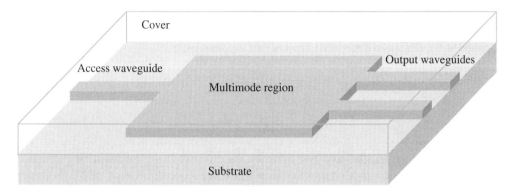

Figure 5.25 Scheme of a multimode interference (MMI) device. It consists of one (or several) access waveguide, a wide region supporting several transversal modes and several output waveguides

example of a 1×2 MMI device, where the most relevant parts are signalled. The performance of the MMI devices can be modelled by 3D-BPM. Nevertheless, as the waveguides are usually single-mode in the in-depth direction, the 3D-structure MMI device can be transferred to its two-dimensional counterpart using the Effective Index Method (EIM) [9] if the strong-guiding limit is valid. In this way, the counterpart 2D-structure can be simulated using 2D-BPM, which allows more efficient calculations.

MMI-couplers are fabrication tolerant, small sized components, which makes them highly attractive for use in a wide variety of integrated photonics devices. Their functionality is based upon the propagation of a certain number of modes, which are excited by the input field. Due to interference between the various modes in the multimode region, single or multiple images of the input field, the so-called self-images, appear at different positions along the MMI-coupler.

5.4.2 *Multimode Interference and Self-Imaging*

The self-imaging property of multimode waveguides, like MMI-couplers, was first suggested by Bryngdahl [10] and extensively described by Ulrich [11]. The functionality of the MMI-coupler can be described by beam propagation type methods [12] or modal propagation analysis (MPA) [13]. This latter method provides the most illustrative description and will be used to explain some of the properties of the MMI-coupler. Nevertheless, for accurate description of the MMI devices BPM analysis is required.

Let us assume that the MMI coupler (multimode region) is invariant along the z-direction. In the modal analysis, the field profile at a distance z from the entrance plane in the MMI coupler is written as the superposition of all propagating m guided modes supported by the MMI-coupler:

$$\Psi(x,z) = \sum_{\nu=0}^{m-1} c_\nu f_\nu(x) e^{-i\beta_\nu z} \tag{5.10}$$

in which $f_\nu(x)$ represents the transversal modal field distribution of the ν-th mode and β_ν its propagation constant. This expression holds under the assumption that no power is lost due to the excitation of radiation modes, which is fulfilled if the spatial profile of the input field

$\Psi(x,0)$ is narrow enough. The different field excitation coefficients c_ν are determined by the overlap between the various guided modes of the MMI-coupler and the optical field $\Psi(x,0)$ applied at the entrance of the MMI coupler (see Appendix C):

$$c_\nu = \frac{\int \Psi(x,0)f_\nu^*(x)dx}{\int |f_\nu(x)|^2 dx}. \tag{5.11}$$

As the MMI-coupler is assumed to be z-invariant, the different guided modes in the multimode region are orthogonal, which implies that no power transfer occurs between the MMI-coupler modes (that is, $\partial c_\nu / \partial z = 0$).

Equation (5.10) can be more conveniently rewritten as:

$$\Psi(x,z) = e^{-i\beta_0 z}\sum_{\nu=0}^{m-1} c_\nu f_\nu(x)e^{i(\beta_0-\beta_\nu)z}. \tag{5.12}$$

Within the strong guiding approximation (which implies a large number of guided modes in the MMI coupler) the propagation constants difference of the fundamental and a generic νth mode can be approximated by [13]:

$$\beta_0 - \beta_\nu \approx \frac{\nu(\nu+2)\pi}{3L_\pi} \tag{5.13}$$

in which L_π represents the beat length of the two lowest order modes, defined by:

$$L_\pi \equiv \frac{\pi}{\beta_0 - \beta_1}. \tag{5.14}$$

The beat length is the most important parameter defining a MMI coupler. By substituting relation (5.13) into (5.12), the phase factor of the field profile of the ν-th mode at the end of a MMI-coupler with length L is found to be:

$$e^{i\frac{\nu(\nu+2)\pi}{3L_\pi}L}. \tag{5.15}$$

If this phase factor equals an even integer multiple of 2π at $z = L$, all modes will interfere positively with the same relative phase as at $z = 0$. In this case the input field will be reproduced at the end of the MMI-coupler. Also, when the phase factor equals an odd multiple of π at $z = L$, an image of the input field, which is mirrored with respect to the MMI-coupler longitudinal axis, will appear at the end of the MMI-coupler (Figure 5.26). The requirement for these single images is fulfilled if:

$$L = p(3L_\pi), \quad \text{with } p = 0, 1, 2, \dots \tag{5.16}$$

Direct and mirrored images of the input are formed at the end on the coupler for p even and p odd, respectively.

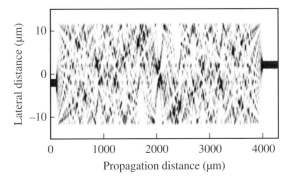

Figure 5.26 Light intensity pattern corresponding to general interference mechanisms in a multimode waveguide, using wide-angle BPM simulation, leading to a mirrored single image at $z = 3L_\pi$. Also, multi-fold images at intermediate distances, non-equally spaced along the lateral axis, are formed

Figure 5.26 shows the TE-light propagation at $\lambda = 1\,\mu\text{m}$ along a wide MMI coupler, computed by wide-angle BPM. The coupler consists of a symmetric planar waveguide with $n_{core} = 1.65$ and $n_{cladding} = 1.45$ and a width of 24 μm. The effective indices of the fundamental and first TE-modes of the coupler are 1.6498728 and 1.6494911, respectively, which according with formula (5.14) gives a coupling length of $L_\pi = 1310\,\mu\text{m}$. For a coupler length of $3L_\pi$ (i.e. 3930 μm), a mirrored image of the input is formed, as can be seen in the example provided in Figure 5.26.

Besides single images, multiple imaging is also found at specific places along the MMI coupler. For instance, an MMI-coupler length of:

$$L = \frac{p}{2}(3L_\pi), \qquad \text{with } p = 1, 3, 5, \ldots. \tag{5.17}$$

gives rise to the presence of two images of the input field at the end of the MMI-coupler. The images have a $\pi/2$ phase difference and each image holds half the power of the input power. This result demonstrates the possibility of using the MMI-coupler as a 3 dB power splitter.

A reduction of the size of the MMI-coupler can be obtained when so-called restricted interference is used. In the absence of every third mode ($\nu = 2, 5, 8,\ldots$), the length of the MMI-coupler can be reduced by a factor of 3 [13]. This case of restricted interference occurs if the input and output waveguides are placed at $x = \pm W_{MMI}/6$, W_{MMI} being the width of the MMI-coupler, which causes the overlap integral of the input field and every third mode to vanish (Figure 5.27). In addition to the mirrored image at a distance from the coupler entrance of $z = L_\pi$ (1310 μm in this example), N-fold images (non-equally spaced in general) are formed at positions (measured form the coupler entrance):

$$z = \frac{L_\pi}{N}. \tag{5.18}$$

A second case of restricted interference is obtained by placing a single monomode input waveguide at $x = 0$ (symmetric interference) (see Figure 5.28). In this way, only the even symmetric

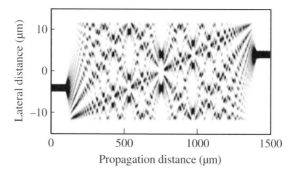

Figure 5.27 Light intensity pattern corresponding to restricted interference mechanisms, obtained by positioning the input waveguide at $x = W_{\mathrm{MMI}}/6$ leading to a mirrored single image at $z = z_0 + L_\pi$ and N-fold images located at $z = z_0 + L_\pi/N$ (being z_0 the position of the coupler entrance)

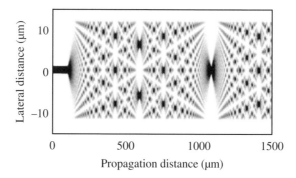

Figure 5.28 Light intensity pattern corresponding to symmetric excitation. N-fold images, equally spaced, are formed at positions $z = (p/N)(3L_\pi/4)$, measured from the multimode region entrance

modes in the MMI-coupler can be excited. By employing this symmetric interference, the length of a 3 dB splitter/coupler can be reduced by a factor of 4 with respect to the non-restricted interference mechanism. In general, N-fold images, equally spaced, are formed at distances from the coupler entrance given by:

$$L = \frac{p}{N}\left(\frac{3L_\pi}{4}\right), \tag{5.19}$$

p being an integer number. An important drawback of symmetric interference couplers/splitters is the presence of only a single input waveguide, which restricts its use to power splitters and combiners.

The parameters of the MMI-couplers, such as dimensions of the multimode region and locations of the input and output waveguides, are obtained in a first approximation by the MPA. Once these parameters are calculated, they should be optimized by means of BPM simulations to obtain a device with low insertion loss and crosstalk level.

5.4.3   1×N Power Splitter Based on MMI Devices

One of the most popular uses of MMI devices is as power splitters. Figure 5.29 shows the 3D structure of a 1 × 3 MMI-power splitter based on Zr-doped sol-gel technology [14]. The core of the input and output waveguides, as well as the multimode region, are made of a homogeneous medium with refractive index of $n_f = 1.492$, whereas the surrounding media (cover and substrate) has an index of $n_s = 1.488$, measured at $\lambda = 1.55$ μm. The high index region, including the input and output waveguides core, has a thickness of $t = 5$ μm. The access and output guides have a rectangular cross section with a width of $W = 6$ μm and the multimode region has a width of $W_{MMI} = 75$ μm.

Figure 5.30 shows the transverse intensity distribution of the mode supported by the input (or output) rectangular buried waveguide, obtained by using semi-vectorial-BPM propagation along the imaginary axis, for quasi-TE polarized light. This field will be used to perform the simulation of the MMI.

The beat length L_π of the two lowest order modes of the multimode region is now calculated by the imaginary distance BPM. The simulation, performed at $\lambda = 1.55$ μm using semi-vectorial-BPM quasi-TE polarized light, gives $N_0 = 1.4899655$ and $N_1 = 1.4898726$ for the effective

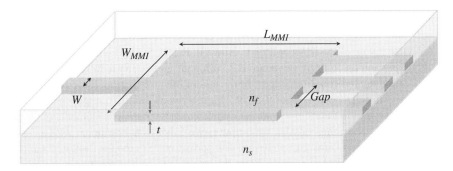

Figure 5.29   MMI device acting as a 1 × 3 power splitter

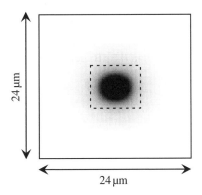

Figure 5.30   Transverse intensity distribution of the quasi-TE mode in a rectangular monomode waveguide obtained by imaginary-distance BPM simulation at $\lambda = 1.55$ μm

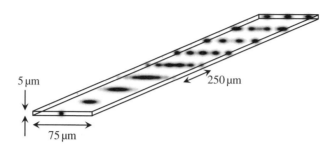

5 µm

250 µm

75 µm

Figure 5.31 Light intensity patterns, taken at 250 µm intervals, of semi-vectorial 3D-BPM propagation at $\lambda = 1.55$ µm in a multimode region of 5×75 µm². The input, centred at $x = 0$, corresponds to the fundamental TE-mode of a rectangular waveguide of 5×6 µm²

indices of the two lowest order propagating modes. Now, according with formula (5.14), a value of $L_\pi = 8342$ µm is obtained. The length of the MMI coupler to produce a three folded image from a centred input, working under the symmetric restricted interference regime, is given by formula (5.19): for $p = 1$ and $N = 3$, this results on a value of $L_{MMI} = 2085$ µm. Figure 5.31 shows the field pattern evolution of the quasi-TE input mode coming from the input waveguide along the 75 µm wide multimode region using semi-vectorial 3D-BPM. The transversal and longitudinal steps used in the simulation are $\Delta x = 0.5$, $\Delta y = 0.25$ and $\Delta z = 0.5$ µm.

The top-view layout of the MMI splitter is drawn in Figure 5.32. The input waveguide has a length of 500 µm and the multimode region is 75 µm wide and 2085 µm long, accordingly with the previously calculated coupling length. At the end of the wide region, three equispaced output guides are connected, with a separation of 27 µm between them. To improve the light coupling between the multimode region and the output guides, three taper elements are included [15]. Also, to avoid coupling between the outputs, two S-bends are designed with a curvature radius of 4000 µm and a lateral shift of 40 µm. The performance of the 1×3 splitter is simulated by 2D-BPM, where the EIM is used to obtain the two-dimensional refractive index map from the 3D original structure. The EIM analysis gives a core refractive index of 1.4899945, surrounded by a substrate index of 1.488. Figure 5.32 (bottom) shows the BPM simulation along the MMI, where the input field corresponds to the guided mode of the input waveguide. Although some losses are produced in the multimode region, the main contribution of the losses comes from the coupling between the field at the multimode region and the output waveguides. In the absence of the tapered regions, 91% of the input power is in the output waveguides, which gives an insertion loss lower than 0.5 dB, defined as:

$$IL = 10 \, \log_{10}(P_{IN}/P_{OUT}). \tag{5.20}$$

If tapers are used (18 µm wide and 400 µm long), a 96% of the total power can be achieved in the output waveguides, which means that the insertion losses are reduced to 0.2 dB.

5.4.4   Demultiplexer Based on MMI

MMI devices can be also used as demultiplexers, in which one input containing two wavelengths split them into separated outputs for each wavelength. In particular, MMI

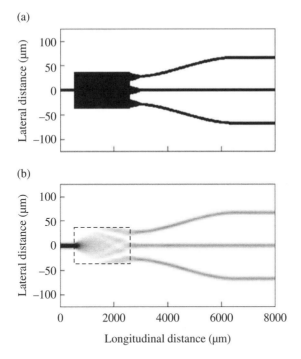

(a)

Lateral distance (μm)

(b)

Lateral distance (μm)

Longitudinal distance (μm)

Figure 5.32 (a) Refractive index map of the 1 × 3 MMI power splitter, where the output guides are connected via taper elements and two curved waveguides are designed to avoid coupling between the ports. (b) Light intensity pattern of 2D-BPM propagation at $\lambda = 1.55\,\mu m$

demultiplexers can be used to separate the pump and signal beams in an optical amplifier. In the general resonance mechanism, at a coupler length of $L_{MMI} = 3pL_{\pi}$, a direct or mirrored image of the input field is reproduced if p is an even or odd integer, respectively. These direct and mirror imaging correspond to bar and cross-coupling states, respectively. An MMI coupler can perform dual-channel wavelength demultiplexing when it is a bar-coupled for one wavelength and cross-coupled for the other. The beat length ratio of the two wavelengths λ_p (pump) and λ_s (signal) is then given by:

$$\frac{L_{\pi}^p}{L_{\pi}^s} = \frac{p+q}{p}, \tag{5.21}$$

where p is a positive integer, q is an odd integer and L_{π}^p and L_{π}^s are the beat lengths at the pump and signal wavelengths, respectively.

Figure 5.33 shows the structure of an MMI device designed to separate two beams at $0.98\,\mu m$ (pump) and $1.55\,\mu m$ (signal) [16]. The substrate and the cladding are made of amorphous SiO_2, with a refractive index of $n_{subs} = 1.46$. The core, made of SiON, is a high index region with $n_{core} = 1.565$ at $\lambda = 0.98\,\mu m$ and 1.561 at $\lambda = 1.55\,\mu m$. The input and output guides have a rib structure as detailed in Figure 5.33, with $h_1 = 1.2\,\mu m$, $h_2 = 1.35\,\mu m$ and $t = 1\,\mu m$, while the width of the rib waveguide is $d = 2.8\,\mu m$. This rib waveguide is monomode at both pump and signal wavelengths.

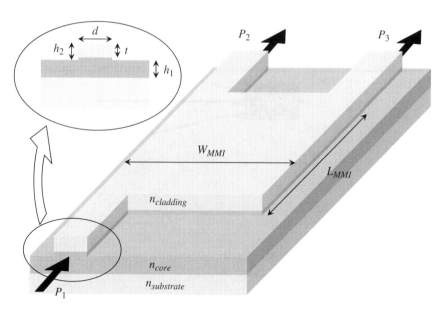

Figure 5.33 Design of an MMI device with one input port and two output ports, to be used as a demultiplexer

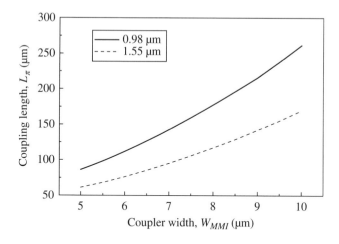

Figure 5.34 Coupling length of the MMI device drawn in Figure 5.33, for two different operating wavelengths. Continuous line: $\lambda_p = 0.98\ \mu m$. Dashed line: $\lambda_s = 1.55\ \mu m$

Using 3D-semi-vectorial BPM, the effective indices of the two lowest TE modes of the multilayer structure in the coupler region are calculated as a function of the coupler width W_{MMI}. With these data the beat length L_π is obtained by formula (5.14) for $\lambda_p = 0.98\ \mu m$ and $\lambda_s = 1.55\ \mu m$ (Figure 5.34). A compact MMI demultiplexer can be designed with a

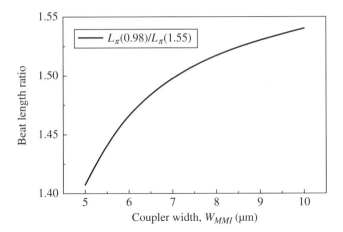

Figure 5.35 Beat length ratio for 0.98 and 1.55 μm as a function of the coupler width, obtained from data in Figure 5.34. The beat ratio 3/2 corresponds to a MMI width of $W_{MMI} = 7.0$ μm

beat length ratio of 1.5, which can be achieved with the natural number $p = 2$ and the odd integer $q = 1$. In this way the coupler acts as a bar-coupler at 0.98 μm and it works as a cross-coupler at 1.55 μm. Figure 5.35 presents the beat length ratio as a function of the coupler width and indicates that a beat length ratio of 1.5 corresponds to a coupler width of 7.0 μm. Taking into account this value, the length of the MMI coupler is given by $L_{MMI} = 2\left(3L_\pi^{0.98}\right) = 3\left(3L_\pi^{1.55}\right) = 852$ μm.

Figure 5.36 shows the top-view of the practical design of the demultiplexer. The input guide is shifted 2.1 μm with respect to the centre of the MMI coupler, which starts at $z_0 = 100$ μm. The exit waveguides are separated a distance of 3.8 μm (from their centres) and separate gradually by two S-bends to avoid crosstalk between them. The S-bend structure has a length of 500 μm and allows a final separation between the guides of 13.8 μm. As the output guides are close at the exit of the wide region coupler, some coupling still persists at the beginning of the S-bends. For that reason, the coupler length must be shorter than the value previously calculated. A coupler length of $L_{MMI} = 820$ μm is chosen in the final MMI design. The simulation of the light propagation along the MMI device is performed at $\lambda_p = 0.98$ μm and $\lambda_s = 1.55$ μm using semi-vectorial-TE-BPM. The eigenmode of the access waveguide is chosen as the input field. The average of the effective indices at the two lowest modes is used as the reference index in the BPM simulations.

The BPM simulations of the light propagation along the MMI device are shown in Figure 5.36 (central and right parts), for a coupler length of $L_{MMI} = 820$ μm, where the transversal intensity maps are plotted at 100 μm intervals in the longitudinal direction. We see that the MMI acts as a bar-coupler for $\lambda_p = 0.98$ μm and for $\lambda_s = 1.55$ μm the output light is in the cross-guide. Contrast and insertion loss are used as measures of performance of the device. The contrast at 980 nm is defined by:

$$Contrast_p = 10 \log_{10}\left(P_2^{\lambda_p}/P_3^{\lambda_p}\right), \tag{5.22}$$

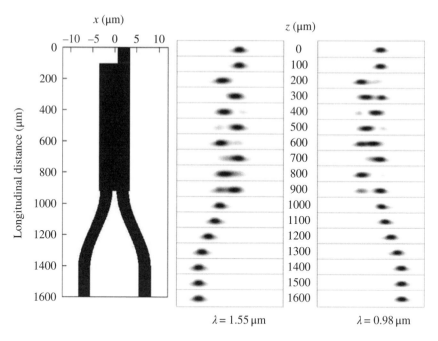

Figure 5.36 Left: top-view of the multiplexer. Centre: snap-shots at 100 μm intervals of the transversal intensity at $\lambda = 0.98$ μm. Right: transversal intensity at $\lambda = 0.98$ μm

and it measures the ability of the device for separating the two wavelengths. On the other hand, the insertion loss (*IL*) at 980 nm is defined by:

$$IL_p = 10 \log_{10}\left(P_1^{\lambda_p}/P_2^{\lambda_p}\right), \tag{5.23}$$

which measures the losses induced at the input-waveguide/coupler, coupler/output-guide and at the S-bend. These losses are due to coupling to radiation modes. Also, the insertion loss takes into account the power lost due to the fraction of light coupled to the other output waveguide. Similar expressions are defined for $\lambda_s = 1.55$ μm, just by exchanging subscripts 2 and 3. The MMI demultiplexer shows contrasts of 17 and 12 dB for the pump and signal wavelengths, respectively. Also, similar insertion losses of $IL = 0.40$ dB are obtained in the BPM simulations for both wavelengths.

5.5 Waveguide Gratings

One of the most versatile elements in integrated optics is the periodic grating structures, which facilitate the interaction between waveguides modes and/or radiation modes [17]. The gratings find applications in integrated optics devices such as filters, couplers, mode converters, distributed feedback oscillators and lasers or demultiplexers, among others. Here we present the use of BPM in simulating some integrated optic elements based on grating structures: mode converters, grating assistant couplers and waveguide reflectors.

5.5.1 *Modal Conversion Using Corrugated Waveguide Grating*

A periodically corrugated surface constitutes a surface relief grating which can be used as modal conversion in waveguides (Figure 5.37) [18]. The presence of the grating provides the extra wavevector in the propagation direction to obtain efficient conversion by phase matching condition. The phase matching condition between two modes μ and ν under the presence of a periodic structure is given by [9]:

$$\beta_\mu = \beta_\nu + \frac{2\pi}{\Lambda}, \tag{5.24}$$

where β_μ and β_ν are the propagation constants of the modes and Λ is the period of the grating.

Conventional BPM can be used to model this energy transfer between modes, providing that the teeth of the corrugated grating do not introduce appreciable back-reflection radiation, which is true for smooth relief gratings where the groove depth is much lower than the waveguide core dimension ($h \ll d$).

Figure 5.38 shows a symmetric step index planar waveguide, where a sinusoidal relief grating is produced at the upper interface, with a total grating length of $L = 2560$ μm. For a working wavelength of $\lambda = 1.55$ μm, the waveguide is multimode and the effective refractive indices of the fundamental and first order modes are $N_0 = 2.22669$ and $N_1 = 2.21712$, respectively. According with Eq. (5.24), the grating period for efficient mode coupling should be:

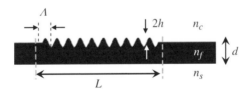

Figure 5.37 Geometry and relevant parameters of a waveguide grating based on a sinusoidal corrugated surface

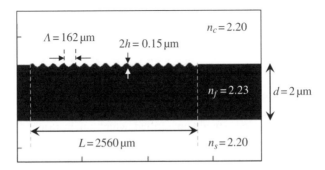

Figure 5.38 Example of a corrugated waveguide grating designed for efficient mode conversion at $\lambda = 1.55$ μm between the two lowest order modes

$$\Lambda = \frac{\lambda}{N_{eff,0} - N_{eff,1}}, \tag{5.25}$$

giving a value of $\Lambda = 162.0\,\mu m$.

On the other hand, for co-directional coupling, maximum conversion of 100% can be obtained theoretically and this can be achieved by using a grating length given by $L = \pi/2\kappa$ [9], where κ denotes the coupling coefficient between the modes involved. Under the approximation of well confined modes ($h \ll d$), the coupling coefficient between TE modes can be expressed as a function of the waveguide parameters and the mode fields at the upper interface $E(0)$ [9]:

$$\kappa = \frac{\pi h}{2\lambda} \frac{n_f^2 - n_c^2}{\sqrt{N_\mu N_\nu}} \frac{E_\mu^*(0) E_\nu(0)}{\sqrt{\int\limits_{-\infty}^{\infty} |E_\mu|^2 dx \int\limits_{-\infty}^{\infty} |E_\nu|^2 dx}}. \tag{5.26}$$

Using this formula, one obtains a coupling coefficient of $\kappa = 6.06 \times 10^{-4}\,\mu m^{-1}$, which corresponds to a coupling length of $L = \pi/2\kappa = 2590\,\mu m$.

The performance of the corrugated grating for mode conversion in waveguides can be conveniently simulated by BPM. Figure 5.39a shows the intensity pattern of the light propagation along the waveguide example described previously, where the grating starts at $z = 200\,\mu m$ and has a length of $L = 2560\,\mu m$. The input field is the fundamental TE-mode of the unperturbed waveguide.

The bottom graph in Figure 5.39 represents the mode power of the first three-order modes as the light propagates along the structure. The power of the fundamental mode (mode-0, continuous line) remains unchanged until the light reaches the grating region, located at $z = 200\,\mu m$. Once the light enters the perturbed region of the grating, its power monotonically decreases and the power corresponding to the mode 1 (dashed line) increases in a complementary manner. At the end of the grating, all the initial power in mode 0 is transferred to mode 1 and therefore the waveguide grating allows a $\sim$100% of power conversion. It is worth noting that the coupling length obtained by means of the coupling coefficient κ is in excellent agreement with the result provided by BPM numerical simulation. Finally, the graph also plots the power evolution of mode 2 (dotted line), which is negligible (<0.2%) with respect to the total power, indicating that if phase matching condition is not fulfilled, very little coupling can take place between the modes.

5.5.2 Injecting Light Using Relief Gratings

Relief grating can be also used for exciting a waveguide mode by light coming from the external medium (cover region) [19]. For efficient coupling, phase matching condition between the external light and the guided mode should be fulfilled, which can be reached by the extra wavevector provided by the periodic modulation grating (Figure 5.40). In this case, Eq. (5.24) is given by:

$$\beta_{mode} = \frac{2\pi}{\lambda} n_{cover} \sin\theta + \frac{2\pi}{\Lambda}, \tag{5.27}$$

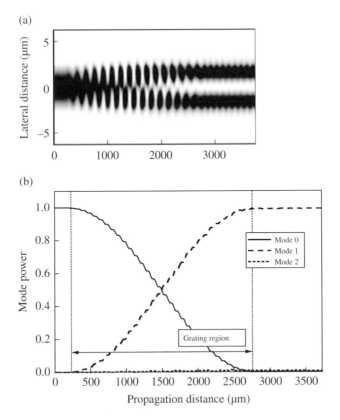

Figure 5.39 (a) BPM simulation along the corrugated waveguide grating of Figure 5.38, for TE-polarized light at a wavelength of 1.55 μm. The input is the fundamental mode of the unperturbed structure. (b) Mode power evolution of the first three guided modes as a function of the propagation distance

where θ is the incident angle of a plane wave into the waveguide surface coming from the cover region. This equation provides the relation between the grating period for a given wavelength, the mode effective index N_{eff} and the external input angle θ:

$$\Lambda = \frac{\lambda}{N_{eff} - n_{cover}\sin\theta}. \tag{5.28}$$

As an example, let us consider the light injection from air into a monomode step-index waveguide at $\lambda = 1.55$ μm, with a substrate index of $n_s = 1.45$ and a film of $d = 3$ μm thickness and refractive index of 1.48. With these waveguide parameters, imaginary-distance BPM gives a mode effective index of 1.46806 for TE polarization. Using these values, and for a grating with a period of $\Lambda = 3.2$ μm, Eq. (5.28) indicates that the phase matching condition is fulfilled for an incident angle of $\theta \approx 80°$.

Although the phase matching condition allows us to determine the precise angle for a given grating, it does not permit to evaluate the efficiency of the coupling. Here, BPM simulations can be used to evaluate the grating performance in terms of the input beam parameters as well as

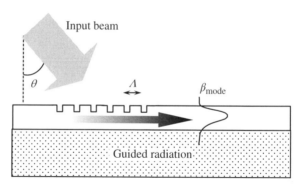

Figure 5.40 Geometry and relevant parameters of a corrugated waveguide grating for guided mode excitation from an external beam

grating design. To see the utility of the BPM in the optimization of grating parameters, let us study the coupling efficiency of a sinusoidal grating when a 15 µm wide Gaussian beam is incident onto the waveguide grating. Wide-angle BPM should be used in the simulations, as wide range of angles is involved in the problem, as well as several media with quite different refractive indices. When using Padé (2,2) wide-angle BPM, there is still a little discrepancy with the grating period to obtain maximum efficiency at 80° ($\Lambda = 3.1$ µm, instead of 3.2 µm previously calculated). Figure 5.41 shows the structure of the waveguide grating described and the BPM simulations in three cases. In the absence of grating (Figure 5.41b), the incident Gaussian beam generates a reflected and a transmitted beam, but light does not couple into the waveguide. Also, if the excitation is not performed under phase-matching conditions, light cannot be coupled as waveguide mode. Finally, for the phase-matched case (Figure 5.41d) the incident beam couples into the waveguide.

Based on BPM simulations, Figure 5.42 show the coupling efficiency, defined as the fraction of power in a waveguide mode relative to the incident power, as a function of the incident angle when an external 15 µm wide Gaussian beam is incident in a waveguide grating of length $L = 145$ µm, a grating depth of $2h = 0.55$ µm and a period of $\Lambda = 3.1$ µm. For this particular incident beam, the waveguide grating allows a maximum transfer efficiency to the guided mode of ∼11%, corresponding to an incident angle of 80°.

The coupling efficiency depends also on the grating depth $2h$. Figure 5.43 presents this analysis, where a maximum coupling is found for a grating depth around 0.55 µm. Also, BPM allows us to study the grating length necessary to achieve the maximum transfer to the guided mode for a given excitation. For the grating depth previously established, and for a 15 µm width Gaussian beam, wide-angle BPM simulations indicate that a grating length is of ∼150 µm is enough to achieve optimum coupling efficiency (Figure 5.44). If fact, for low efficiency gratings, the length of the grating can be estimated by the extension covered by the input beam, given by ∼$2W/\sin\theta$ (∼173 µm).

5.5.3 Waveguide Reflector Using Modulation Index Grating

A third example of BPM application on waveguide grating is presented, which consists of a two-dimensional high index contrast waveguide grating. The structure is composed by alternate

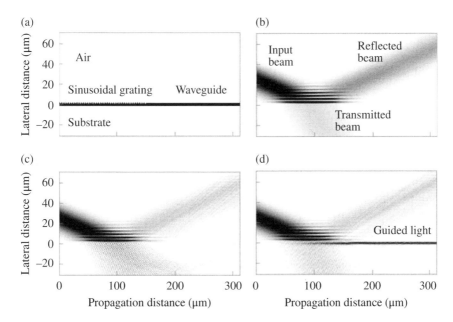

Figure 5.41 BPM simulations of a Gaussian beam incident to a waveguide relief grating. (a) Geometry of the structure. (b) Wide angle BPM simulation of the structure without grating. (c) Simulation of the light propagation for a non-phase-matched grating. (d) BPM simulation for the phase-matched waveguide grating

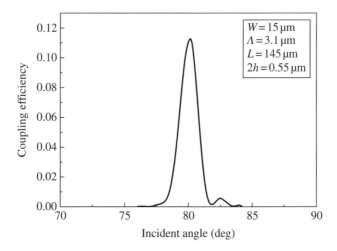

Figure 5.42 Efficiency for the excitation of the fundamental guided mode by an external Gaussian beam, using a sinusoidal relief grating, as a function of the incident angle

layers of low and high index along a planar step-index waveguide (Figure 5.45) and acts as a distributed feedback reflector. The two-dimensional waveguide to be analysed is a symmetric structure with a core layer of refractive index $n_{co1} = 1.55$ and thickness $W = 0.3$ μm, surrounded

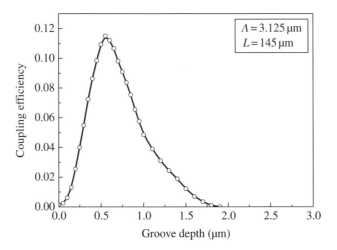

Figure 5.43 Efficiency for the excitation of the fundamental guided mode by an external Gaussian beam, using a sinusoidal relief grating, as a function of the grating groove depth

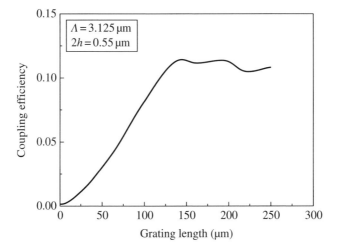

Figure 5.44 Coupling efficiency for the excitation of the fundamental guided mode by an external Gaussian beam, using a sinusoidal relief grating, as a function of the grating length

by a cladding of refractive index $n_{cl} = 1.00$. On the other hand, the grating is composed of eight alternate layers of material with $n_{co1} = 1.55$ and $n_{co2} = 1.45$. The thicknesses Λ_1 and Λ_2 of the layers have to be determined to add reflections constructively [20] following the relation:

$$\Lambda_1 = \frac{\lambda_0}{4N_{eff,1}}, \quad \Lambda_2 = \frac{\lambda_0}{4N_{eff,2}}, \tag{5.29}$$

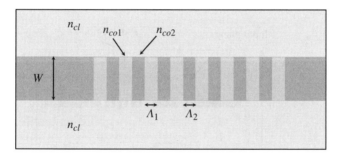

Figure 5.45   Step index waveguide grating used as distributed feedback reflector

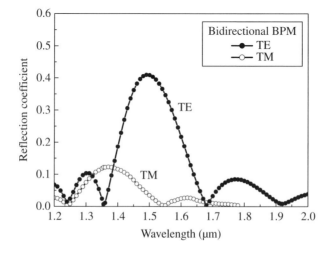

Figure 5.46   Reflection coefficient for TE (full circles) and TM (open circles) propagation for the high index contrast waveguide grating depicted in Figure 5.45 as a function of the operating wavelength

where $N_{eff,1}$ and $N_{eff,2}$ are the effective indices of the waveguide in the high and low index regions (with core of refractive indices n_{co1} and n_{co2}, respectively) and λ_0 is the operation wavelength. With $N_{eff,1} = 1.2088$ and $N_{eff,2} = 1.1447$ obtained by TE polarization at the phase-matching wavelength $\lambda_0 = 1.5\ \mu m$, it gives $\Lambda_1 = 0.328\ \mu m$ and $\Lambda_2 = 0.310\ \mu m$ for the thicknesses of the layers.

As result of the high index contrast structure, the waveguide grating exhibits strong polarization dependence. Figure 5.46 shows the reflection coefficient r of the fundamental mode of this waveguide grating as a function of the wavelength, obtained by bidirectional-BPM. The discretization steps are $\Delta x = 0.01\ \mu m$ and $\Delta z = 0.05\ \mu m$ and window dimensions of $L_x = 5.12\ \mu m$ and $L_z = 7.5\ \mu m$. Five trips are enough for reaching convergence in the iterative Bi-BPM, using a damping factor of $\gamma = 0.1$.

The waveguide reflector is also examined by FDTD (finite difference time domain). This numerical method is well suited to this type of short length structure with high index contrast that induces high reflection. For FDTD analysis of the structure, the fundamental waveguide

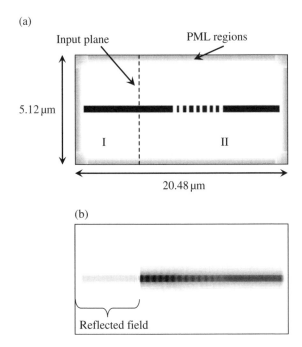

Figure 5.47 (a) Computational window for performing FDTD simulations of a step-index grating inserted along a symmetric step-index waveguide, surrounded by PML regions (grey regions). The input plane is indicated by a vertical dashed line. (b) Map of the averaged energy corresponding to a CW (continuous wave) excitation at 1500 nm for TE polarization

mode is used as the input source, both for TE and TM propagation, where the TF/SF (total-field/scattered-field) formulation has been implemented, following the scheme indicated in Figure 5.47a. The computational window has 128×128 cells, with $\Delta x = \Delta z = 40$ nm and it is surrounded by PML (perfectly matched layer) regions. To illustrate the performance of the TF/SF formulation, Figure 5.47b shows the map of the averaged energy corresponding to a TE-polarized sinusoidal source with a frequency of $\nu = 2 \times 10^{14}$ s^{-1} ($\lambda_0 = 1.5$ μm). Beyond the grating, only transmitted field is present. In the region between the grating and the input plane some pattern is generated, due to the interference between the incident field and the reflected field from the grating. On the contrary, in region I (left from the input plane) only pure reflected field is present, which permits a correct analysis of the waveguide grating reflectivity.

To calculate the reflectivity of the waveguide grating as a function of the wavelength by FDTD, a pulse excitation located at the input plane is used. The data are obtained by FFT (fast Fourier transform) of the temporal history of the reflected field, being the excitation source a Gaussian pulse modulated sine wave with a reference oscillation frequency source centred at $\omega_0 = 1.25 \times 10^{15}$ s^{-1}, with a pulse width of $\tau = 2$ fs. The simulation time step is $\Delta t = 0.08$ fs and the total simulation time is $T = 1310.72$ fs (2^{14} steps). That gives a frequency resolution $\Delta\nu$ in the FFT spectra of $1/T = 7.6 \times 10^{11}$ s^{-1} ($\Delta\lambda \approx 8$ nm in the range of interest). The spectral response, obtained by the FFT of the ratio between the source at the input plane and the reflected field located at the scattered region (region I), is plotted in Figure 5.48 for both TE (full circles) and TM (open circles) polarization. The most salient feature of the spectra

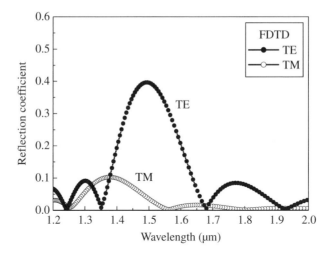

Figure 5.48 Reflection coefficient for a step-index grating in a symmetric step-index waveguide calculated by FDTD, for TE (full circles) and TM (open circles) propagation

is its strong polarization dependence, due to the high index contrast of the structure. Also, as expected, the data supplied by Bi-BPM and FDTD are in good agreement.

5.6   Arrayed Waveguide Grating Demultiplexer

Wavelength Division Multiplexing (WDM) is an efficient method of increasing the bandwidth in an optical communication network, where several channels, each carried by a different wavelength, are transmitted through a single optical fibre. One of the key optical elements in WDM are phased-array (PHASAR) based devices (also called arrayed waveguide gratings, AWGs). AWGs have become increasingly popular as optical wavelength (de)multiplexers for WDM applications, as they have proven to be capable of precise demultiplexing of a large number of channels with relative low losses [21–23].

Here, the operation principle and relevant parameters of AWG used as demultiplexer are described, besides some simple design rules for this device. Also, one example of AWG-demultiplexer is presented and its performance is then simulated by means of BPM. As usually these devices are very large (several millimetres), for the BPM simulations, in practice, the AWG-demultiplexer is break into three main parts and each part is simulated separately, with the proper start-field for each simulation. The spectral response of the demultiplexer is then studied by performing several BPM-runs as a function of wavelength and recording the modal power for each output channel.

5.6.1   Description of the AWG Demultiplexer

Figure 5.49 shows a schematic representation of the PHASAR demultiplexer. The device consists of two slab waveguide star couplers (free propagation regions, FPRs), connected by a dispersive waveguide array. The operation principle is as follows. Light propagating

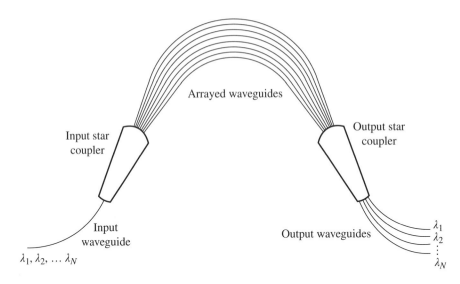

Figure 5.49 Scheme of the AWG-demultiplexer, showing the relevant five parts: input waveguide, input star coupler, arrayed waveguides, output star coupler and output waveguides

in the input waveguide will be coupled into the array via the first star coupler. The array has been designed such that (for the central wavelength of the demultiplexer) the optical path length difference between adjacent array arms equals an integer multiple of the central wavelength of the demultiplexer. As a consequence, the field distribution at the input aperture will be reproduced at the output aperture. Therefore, at this wavelength, the light will focus at the centre of the image plane (provided that the input waveguide is centred in the input plane) of the output star coupler.

If the input wavelength is detuned from this central wavelength, phase changes will occur in the array branches. Due to the constant path length difference between adjacent waveguides, this phase change will increase linearly from the inner to outer array waveguides, which will cause the wavefront to be tilted at the output aperture. Consequently, the focal point in the image plane will be shifted away from the centre. The positioning of the output waveguides in the image plane allows the spatial separation of the different wavelengths (channels). Thus, the wavelength dependent shift (dispersion) of the focal point in the image plane is the key feature of the demultiplexer.

5.6.1.1 Relative Dispersion of the Demultiplexer

To obtain the lateral displacement of the focal spot along the image plane of the demultiplexer as a function of the wavelength (dispersion), first it is necessary to calculate the wavefront tilting angle $d\theta$ due to a phase difference $d\phi$ between adjacent arrayed waveguides using the parameters defined in Figure 5.50.

This phase difference can be expressed as:

$$d\phi = d\beta_g \Delta l, \tag{5.30}$$

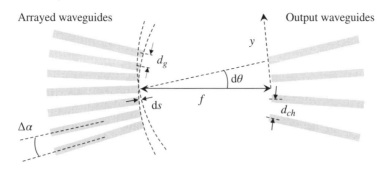

Figure 5.50   Geometry of the output star coupler, showing some relevant parameters for its analysis

where β_g is the propagation constant in the arrayed waveguides and Δl is the path length difference between adjacent array arms. The phase difference can also be written as:

$$d\phi = \beta_f ds, \qquad (5.31)$$

with β_f the propagation constant in the slab waveguides and ds denotes the advance of the wavefront (see Figure 5.50). From that figure and for small angles, it holds:

$$ds \approx d_g d\theta, \qquad (5.32)$$

d_g being the separation between arrayed guides at the entrance of the output-star coupler. Substituting Eq. (5.32) into Eq. (5.31), we obtain:

$$d\theta \approx \frac{d\varphi}{\beta_f d_g}. \qquad (5.33)$$

On the other hand, the relation between the propagation constant β_g and the effective index N_{eff} of the array waveguide is:

$$\beta_g = \frac{2\pi}{\lambda} N_{eff}, \qquad (5.34)$$

which is evaluated at the central wavelength of the demultiplexer $\lambda = \lambda_c$. Differencing this formula, it yields:

$$d\beta_g = \beta_g \left[\frac{dN_{eff}}{N_{eff}} - \frac{d\lambda}{\lambda} \right] \approx -\beta_g \frac{d\lambda}{\lambda}, \qquad (5.35)$$

where the term dN_{eff}/N_{eff} (which includes both material and waveguide dispersion) has been neglected for the approximation. Combining Eqs. (5.30), (5.33) and (5.35), the wavefront tilting $d\theta$ due to wavelength variation results in:

$$\frac{d\theta}{d\lambda} = -\frac{rm}{N_{eff} d_g}, \qquad (5.36)$$

with $r \equiv \beta_g/\beta_f \, (\approx 1)$ and m the diffraction order of the demultiplexer defined as:

$$m \equiv \frac{\Delta l}{\lambda/N_{eff}} = \frac{\Delta l}{\lambda_g}, \tag{5.37}$$

where we have introduced the central wavelength measured in the arrayed waveguides $\lambda_g = \lambda/N_{eff}$ and evaluated at $\lambda = \lambda_c$.

On the other hand, the dispersion D of the AWG demultiplexer is defined by:

$$D \equiv \frac{dy}{d\lambda}, \tag{5.38}$$

where dy is the shift of the focal point in the image plane (see Figure 5.50). From the geometry of the figure it also holds that:

$$dy = f \, d\theta, \tag{5.39}$$

f being the focal length of the star-coupler. Combining these equations, the dispersion of the demultiplexer can be expressed as:

$$D = -\frac{rmf}{N_{eff} d_g}. \tag{5.40}$$

5.6.1.2 Free Spectral Range

Another important parameter of the AWG is the Free Spectral Range ($\Delta\lambda_{FSR}$, FSR), also known as the demultiplexer periodicity. This periodicity is due to the fact that constructive interference at the output star coupler can occur for a number of wavelengths and it is easily calculated by specifying:

$$\Delta l = m \frac{\lambda}{N_{eff}(\lambda)} = (m-1) \frac{\lambda + \Delta\lambda_{FSR}}{N_{eff}(\lambda + \Delta\lambda_{FSR})}. \tag{5.41}$$

From that condition (and again ignoring material and waveguide dispersion) one obtains:

$$\Delta\lambda_{FSR} \approx \frac{\lambda}{m}, \tag{5.42}$$

where the approximation $m \approx m - 1$ has been made and is evaluated at $\lambda = \lambda_c$.

5.6.1.3 Arrayed Waveguides

The critical part of the AWG demultiplexer is the arrayed waveguides, which have to be designed correctly. Here, there is a great freedom in designing the array branch geometry and one of the most common designs [24] consists of a curved waveguide connected to a straight waveguide on either side of the curve, as is drawn in Figure 5.51. Each array branch

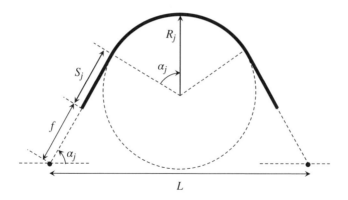

Figure 5.51   Schematic representation of the array branch geometry

has circles of different curvature radii and different centre positioning, as well as different lengths of straight sections.

From the geometry of the figure, it is easily seen that the straight section length S_j and radius of curvature R_j corresponding to the j-th array guide (for $j = 1$ to n_g, being n_g the number of arrayed guides) are related to the input to output plane spacing length L by:

$$L = 2R_j \sin \alpha_j + 2(S_j + f) \cos \alpha_j, \tag{5.43}$$

where α_j is the angle of the j-th array, which is given by:

$$\alpha_j = \alpha_1 + (j-1)\Delta\alpha, \tag{5.44}$$

$\Delta\alpha$ being the angular separation of the straight sections of adjacent waveguides.

On the other hand, the path length l_j of the j-th arrayed guide is:

$$l_j = 2(S_j + R_j \alpha_j). \tag{5.45}$$

The array must be designed such that, for the central wavelength of the demultiplexer, the optical path length difference between adjacent array arms equals an integer multiple of the central wavelength of the demultiplexer. Therefore, the path length l_j of the j-th element has to satisfy the condition:

$$l_j = l_1 + m(j-1)\lambda_g. \tag{5.46}$$

This requirement can be fulfilled by different design parameters, which can be used to optimize the AWG demultiplexer in terms of its size, performance and tolerances.

5.6.1.4   Setting the AWG Design Parameters

Following the relationships between the different parameters and the properties of the AWG-demultiplexer described here, the design strategy of the device can be now established. The first step for designing the AWG is to choose the spacing d_{ch} of the output waveguides in the image plane of the output star. The output channels should have low insertion loss and low crosstalk between them. Low crosstalk implies sufficient isolation between neighbour output guides, thus the gap between them should be sufficiently large. On the contrary, to achieve low insertion losses the output guides should be situated as close as possible. As these two requirements cannot be fulfilled at the same time, a compromise must be taken and as a rule of thumb, the gap between adjacent output channels should be chosen to be twice the width of the waveguide.

Once the output waveguide spacing d_{ch} has been fixed in the focal plane (see Figure 5.50), from Eq. (5.38) the relative dispersion D is simply given by:

$$D = \frac{d_{ch}}{\Delta\lambda},\tag{5.47}$$

where $\Delta\lambda$ is the wavelength channel spacing of the demultiplexer.

For a fixed free spectral range, the array order m can be obtained using expression (5.42):

$$m = ROUND\left[\frac{\lambda_c}{\Delta\lambda_{FSR}}\right].\tag{5.48}$$

The rounding to the nearest integer in the brackets implies a slight correction of the FSR. Once the array order has been fixed, the focal length f can be calculated, using expression (5.40), to be:

$$f = \frac{d_g N_{eff} D}{rm}.\tag{5.49}$$

From Figure 5.50, the angular increment $\Delta\alpha$ of the array branch angle simply follows as:

$$\Delta\alpha = \frac{d_g}{f}.\tag{5.50}$$

Now, given L and fixing α_1 and l_1, the value of the angle α_j is obtained by using Eq. (5.44), while the radius of curvature R_j and the length of the straight section S_j of the array branches can be calculated by:

$$R_j = \frac{L - [2f + l_1 + m(j-1)\lambda_g]\cos\alpha_j}{2(\sin\alpha_j - \alpha_j\cos\alpha_j)};\tag{5.51a}$$

$$S_j = \frac{l_1 + m(j-1)\lambda_g}{2} - R_j\alpha_j.\tag{5.51b}$$

The path lengths l_j of the waveguides in the array are simply given by Eq. (5.46). The input to output plane spacing L, the angle α_1 and the length l_1 are free parameters that can be used to optimize the AWG design. In fact, several optimization procedures have been devised in order to arrive to an optimum layout [24, 25].

The last thing that needs to be decided is the number of array waveguides (n_g), which has to be sufficiently large, such that almost all the light diffracted into the input-star coupler is collected by the array aperture. Also, as for a conventional diffraction grating, the full width at half-maximum of the converged beam is at best $\lambda f/n_0 n_g d_g$ because of the diffraction limit (being n_0 the refractive of the medium) and then, to resolve efficiently the wavelength channels (and taking into account Eq. (5.40)) the number of arrayed guides must meet:

$$n_g > \frac{\lambda}{m\Delta\lambda}. \tag{5.52}$$

In practice, this relation is usually fulfilled by a factor of 4.

5.6.2   Simulation of the AWG

The performance of the AWG demultiplexer can be simulated by using BPM. For doing that efficiently, it is recommended to break the demultiplexer into three parts: the input star coupler, the array of decoupled waveguides and the output star coupler. The input-star coupler is simulated with wide angle BPM for the central wavelength of the AWG only, since the star coupler performance is very insensitive to wavelength. For each of the array waveguides the power and phase are determined at a location where the waveguides are sufficiently decoupled. Figure 5.52 shows a schematic representation of the configuration used for the simulation of the first star coupler. The eigenmode of the input waveguide is used as the start-field of this simulation, which can be obtained by imaginary-distance BPM. The relevant output from this simulation is the modal field power carried by the individual arrayed guides, as well as their relative phase. Let us note that for a given longitudinal distance, the light in each waveguide has propagated a different path.

Then, the phase change in each array waveguide $\beta_g l_j$ is calculated by taking into account the optical path length of the waveguide. Although the propagation constant in the circled

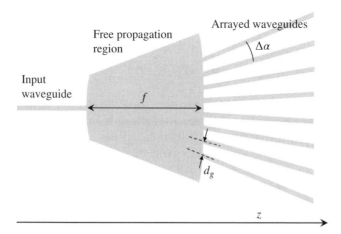

Figure 5.52   Input star coupler: geometry and relevant parameters

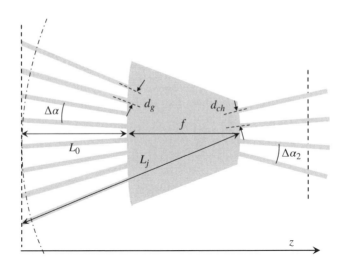

Figure 5.53   Schematic representation of the output-star coupler for BPM simulations

section of the waveguides (and in some tapered sections, if any) differs from that the straight sections, this difference can be neglected. Also, for sufficient large curvature radius, the power loss in the circled section can be ignored. The dispersive effect of the waveguide array is determined by calculating the phase change $\beta_g l_j$ as a function of the wavelength.

For the BPM simulation of the second star-coupler, the configuration represented in Figure 5.53 is used. The start field for this simulation consists of the eigenmodes of each of the array waveguides, taking into account the proper power and phase. As the input field is launched at a fixed longitudinal position (e.g. $z = 0$), the phase of the modal field for each arrayed waveguide must be corrected. The phase correction with respect to the position on a circle of radius $(L_0 + f)$ centred at $z = L_0 + f$ (see Figure 5.53, dashed-dotted line) can be easily calculated to be:

$$d\varphi_j = \beta_g \Delta L_j, \tag{5.53}$$

where ΔL_j is:

$$\Delta L_j = L_j - (f + L_0). \tag{5.54}$$

If the demultiplexer is adequately designed, for a particular channel wavelength the multiple inputs interfere in the free region and produce a focused beam at the corresponding channel guide. The modal power in the output channel is then computed by the overlapping between the numerically simulated field and the corresponding eigenmode, which is previously calculated and stored. For a correct eigenmode calculation, the tilting of the output guide must be considered. Finally, the spectral response of the demultiplexer is analysed by computing the modal power in each output channel waveguide as a function of the wavelength.

The AWG-demultiplexer example here presented is designed for demultiplexing eight channels, with a channel spacing of $\Delta\lambda = 3.4$ nm and to operate around a central wavelength of $\lambda_c = 1.55$ µm. To avoid overlapping between the outputs, a free spectral range of $\Delta\lambda_{FSR} = 28$ nm is at least required. For compatibility with silica optical fibres, the waveguide refractive indexes of the core and substrate are chosen to be $n_1 = 1.455$ and $n_0 = 1.450$, respectively. To assure mono-mode TE propagation, the input and output waveguides, as well as the arrayed waveguides, have a width of $w_g = 6 \, \mu m$, which gives a modal effective index of $N_{eff} = 1.453096$, measured at λ_c.

Taking into account the waveguide width, the separation between output waveguides has been taken as $d_{ch} = 20 \, \mu m$, which assures sufficient isolation between neighbour output wave-guides and thus low crosstalk between channels. With these data, the dispersion of the demul-tiplexer can be now calculated by Eq. (5.47), giving a value of $D = 5882$. Also, the array order m is obtained using expression (5.48), resulting in $m = 55$ and the correction on FSR is $\Delta \lambda_{FSR} = 28.2 \, nm$.

Choosing a separation between input waveguides in the output star of $d_g = 15 \, \mu m$, the output coupler star focus is calculated by means of Eq. (5.49) and gives a value of $f = 2334 \, \mu m$, where the central wavelength measured in the waveguide has been used ($\lambda_g = 1.0667 \, \mu m$).

Using the focal of the output star and the separation between input waveguides, the angular increment of the array branch angle can be calculated according to Eq. (5.50), resulting in $\Delta \alpha = 0.368°$. Also, from Figure 5.53, the angular separation between the output channels can be obtained, giving $\Delta \alpha_2 = 0.491°$.

The number of arrayed guides is set to $n_g = 32$ (four times the number of channels). The dimensions of the arrayed guides are obtained by choosing an input to output plane spacing of $L = 20 \, mm$, the angle $\alpha_1 = 45°$ and the length $l_1 = 20 \, mm$. With these data, it results on a minimum radius of curvature of $R_1 = 8.46 \, mm$ (curvature losses of 0.15 dB/cm) and a maxi-mum of $R_{32} = 9.29 \, mm$ (losses of 0.065 dB/cm). The length of the straight sections ranges from $S_1 = 3.35 \, mm$ to $S_{32} = 1.75 \, mm$. The longest arrayed waveguide has a total length of $l_{32} = 21.82 \, mm$ and an angle of $\alpha_{32} = 56.45°$.

Finally, for improving the insertion losses, linear tapers are designed at the input and output of the arrayed waveguides, with a taper width of $w_t = 14 \, \mu m$ and a taper semi-angle of $0.4°$. Also, the output channels are connected via tapered waveguides having a width of $w_{ch} = 10 \, \mu m$ and a semi-angle taper of $0.4°$.

The final design of the demultiplexer is depicted in Figure 5.54 and Table 5.1 lists its para-meters, as well as other relevant parameters used in the simulations.

Based on the design described here, the geometry of the input star coupler for BPM simu-lation is shown in Figure 5.55. For the simulation, a computational window of 6400 points in the x-direction is used, with a transversal step size of $\Delta x = 0.125 \, \mu m$. Figure 5.56 shows the

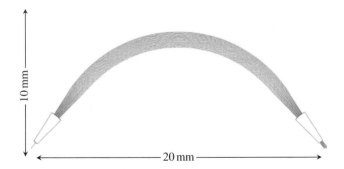

Figure 5.54 PHASAR layout following the parameters indicated in the text

Table 5.1   Parameters of the AWG-demultiplexer, besides the parameters used in the simulations

Variable	Symbol	Value
General parameters of the AWG		
Number of channels	n_{ch}	8
Central wavelength	λ_c	1.55 µm
Channel separation	$\Delta\lambda$	3.4 nm
Free spectral range	$\Delta\lambda_{FSR}$	28 nm
Waveguide parameters		
Refractive index of the core region	n_1	1.455
Refractive index of the cladding region	n_0	1.450
Waveguide width	w_g	6 µm
Waveguide effective index	N_{eff}	1.453096
Input star combiner		
Access guide length	—	500 µm
FPR entrance width	—	150 µm
FPR exit width	—	750 µm
FPR length	—	2334 µm
Arrayed waveguides		
Number of waveguides	n_g	32
Angular increment of array branch angle	$\Delta\alpha$	0.368°
Total length of the innermost array branch	l_1	20 mm
Angle of the innermost guide	α_1	45°
Curvature radius of the innermost guide	R_1	8.46 mm
Straight section length of the innermost guide	S_1	3.35 mm
Taper width	w_t	14 µm
Taper semi-angle	α_t	0.4°
Output star combiner		
Separation between input waveguides	d_g	15 µm
Separation between output waveguides	d_{ch}	20 µm
Angular separation between output channels	$\Delta\alpha_2$	0.491°
Taper width of output channels	w_{ch}	10 µm
Taper semi-angle of output channels	α_{ch}	0.4°
Focus	f	2334 µm
Dispersion of the demultiplexer	D	5882
Array order	m	55
Length of central access guide	—	500 µm
FPR entrance width	—	600 µm
FPR exit width	—	300 µm
Other parameters		
Central wavelength measured in the waveguide	λ_g	1.0667 µm
Ratio of effective indices guides/material	r	1.0013
Input to output plane spacing	L	20 mm
Length of the arrayed waveguides	l_j	20–21.82 mm
Angles of the straight section of the arrayed waveguides	α_j	45–56.45°
Simulation parameters		
2D Wide angle TE-BPM		
Transversal grid spacing	Δx	0.125 µm
Number of transversal points	N_x	6400

Table 5.1   (*continued*)

Variable	Symbol	Value
Longitudinal step	Δz	2 µm
Total propagation length	L_z	4000 µm
Reference index	N_{ref}	1.453

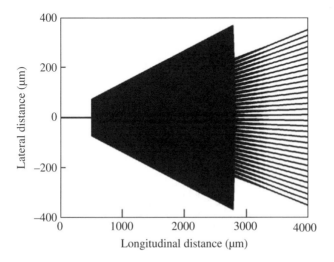

Figure 5.55   Refractive index map of the input star coupler. The free propagation region starts at 500 µm and has a length of 2334 µm. The exit connects to 32 tapered waveguides, with a taper angle of 0.4° and 14 µm of taper width. The angular separation between adjacent exit guides is 0.368°

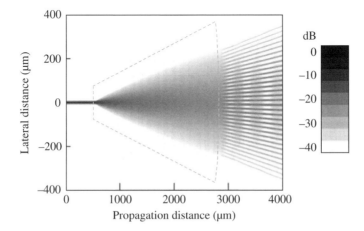

Figure 5.56   Field intensity distribution (log-scale) in the input-star coupler simulated by wide-angle BPM at $\lambda = 1.55$ µm

intensity distribution (in log-scale) of the input star, simulated using wide angle BPM algorithm with a propagation step size of $\Delta z = 2\,\mu m$. The light injection in the input waveguide corresponds to its fundamental mode obtained by imaginary-distance BPM. Once the light reaches the free propagating region, it diffracts, feeding the 32 tapered arrayed waveguides.

Figure 5.57 shows the transversal field intensity, measured at $970\,\mu m$ from the exit of the input star coupler, where it can be observed that the light has been correctly accommodated to the arrayed waveguides. The power in the guided mode of each array waveguide, computed as a fraction of the launched power in the entrance waveguide, is presented in Figure 5.58. The sum of the modal power carried by each individual guide gives the total power coupled to

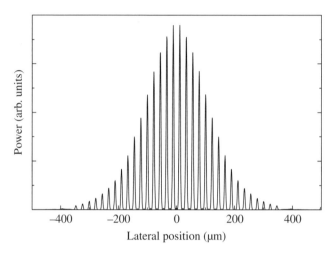

Figure 5.57   Distribution of the power at a distance of $970\,\mu m$ from the exit of the input star

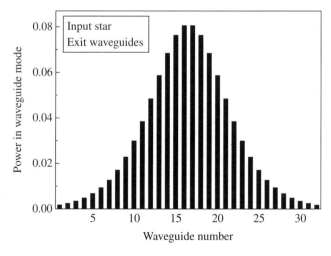

Figure 5.58   Modal power carried by each of the 32 exit waveguides

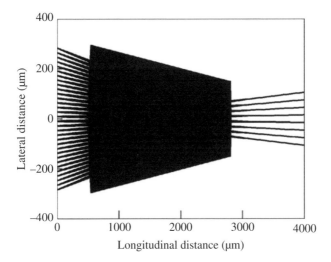

Figure 5.59 Layout of the output star coupler used for the simulations, with 32 input arrayed waveguides and 8 output channels

the array, resulting in a 97% with respect to the input power. This amounts to an insertion loss of $IL = -10 \log(0.97) = 0.13$ dB. If tapered waveguides are not used, the losses would increase up to 1.0 dB.

The geometry of the output star coupler, following the design parameters settled previously, is drawn in Figure 5.59. The arrayed waveguides enter the free region at the position $z = 500\,\mu\text{m}$. At the end of the start coupler, located at $z = 2834\,\mu\text{m}$, eight output tapered waveguides are positioned as equally spaced, which correspond to the eight channels of the AWG demultiplexer.

The start field of the output star simulation at $z = 0$ is the superposition of the eigenmodes of the individual array waveguides with the proper power and phase corrections. The tilting of each waveguide has also been taken into account for the correct simulation of the start field. Figure 5.60 shows the field intensity distribution (in the log-scale) for a wavelength of 1548.3 nm. Proper focusing of the output beam at the image plane can be observed for the output channel number 4, as would expected based on the central wavelength of 1550 nm and a channel spacing of 3.4 nm, fulfilling:

$$\lambda_i = \lambda_c + \left[i - \frac{n_{ch} + 1}{2} \right] \Delta\lambda. \tag{5.55}$$

The adjacent diffraction orders are also clearly visible. The total power (as a fraction of the start-field power) in the output channel guide is found to be 97%. This amounts to an insertion loss of $-10^{*}\log(0.97) = 0.13$ dB.

Finally, the performance of the AWG is evaluated by plotting the modal power in each output channel as a function of the wavelength. The results of this spectral response are presented in Figure 5.61. The simulations were taken at wavelength intervals of 0.5 nm, covering the 1535–1565 nm range. While the insertion loss for the central channels is low, the outer channels show the higher insertion losses and the numerical simulations give a value of 1.9 dB.

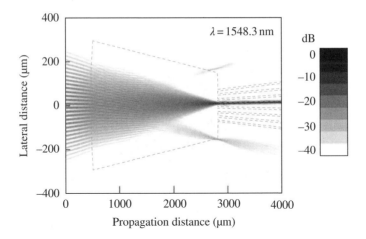

Figure 5.60 Field intensity distribution (log-scale) of the output star coupler simulation for an input wavelength of 1548.3 nm

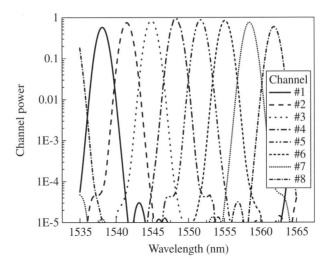

Figure 5.61 Spectral response of the AWG-demultiplexer

In addition, the crosstalk between channels, caused by extraneous light power from adjacent channels, is found to be better than −40 dB.

5.7   Mach-Zehnder Interferometer as Intensity Modulator

A Mach–Zehnder interferometer (MZI) is a device that measures phase shifts of a sample placed in one arm of the interferometer. In a Mach–Zehnder (MZ) modulator a phase shift is intentionally introduced to obtain an intensity modulation at the output. In integrated optic

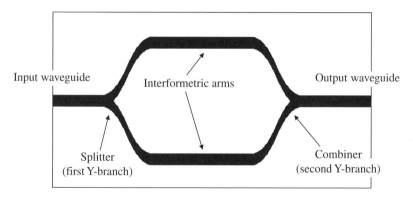

Figure 5.62   Scheme of a Mach–Zehnder interferometer using Y-junctions

devices, the interferometer consists of single mode optical waveguides fabricated from electro-optic materials (e.g. polymers, dielectric crystals or semiconductors). The simplest structure of a MZI has a single input waveguide that splits into two branches (interferometric arms) of straight waveguides of length L (where the phase changes will take place) and recombines again to a single output waveguide. The split of the input beam and the recombination of the light propagating along the two arms can be achieved either by MMI devices or using Y-branches [26, 27]. Figure 5.62 shows the layout of a MZI based on Y-junctions as splitter/combiner elements.

For the MZI configuration with symmetric Y-branches at the input and the output, zero phase difference between the arms equals a constructive interference, giving the ON state for intensity modulation. By contrast, a phase difference of π rad between both arms results in destructive interference with no light being coupled to the output waveguide (OFF state). For a symmetric MZI, the relation between the output and input power is a function of the phase difference $\Delta\varphi$ between both interferometer arms and is given by [28]:

$$P = \frac{P_0}{2}(1 + \cos\Delta\phi). \tag{5.56}$$

In an electro-optic MZI intensity modulator, the phase difference is induced by the application of a drive voltage to two parallel metallic electrodes in both sides of one interferometer arm as shown in Figure 5.63 (although other electrode configurations can be used, see Appendix K). The electric field across the optical waveguide, formed on an electro-optic substrate (for instance lithium niobate), induces a change in the effective index of the mode, which causes a phase shift proportional to the applied voltage and the electrode length (see Appendix K). The induced phase change can be expressed as:

$$\Delta\phi = \frac{V}{V_\pi}\pi, \tag{5.57}$$

where V is the applied voltage and V_π is the so-called half-wave voltage of the modulator, which is defined by the required voltage to induce π rad of phase shift between arms. This parameter

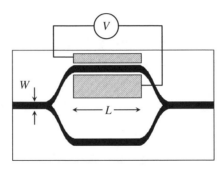

Figure 5.63 Scheme of an integrated electro-optic modulator based on MZI

depends on the electrodes configuration and the overlap between the induced electric field and the optical mode (see Appendix K).

The relationship between the modal effective index change ΔN_{eff} and the applied voltage V is expressed as:

$$\Delta N_{eff} = n^3 r \frac{V}{2G} \Gamma, \tag{5.58}$$

where n is the refractive index of the substrate, r is the relevant electro-optic coefficient, G is the gap between electrodes and Γ is the overlap between the externally applied electric field and the mode profile.

The behaviour of the intensity modulator based on MZI can be simulated by means of BPM. As an example, let us consider a MZI fabricated on a c-cut LiNbO$_3$ substrate, consisting on two opposite Y-junctions connected by a pair of parallel straight waveguides. We also assume that the channel waveguides forming the MZI structure has a Gaussian profile in the horizontal direction and a semi-Gaussian profile in the depth direction, typical of Ti-in diffusion waveguides fabricated on LiNbO$_3$ [29, 30]:

$$n(x,y) = n_s + \Delta n \cdot e^{-\left(\frac{x-x_0}{\sigma_x}\right)^2} e^{-\left(\frac{y}{\sigma_y}\right)^2} \quad (y < 0); \tag{5.59a}$$

$$n(x,y) = n_c \quad (y > 0). \tag{5.59b}$$

For the simulations, we will use $n_s = 2.203$, $\Delta n = 5 \times 10^{-3}$, $n_c = 1$, $\sigma_x = 2.5\,\mu\text{m}$ and $\sigma_y = 2.5\,\mu\text{m}$. For these parameters the waveguide is single moded for quasi-TM propagation at the operation wavelength of $\lambda = 0.633\,\mu\text{m}$. The TM-mode intensity profile can be approximated to a Gaussian–Hermite function with $w_x = 1.6\,\mu\text{m}$ and $w_y = 1.8\,\mu\text{m}$. Using a symmetric CS (coplanar strip) configuration for the electrodes with a gap of $G = 8\,\mu\text{m}$, an overlap integral of $\Gamma \approx 0.5$ is found (see Appendix K).

A first requirement for the MZI to work properly is that the light in the two parallel arms must propagate independently, that is, the parallel waveguides must be separated a distance long enough to avoid coupling between them. Figure 5.64 shows the coupling length L_c between two parallel channel waveguides having the parameters indicated previously, as a function of their separation distance, S, computed by semi-vectorial 3D-BPM. The coupling length

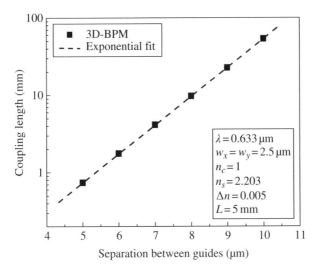

Figure 5.64   Coupling length between parallel waveguides as a function of their separation, computed by 3D-BPM

obtained by BPM simulations follows an exponential behaviour with the distance between guides:

$$L_c = A e^{S/p}, \tag{5.60}$$

with $A = 0.0102$ mm and $p = 1.17\,\mu$m. For a propagation length of 5 mm, which is adequate for having low drive voltage in the MZI modulator, the coupling is negligible when the distance between the channel waveguides is greater than $\sim 12\,\mu$m. According to this result, we set the separation of waveguides to 20 μm, which assures independent waveguides and it also allows enough space for the placement of metallic electrodes for electro-optic modulation.

Once the separation between the waveguides has been established, the Y-branch design can be accomplished. The S-bends of the Y-junctions should be as short as possible to build a compact device, but they must be designed to avoid excessive losses. Figure 5.65 plots the power carried by the waveguide mode at the end of an S-bend made by two opposite arcs of circumference, as a function of the length of the S curve for a lateral shift of 10 μm. Beyond an S-bend length of 1000 μm the losses are low enough and therefore Y-junctions with a length of 2000 μm will assure a good performance in terms of losses. This design corresponds to a radius of curvature of 100 mm.

Following the previously mentioned design parameters, Figure 5.66 presents the top view layout of the MZI for electro-optic modulation. The first Y-junction starts at 1 mm and the straight section is 5 mm long. The electrode's placement with respect to the interferometer is also shown in the figure (grey rectangles).

The performance of the MZI modulator is now determined by evaluating the fraction of power carried in the fundamental mode at the straight section end of the interferometer. The simulation starts by launching the TM-fundamental mode at $z = 0$, obtained by

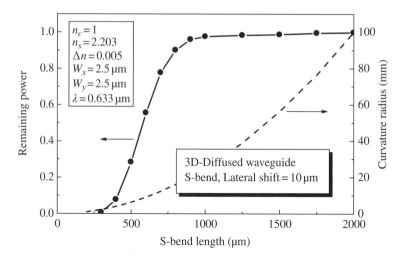

Figure 5.65 Relative output power carried in the waveguide mode as a function of the S-bend length, for a fixed lateral shift of 10 μm, computed by 3D-BPM. On the right scale the corresponding curvature radii of the bends are indicated

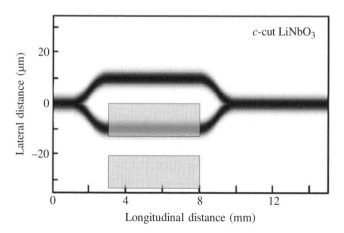

Figure 5.66 Top view layout of the electro-optic Mach–Zehnder intensity modulator. The CS electrodes (12 μm wide, separated 8 μm) are along one side of the straight sections and provide a vertical field to induce phase shift trough the r_{33} electro-optic coefficient of the c-cut LiNbO$_3$ substrate

imaginary-distance BPM. The output power as a function of the induced index change in one of the straight sections of the MZI is plotted in Figure 5.67 (squares).

The phase difference $\Delta\varphi$ of the optical modes at the end of the two arms of length L is related by the effective index change of the waveguide mode in one arm:

$$\Delta\phi = 2\pi\Delta NL/\lambda. \tag{5.61}$$

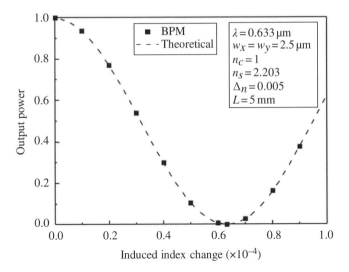

Figure 5.67 Performance of the electro-optic MZI modulator, analysed by 3D-BPM. The optical power at the end of the MZI structure is numerically evaluated as a function of the index change in one of the interferometers arm induced by an external electric field

At the wavelength of $\lambda = 0.633$ μm and for an electrode length of $L = 5$ mm, it indicates that an index change of $\Delta N = 0.633 \times 10^{-4}$ is required to induce a phase difference of π rad between the two straight sections of the interferometer ($\Delta N = \lambda/2L$). From Eqs. (5.61) to (5.56), the theoretical response of the MZI as a function of the effective index can be expressed by:

$$P = P_0 \cos^2(\pi \Delta N L / \lambda). \tag{5.62}$$

The dashed line in Figure 5.67 represents the theoretical curve expressed by this equation with $\Delta N = 0.633 \times 10^{-4}$. As can be seen, the results provided by 3D-BPM simulations (squares) are in excellent agreement with the expected theoretical values. Finally, the required voltage to induce that effective index change can be calculated as:

$$V_\pi = \frac{2G\Delta N}{n_e^3 r_{33} \Gamma}. \tag{5.63}$$

Using the these quoted values, this results in a half-wave voltage of $V_\pi = 6.3$ V.

The behaviour of the light along the integrated MZI can be conveniently visualized by 2D-BPM simulations, using the EIM to convert the 3D into the counterpart 2D-structure. Figure 5.68 (top) shows the light propagation along the MZI without external electric field, that is, with no phase shift between the upper and lower interferometric arms. When the fundamental mode is launched at the input straight section of the MZ, it splits symmetrically at the first Y-branch. The in-phase recombination at the second Y-branch recovers the guided mode at the straight section end of the MZ (on-state), with negligible loss power. By contrast, when a voltage of 6.3 V is applied between the metallic electrodes a phase shift of π rad is induced

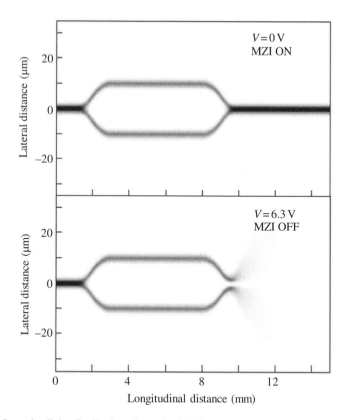

Figure 5.68 Intensity light distribution along the MZ interferometer computed by 2D-BPM. Upper figure: without external applied field (on state). Bottom figure: a voltage of 6.3 V is applied between the electrodes, which induces a phase shift of π rad between arms (off state)

between the parallel waveguides. The out of phase recombination at the second Y-branch avoids the light to be coupled into the guided mode and light spreads out through the substrate (Figure 5.68, bottom). The MZI is said to be on the off-state.

5.8 TE-TM Converters

TE to TM converters are important elements in some integrated optics devices. The polarization rotation can be accomplished using the electro-optic or acousto-optic effects, where the external field (either electric or a surface acoustic wave) induces non-diagonal elements in the permittivity tensor that causes coupling between the two polarized modes [31, 32]. If the two polarizations have different propagation constants, then a phase matching mechanism has to be introduced to synchronize the coupling to achieve maximum rotation. Also, passive TE/TM mode converters can be achieved by using a number of periodically arranged abrupt waveguide interfaces [33]. Very efficient polarization rotation is observed in these passive devices, despite the fact that isotropic materials can be used to fabricate the waveguide structures. Very compact passive devices have also been demonstrated using specially designed cross-sectional wave-guides [34].

In what follows, active and passive TE-TM mode converters are analysed using 3D-BPM. A full-vectorial approach is required as the performance of the devices is based on the different properties of the polarized modes and the fact that coupling between them is essential for allowing polarization conversion. Also, anisotropic treatment is required in analysing the electro-optic TE-TM converter, as it is based on anisotropic materials. The finite difference approach of 3D-anisotropic BPM is deduced in Appendix L. Where possible, a comparison between numerical results and theoretical results is made.

5.8.1   Electro-Optical TE-TM Converter

The TE-TM polarization converter example considered here consists of a LiNbO$_3$ channel waveguide in which an input TE mode is coupled to a TM mode by means of an electrode array that periodically modulates the off-diagonal elements of the permittivity tensor along the wave-guide axis [31] (see Figure 5.69). The coupling is due to the off-diagonal element induced by EO-effect (EO, electro-optic). To achieve total coupling the electrode period must be adjusted properly to match the phase since the propagation constants of the TE and TM modes are different. This difference is caused by both the material anisotropy and the vectorial property of the modal fields.

Using a c-cut LiNbO$_3$ substrate the off-diagonal element ε_{xz} induced by the linear electro-optic effect under an x-oriented external electric field E is given by (Appendix H):

$$\varepsilon_{xz} = -\varepsilon_0 \frac{n_o^2 n_e^2 r_{51} E}{1 - (n_o n_e r_{51} E)^2} \approx -\varepsilon_0 n_o^2 n_e^2 r_{51} E, \tag{5.64}$$

where n_o and n_e are the ordinary and extraordinary refractive indices and r_{51} is the electro-optic coefficient.

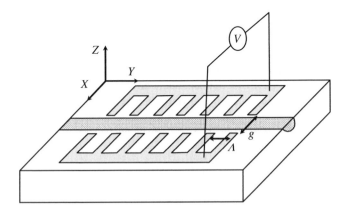

Figure 5.69   Schematic portray of the channel waveguide TE-TM electro-optic converter based on a c-cut LiNbO$_3$. The external electric field is directed along the x-axis, giving rise to the appearance of the off-diagonal element ε_{xz} in the permittivity tensor through the r_{51} electro-optic coefficient

To utilize directly the BPM equations in anisotropic media presented in Section 4.6 (and developed in Appendix L), we transform the coordinates as: $x \rightarrow y$, $y \rightarrow z$ and $z \rightarrow x$. Within this coordinate system, we have $n_{xx} = n_e = 2.2022$ (extraordinary index) and $n_{yy} = n_o = 2.2864$ (ordinary index) at $\lambda = 0.633 \, \mu m$. Also the off-diagonal elements n_{xy}^2 and n_{yx}^2 are calculated using formula (5.64). For a homogeneous electric field of $E = 1.4 \, V/\mu m$ and taking into account the value of $r_{51} = 28 \, pm/V$, we obtain $n_{xy}^2 = n_{yx}^2 = 0.001$.

To simulate the TE-TM waveguide converter we consider that the channel waveguide consists of a graded index profile, typical of Ti-in diffusion waveguides fabricated on LiNbO$_3$. The channel waveguide is modelled by a Gaussian index profile in the horizontal direction and a semi-Gaussian index profile in the depth direction [29, 30]:

$$n(x,y) = n_s + \Delta n \cdot e^{-(x/\sigma_x)^2} e^{-(y/\sigma_y)^2} \qquad (y < 0), \qquad (5.65a)$$

$$n(x,y) = n_c \qquad\qquad\qquad\qquad (y > 0), \qquad (5.65b)$$

with $\Delta n = 5 \times 10^{-3}$, $\sigma_x = 3 \, \mu m$, $\sigma_y = 4 \, \mu m$ and $n_s = 2.2022$ and $n_s = 2.2864$, for n_{xx} and n_{yy}, respectively. The refractive index along the propagation direction corresponds also to the ordinary axis and then $n_{zz} = 2.2864$. The waveguide is monomode at the working wavelength of $\lambda = 0.633 \, \mu m$. We also assume that the off-diagonal elements $n_{xy}^2 (= n_{yx}^2)$ follows a step-like function between 0 and +0.001 along the waveguide axis, with the period of the electrode grating. This period, Λ, is determined by the phase-matching condition [35]:

$$\Lambda = \frac{\lambda}{|N_{TE} - N_{TM}|}, \qquad (5.66)$$

where N_{TE} and N_{TM} are the effective indices of the (quasi-) TE and TM polarized modes, respectively, which are calculated by the imaginary-distance BPM. At the operating wavelength, the effective indices of the TE and TM modes are 2.20381 and 2.28806, respectively, which correspond to an electrode grating period of $\Lambda = 7.51 \, \mu m$.

Figure 5.70 shows the BPM results of the TE and TM powers as a function of the propagation distance, using a step size of $\Delta z = 0.5 \, \mu m$. For the reference index we used an average between the ordinary and extraordinary refractive indices of the substrate. A complete power conversion from the input TE mode to the TM mode is achieved after a propagation distance of 2250 μm. At the coupling length, only 4×10^{-5} of the incident power remains in the incident polarization. The efficiency of the polarization converter (ratio between the TM power with respect to the incident TE power) at the phase matching condition is accurately fitted by the expected sin^2 function:

$$\eta = \sin^2(\kappa L), \qquad (5.67)$$

where κ is the coupling coefficient. It is worth commenting that power is conserved during the process of TE-TM conversion.

For a fixed electrode period, the phase matched condition (5.66) is fulfilled only at a particular wavelength. This fact can be utilized to perform wavelength filtering with the EO-mode converter. The spectral response of this device, analysed by anisotropic-BPM, is

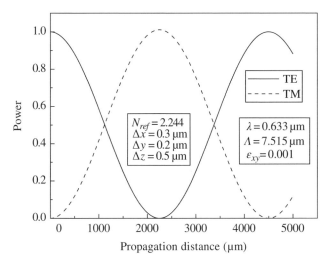

Figure 5.70 Power evolution of the TE and TM modes in a waveguide electro-optic polarization mode converter

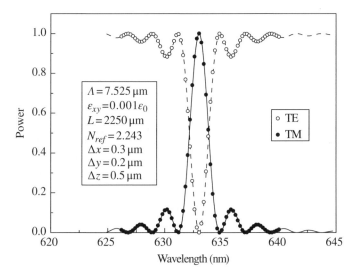

Figure 5.71 Spectral response of the EO-polarization mode converter calculated by 3D-anisotropic-BPM. Continuous and dashed lines represent the theoretical predictions

presented in Figure 5.71 (circles). The theoretical efficiency conversion of the device is expressed as [36]:

$$\eta = \frac{\sin^2\left\{\kappa L\sqrt{1+(\delta/\kappa)^2}\right\}}{1+(\delta/\kappa)^2}, \tag{5.68}$$

where:

$$\delta = -\frac{\pi}{\Lambda}\frac{\Delta\lambda}{\lambda_0 + \Delta\lambda}, \tag{5.69}$$

λ_0 being the wavelength corresponding to the phase matched condition. The continuous line is the plot of the function (5.68) using the parameters used in the BPM simulations, showing excellent accordance with the numerical results provided by the anisotropic BPM analysis.

Also, a comparison of the results provided by anisotropic-3D-BPM can be made if the EO coefficient varies sinusoidally along the propagation direction. In this case, the BPM simulation gives a coupling length of $L_c = 1430$ μm. On the other hand, for a sinusoidally varying EO coefficient along the propagation axis, the coupled mode theory provides us an expression for the coupling coefficient [35, 37], given by:

$$\kappa = \alpha(\pi/2\lambda)n^3 r_{51} E, \tag{5.70}$$

where n is the substrate refractive index (geometric average of ordinary and extraordinary refractive indices) and α is a parameter that measures the overlap between the external electric field and the TE and TM waveguide modes. Using the values previously quoted (and assuming $\alpha = 1$), it gives a coupling efficiency value of $\kappa = 1100$ m^{-1} and therefore a coupling length of is $L_c = \pi/2\kappa = 1429$ μm. This value agrees favourably with the anisotropic-BPM numerical simulation.

5.8.2   Rib Loaded Waveguide as Polarization Converter

A periodic loaded rib waveguide, as shown in Figure 5.72, can be also used to achieve polarization conversion [33, 38]. At variance with the previous example of EO-TE/TM converter, which requires the existence of non-diagonal elements on the permittivity tensor, the polarization rotation of the asymmetrically loaded rib waveguide is purely due to the vectorial properties of the guided modes in the structure. In fact, the device can be fabricated using isotropic materials. Although an analysis based on coupled-mode theory can be performed [39], 3D-BPM simulations provide further insights of the device performance; in particular, it can predict the scattering losses at the junctions [40, 41].

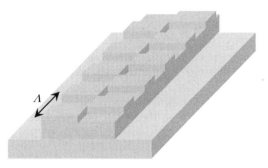

Figure 5.72   Schematic picture of the asymmetric periodic loaded rib waveguide, to be used as TE-TM polarization converter

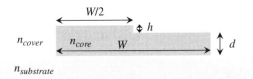

Figure 5.73   Cross section of the asymmetric loaded rib waveguide

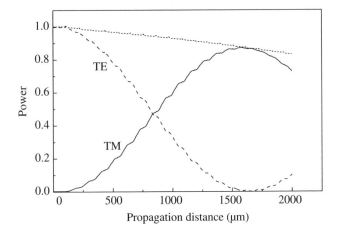

Figure 5.74   Power of the quasi-TE (dashed line) and quasi-TM guided (continuous line) modes as a function of the propagation distance. The total power in the rib waveguide is also shown by dotted line

The cross-section of the structure to be analysed by vectorial 3D-BPM is indicated in Figure 5.73. The waveguide consists of a homogeneous substrate with refractive index $n_{subs} = 1.45$ and a core refractive index $n_{core} = 1.54$, where the cover region is air ($n_{cover} = 1$). The width and the thickness of the core region are $W = 3.0\,\mu m$ and $d = 1.0\,\mu m$, respectively. The width of the dielectric loading is $W/2 = 1.5\,\mu m$ and the height of the load is $h = 0.1\,\mu m$. The device is examined at a wavelength of $\lambda = 1.3\,\mu m$.

At the working wavelength, the difference of effective indices between the quasi-TE and quasi-TM modes is $\Delta N_{TE\text{-}TM} = 6.08 \times 10^{-3}$, obtained by imaginary-distance BPM. Applying the formula (5.66), the period of the load for maximum polarization conversion is calculated to be $\Lambda = 214\,\mu m$.

The performance of the polarization conversion of the periodic asymmetric loaded rib waveguide is now studied by vectorial 3D-BPM. To do that, the fundamental quasi-TE mode, obtained by imaginary propagation BPM, is launched and the power associated at the quasi-TE and quasi-TM modes are monitored as a function of the propagation distance, using a reference index of 1.473 and a propagation step of $\Delta z = 0.1\,\mu m$. Figure 5.74 shows the results of the numerical simulation, which indicates that at a distance of $1660\,\mu m$ all the remaining power in the rib waveguide corresponds to the quasi-TM mode. Also, as indicated by dotted line in the figure, there is a clear loss of total power. This loss is due to scattering at the abrupt discontinuities of the periodic loaded waveguide along the propagation direction, which induces coupling from guided- to radiation-modes. This loss is calculated to be 0.4 dB/mm.

5.9 Waveguide Laser

Due to the light confinement in optical waveguides structures, waveguide lasers generally exhibit high optical gain and a low threshold [42–44]. This is already true for planar waveguide lasers, but particularly in channel waveguides, which provide a two-dimensional light confinement (Figure 5.75). Such waveguide lasers are very promising, as they offer a high potential for miniaturization and can lead to more complex integrated optical devices, featuring multiple active and passive elements on a small chip [45, 46].

Besides semiconductor lasers, crystalline rare-earth (RE) doped dielectric oxides are well suited for the realization of lasers with high frequency stability or high (peak) power, since they possess sharp emission peaks and high peak cross sections. In particular, end-pumped waveguide lasers results in an excellent mode overlap of pump and laser light, which in turn provides compact and high efficient devices.

Here, we present the use of active-BPM for simulating end-pumped waveguide lasers. At variance with the mode-overlap formalism [47, 48] (see Appendix N), the BPM-based method allows us to simulate z-variant structures, such as tapered structures, Mach–Zhender or Y-branch devices [49]. The algorithm is based on an iterative approach, wherein the pump and laser fields are propagated back and forth inside the cavity waveguide until a self-consistent solution is found. Strictly speaking, interference between the forward and backwards fields would lead to longitudinal spatial hole-burning where the holes are a half-wavelength apart. Nevertheless, here we will consider long waveguide cavities ($\lambda \gg L$), where a large number of longitudinal modes can simultaneously oscillate due to small inhomogeneities in the medium. In this case, it is possible to average the intensity along the waveguide in such a way that the intensity at each point is taken as the sum of the intensities of the forward and backward beams [49]. For short cavities (microchip lasers), one must sum the forward and backward fields and then take the square to obtain the intensity at each grid point. This approach (that takes into account the phase of the fields) allows us to describe Fabry–Perot effect in the cavity, such as the selection of a particular resonant longitudinal cavity mode [50].

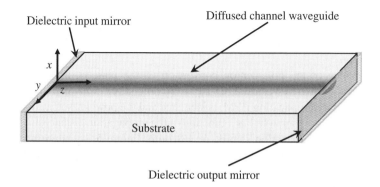

Figure 5.75   Waveguide laser

5.9.1   Simulation of Waveguide Lasers by Active-BPM

A typical end-pumped waveguide laser is formed by a straight RE doped channel waveguide, where two dielectric mirrors are butted to the ends of the waveguide to provide feedback (Figure 5.76). The mirrors M_1 and M_2 have partial reflectances R_{1P}, R_{1L}, R_{2P} and R_{2L} at the pump and laser wavelengths (λ_P and λ_L, respectively). The cavity is considered to be a 'double pass' cavity, so that both pump and laser beams travel forward and backward through the active medium. The analysis of the laser performance by means of BPM is based on the propagation of the forward and backward fields along the longitudinal direction for both pump and laser beams, using the (scalar) Active-BPM presented in Section 4.4, besides the rate equations describing the population dynamics of the active ions (see Appendix M). In contrast to the waveguide amplifier where only forward fields are considered, the waveguide laser modelling in addition requires simulation of the back-propagation of both pump and laser beams. The forward fields, u_P^+ and u_L^+ and the backward fields, u_P^- and u_L^-, evolve according with the dynamic equations of the active ions, the injected pump power and the partial reflectances of the mirrors. The plus and minus signs used in the superscript denote forward and backward fields, respectively, while the subscripts P and L indicate the pump and laser fields, respectively. The method presented here is based on an iterative approach, wherein the pump and laser fields are propagated back and forth inside the cavity waveguide by means of Active-BPM until a self-consistent solution is found [49].

The algorithm for the simulation performance of end-pumped waveguide lasers proceeds following this sequence:

i. The initial forward field for the pump, usually taken as the fundamental waveguide eigenmode, is launched at $z = 0$ (position of the input mirror), with the correct amplitude to take account the absolute input pump power (in Watts units). If the injected pump power in the waveguide is $P_{input,P}$ and the fundamental eigenmode (obtained by imaginary-distance BPM) is $\psi_P(x,y)$, then the input forward field $u_P^+(x,y)$ for the pump is calculated by:

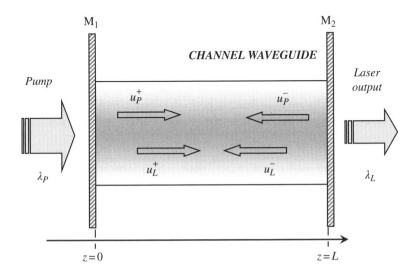

Figure 5.76   Pump and laser fields inside the waveguide laser double-pass cavity

$$u_{input,P}^{+}(x,y) = \sqrt{\frac{P_{input,P}}{\displaystyle\iint |\psi_P(x,y)|^2\,dxdy}}\,\psi_P(x,y). \qquad (5.71)$$

The forward field of the laser radiation u_L^+ is initialized (at $z = 0$) at an arbitrarily low power value (for instance, $P_{initial,L}$ around 1 μW). In any case, the solution at the steady state is independent of this initial value. If the initial field profile corresponds to the fundamental mode $\psi_L(x,y)$ at λ_L, then the initial laser field is calculated as:

$$u_L^+(x,y,0) = \sqrt{\frac{P_{initial,L}}{\displaystyle\iint |\psi_L(x,y)|^2\,dxdy}}\,\psi_L(x,y). \qquad (5.72)$$

The backward fields for the pump and laser beams are initialized to zero.

ii. The induced transition rate $R(x,y,z)$ (absorption process) at the transversal plane z is calculated in terms of the intensity of the (actual) forward pump beam $I_P^+(x,y,z)$ and the (previously stored) backward pump beam $I_P^-(x,y,z)$ at the longitudinal position $z = m\Delta z$, being Δz the propagation step and m an integer number for the longitudinal discretization, using (see Appendix M):

$$R(x,y,z) = \frac{\sigma_{abs}\lambda_P}{hc}\left[I_P^+(x,y,z) + I_P^-(x,y,z)\right], \qquad (5.73)$$

where h is the Planck's constant, c is the speed of light in free-space and σ_{abs} is the absorption cross section corresponding to the pumping transition at λ_P.

In a very similar way, the induced transition rate $W(x,y,z)$ (stimulated emission process) is calculated by (see Appendix M):

$$W(x,y,z) = \frac{\sigma_{emi}\lambda_L}{hc}\left[I_L^+(x,y,z) + I_L^-(x,y,z)\right], \qquad (5.74)$$

where $I_L^+(x,y,z)$ and $I_L^-(x,y,z)$ are the local intensity of the (actual) forward laser beam and the (stored) backward laser beam at position $z = m\Delta z$, respectively, and the spectroscopic parameter σ_{emi} is the emission cross-section corresponding to the laser transition at λ_L.

iii. Taking into account the absorption and emission rates, the populations of the relevant levels are calculated. In the case of Nd^{3+} ions, which form an ideal four-level system, the populations of the fundamental level 1 and the upper laser level 3 are given by (see Appendix M):

$$N_3(x,y,z) = \frac{R(x,y,z)}{R(x,y,z) + W(x,y,z) + 1/\tau_{exp}}N_T(x,y,z); \qquad (5.75a)$$

$$N_1(x,y,z) = N_T(x,y,z) - N_3(x,y,z), \qquad (5.75b)$$

where τ_{exp} is the experimental lifetime of level 3. Let us note that the model includes the possibility of inhomogeneous doping of the active ions, as $N_T(x,y,z)$ is allowed to be position dependent.

iv. The imaginary part of the refractive index distributions, for both pump and laser wavelengths, κ_P and κ_L, respectively, are calculated using:

$$\kappa_P(x,y,z) = \frac{\lambda_P}{4\pi}[\sigma_{abs}N_1(x,y,z) + \tilde{\alpha}_P]; \tag{5.76a}$$

$$\kappa_L(x,y,z) = \frac{\lambda_L}{4\pi}[-\sigma_{emi}N_3(x,y,z) + \tilde{\alpha}_L], \tag{5.76b}$$

where the parameters $\tilde{\alpha}_P$ and $\tilde{\alpha}_L$ (units of m^{-1}) take into account the intrinsic propagation losses of the passive waveguide at the pump and laser wavelengths, respectively.

v. Using Active-BPM, pump and laser fields (u_P^+ and u_L^+) are forward propagated simultaneously from the transversal plane z to the next plane $z + \Delta z$ (from m to $m + 1$) considering complex refractive indices in the wave equation.

vi. Calculate and store the intensity of the forward fields at the actual longitudinal position $z = m\Delta z$, according to:

$$I_P^+(x,y,z) = |u_P^+(x,y,z)|^2; \tag{5.77a}$$

$$I_L^+(x,y,z) = |u_L^+(x,y,z)|^2. \tag{5.77b}$$

Steps (ii–vi) are performed in a loop until the beams reach the output mirror, which is located at $z = L$ (which corresponds to the index longitudinal position $m = L/\Delta z$). The laser output power is then calculated by:

$$P_{Output} = (1 - R_{2,L}) \int\int |u_L^+(x,y,L)|^2 dxdy, \tag{5.78}$$

Now, the backward propagation of both fields is performed.

vii. The backward calculation starts by setting the fields at position $z = L$ ($m = L/\Delta z$) with the following boundary conditions:

$$u_P^-(x,y,L) = \sqrt{R_{2,P}} u_P^+(x,y,L); \tag{5.79a}$$

$$u_L^-(x,y,L) = \sqrt{R_{2,L}} u_L^+(x,y,L). \tag{5.79b}$$

viii. The induced transition rate $R(x,y,z)$ at the transversal plane z is calculated in terms of the intensity of the (actual) backward pump beam $I_P^-(x,y,z)$ and the (previously stored) forward pump beam at position z $I_P^+(x,y,z)$ using Eq. (5.73). In a very similar way, the induced transition rate $W(x,y,z)$ is calculated by Eq. (5.74), where $I_L^-(x,y,z)$ and $I_L^+(x,y,z)$ are now the intensity of the (actual) backward laser beam and the (stored) forward laser beam at position $z = m\Delta z$, respectively.

ix. The populations of the level 1 and level 3 are calculated using Eq. (5.75a–b).

x. The imaginary part of the refractive index distribution, for both pump and laser wave-
    lengths, is calculated by Eq. (5.76a–b).
xi. Using active-BPM, pump and laser fields are backward propagated simultaneously from
    the transversal plane z to the previous plane $z - \Delta z$ (from m to $m - 1$) considering complex
    refractive indices in the wave equation.
xii. Calculate and store the intensity for the backward fields, according to:

$$I_P^-(x,y,z) = \left| u_P^-(x,y,z) \right|^2; \tag{5.80a}$$

$$I_L^-(x,y,z) = \left| u_L^-(x,y,z) \right|^2. \tag{5.80b}$$

Steps (viii–xii) are performed in a loop until the beams reach the input mirror located at
$z = 0$ $(m = 0)$.

xiii. The forward calculations start once again by re-setting the fields at position $z = 0$ $(m = 0)$,
    with the following boundary conditions:

$$u_P^+(x,y,0) = u_{input,P}^+(x,y) + \sqrt{R_{1,P}}\,u_P^-(x,y,0); \tag{5.81a}$$

$$u_L^+(x,y,0) = \sqrt{R_{1,L}}\,u_L^-(x,y,0). \tag{5.81b}$$

The procedure (ii–xiii) iterates until the laser output power differs less than 0.5% between
two consecutive runs. This is accomplished when forward and backward fields are linked
according to the boundary conditions at the mirrors (Eqs. (5.79a–b) and (5.81a–b)). This
method gives a very good approximation, although it is an iterative procedure where the
convergence is fast for high pump powers and it is low for values near the laser threshold.
A flow diagram of the algorithm described here is presented in Figure 5.77.

5.9.2   Performance of a Nd³⁺-Doped LiNbO₃ Waveguide Laser

The end-pumped waveguide laser, modelled here by active-BPM, is based on a longitudinal
invariant channel waveguide made in a LiNbO$_3$ substrate activated with Nd^{3+} RE ions [51,
52]. It is assumed that the refractive index of the waveguide beneath the surface follows Gaussian profiles along the x and y directions, typical of waveguides fabricated by thermal metallic
in-diffusion and is given by:

$$n(x,y) = \begin{cases} n_c & x \geq 0 \\ n_s + \Delta n \cdot \exp\left[-(x/W_x)^2\right] \exp\left[-(y/W_y)^2\right] & x < 0 \end{cases} \tag{5.82}$$

with $n_s = 2.1727$ at $\lambda = 0.81\ \mu\text{m}$ and $n_s = 2.1546$ at $\lambda = 1.08\ \mu\text{m}$ (extraordinary refractive indices), $\Delta n = 0.004$ and $W_x = W_y = 4\ \mu\text{m}$. The cover is assumed to be air ($n_c = 1.000$). This waveguide has been designed to be monomode at 0.81 and 1.08 μm. The refractive index map and
the mode intensity profiles at the pump and laser wavelengths, obtained by scalar Im-Dis-BPM,
are presented in Figure 5.78. The overlap between theses modes is $\Gamma = 0.92$. These fundamental
modes at 0.81 and at 1.08 μm are used to start the propagation in the simulations, where the

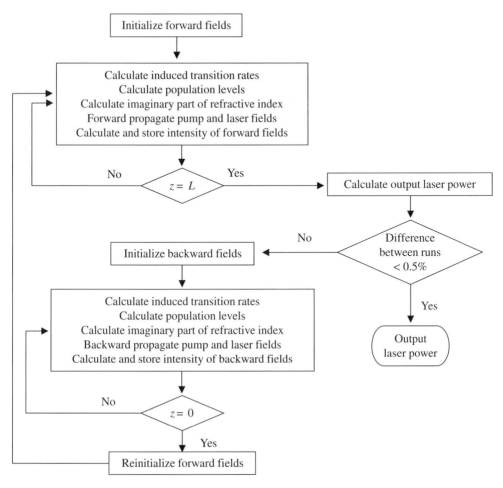

Figure 5.77 Flow diagram used in the simulation of waveguide lasers based on active-BPM

input power for the pump field is conveniently selected at each simulation run. Table 5.2 summarizes the waveguide parameters used in the simulations of the channel waveguide laser.

The LiNbO$_3$ substrate is assumed to be homogeneously doped with a 0.2% of Nd^{3+} ions, which corresponds to a concentration of 3.8×10^{19} ions/cm^3. This concentration is low enough to avoid luminescent quenching effects. Table 5.3 presents the relevant spectroscopic data of the Nd^{3+} ions in LiNbO$_3$ required to evaluate the waveguide laser performance [53], which includes the values of the absorption and emission cross sections (π-polarized absorption and emission) and the lifetime of the upper laser level.

In the simulations here presented, it is assumed that the M$_1$ mirror (input mirror) transmits a 100% at the pump wavelength and it reflects a 100% at the laser wavelength. Also, the M$_2$ mirror (output mirror) reflects a 100% the pump radiation and it is partially reflectant at the laser wavelength. Thus, the laser acts as a double pass cavity, where the pump radiation makes a complete forward–backward trip. A first rough estimation of the cavity length L can be obtained by imposing that a ~90% of the input power is absorbed by the Nd^{3+} ions.

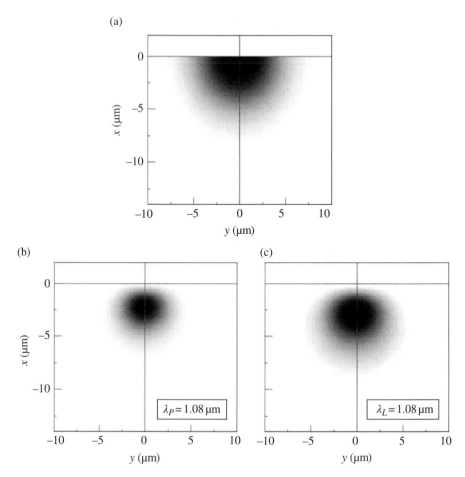

Figure 5.78 (a) Refractive index map of the channel waveguide. Intensity map of the waveguide modes for the pump (b) and laser (c) wavelengths

Table 5.2 Waveguide parameters of the channel waveguide laser

Parameter	Symbol	Value
Waveguide depth in x direction	W_x	4 μm
Waveguide depth in y direction	W_y	4 μm
Waveguide losses	$\tilde{\alpha}_P$, $\tilde{\alpha}_L$	0.3 dB/cm
Cover refractive index	n_c	1 (air)
Substrate refractive index (extraordinary index)	$n_s(\lambda_P)$	2.1727
	$n_s(\lambda_L)$	2.1546
Maximum index increase	Δn	0.004

Table 5.3 Relevant spectroscopic parameters used in the Nd^{3+}-activated waveguide laser in LiNbO$_3$

Parameter	Symbol	Value
Pump wavelength	λ_P	0.81 μm
Laser wavelength	λ_L	1.08 μm
Absorption cross section	σ_{abs}	2.7×10^{-20} cm^2
Emission cross section	σ_{emi}	2.7×10^{-19} cm^2
Lifetime of level (3)	τ_{exp}	90 μs
Active ions/unit volume	N_T	0.2% (3.8×10^{19} cm^{-3})

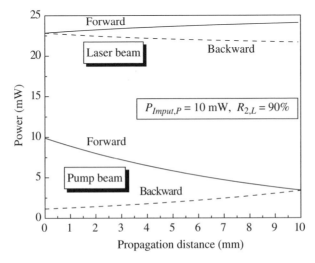

Figure 5.79 Evolution of the forward and backward radiation for the pump and laser fields, for an input pump power of 10 mW, an output mirror reflectance of 0.9 at the laser wavelength and a cavity length of 1 cm

Using the absorption cross section and the concentration of active ions presented in Table 5.3, it gives:

$$1 - e^{-2\sigma_{abs}N_T L} \approx 0.9 \qquad \rightarrow \qquad L \approx 1.1 \text{ cm}$$

Figure 5.79 shows the evolution of the pump and laser powers inside the waveguide laser cavity for an input power of 10 mW and an output mirror reflectance of $R_{2,L} = 0.9$, obtained by the algorithm explained in the previous section for a cavity length of $L = 1$ cm. It can be seen that the forward pump power starts with 10 mW at $z = 0$ (because $R_{1,P} = 0$, Eq. (5.81a)) and decays along the forward trip. Once it reaches the output mirror at $z = 1$ cm, it reflects totally (Eq. (5.79a) with $R_{2,P} = 1$). The backward pump power continues the decay along its backward trip. The pump power decrease is regulated by the induced absorption of the Nd^{3+} ions and by the intrinsic losses of the waveguide. For the parameters indicated in Table 5.2, the fraction of useful absorbed power is 0.76 with respect to the input power. On the other hand, the power of

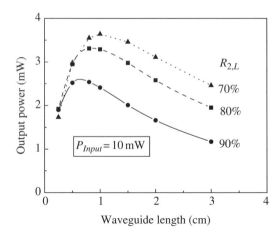

Figure 5.80 Output laser power as a function of the cavity length for an input pump power of 10 mW (absorbed pump power 7.6 mW), for three different output mirror reflectances

the forward laser radiation increases along the forward trip due to the amplification by stimulated emission. After losing a 10% of its power at the output mirror M_2 (due to the 90% reflectance mirror, Eq. (5.79b)), the laser power continues growing along its backward trip. Once the radiation reaches the input mirror (position $z = 0$), the forward and backward laser fields should be coincident and it is imposed by the boundary condition in the simulation (Eq. (5.81b), with $R_{1,L} = 1$), indicating that the iteration has correctly reached the solution.

Although a rough estimation of the cavity length has been obtained on the basis of the fraction of absorbed pump power, the optimum cavity length depends both on the input pump power and on the reflectance of the output mirror (and also on the intrinsic waveguide losses). Figure 5.80 plots the output power as a function of the waveguide length for three different output mirror reflectances. As can be seen, the maximum laser power is reached at different lengths depending on the mirror reflectance. In what follows, we have chosen a cavity length of $L = 1$ cm, close to the optimum length for output mirror reflectances of 70 and 80%.

The waveguide laser performance is fully described by the plot of the laser output power as a function of the absorbed pump power. Figure 5.81 presents the results obtained by three different output mirror reflectances, where each laser curve is characterized by its threshold and its slope efficiency. The threshold is defined as the pump power that needs to be absorbed by the active ions in the waveguide laser to reach oscillation, while the slope efficiency is defined as the slope of the curve beyond the threshold. It is seen from Figure 5.81 that the threshold for laser oscillation decreases as the reflectance increases. On the contrary, the slope efficiency increases as the reflectance of the output mirror decreases. In general, it is desirable to design a waveguide laser with low threshold and high slope efficiency. Nevertheless, the optimum mirror reflectance cannot be determined unless the range of the pump power is established.

This fact can be better visualized by plotting the laser output power as a function of the output mirror reflectance. Figure 5.82 displays these results for three different input pump powers. The plot shows that, for each pump power, there exists an optimum mirror reflectance where the laser output is maximum, indicating that the maxima are reached at different mirror reflectances

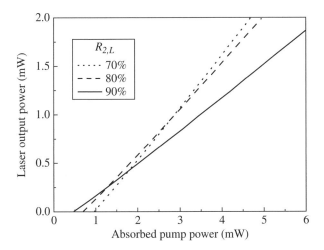

Figure 5.81 Laser output power as a function of the absorbed pump power, for three different output mirror reflectances. Waveguide length: $L = 1$ cm

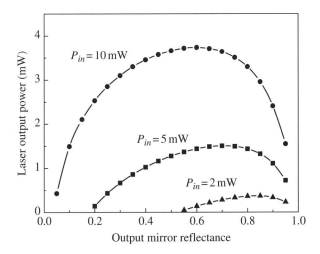

Figure 5.82 Laser output power as a function of the output mirror reflectance, for three different input pump powers. The absorbed pump power is 0.76 of the input power. Waveguide length: $L = 1$ cm

depending on the pump conditions. As the input pump power increases, the optimum mirror reflectance decreases. Thus, for optimizing the waveguide laser performance, the input pump power should be also taken into consideration.

Finally, the threshold and the slope efficiency as a function of the output mirror reflectance are plotted in Figure 5.83, where symbols indicate the results provided by the BPM algorithm. These results can be compared by an approximate analysis based on the mode overlap method [54]. This analysis assumes fixed transversal fields along the propagation (and thus z-invariant waveguides) and can be applied only to four-level atomic systems. As the waveguide laser

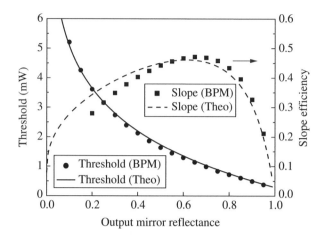

Figure 5.83 Absorbed pump power at the threshold (circles and continuous line) and slope efficiency (squares and dashed line) as a function of the output mirror reflectance. Waveguide length: $L = 1$ cm

example presented here fulfils these assumptions, a comparison between both methods is possible. From the formalism of mode overlap, the threshold can be approximated by [54]:

$$P_{th} = \frac{hc}{\eta_P \lambda_P \sigma_{emi} \tau_{exp}} \frac{\delta}{2} A_{eff},$$ (5.83)

where η_P is the pump quantum efficiency, that is the probability for one absorbed pump photon to create an excited ion in the upper laser level and δ is the round-trip cavity loss exponential factor, which is given by:

$$\delta = 2\widetilde{\alpha}_L L - \ln(R_{1,L} R_{2,L}).$$ (5.84)

The effective pump area A_{eff}, defined as the inverse of the normalized pump and laser distributions overlap integral in the active waveguide, is calculated by:

$$A_{eff}^{-1} = \frac{\iint |\psi_P(x,y)|^2 |\psi_L(x,y)|^2 dxdy}{\iint |\psi_P(x,y)|^2 dxdy \iint |\psi_L(x,y)|^2 dxdy},$$ (5.85)

where $\psi_P(x,y)$ and $\psi_L(x,y)$ are the eigenmodes of the pump and laser fields, respectively.

For the case of Nd^{3+} ions in $LiNbO_3$, the pump quantum efficiency η_P is close to unity, while the effective pump area A_{eff} computed by using the eigenmodes shown in Figure 5.78 gives a value of 40.7 μm^2. With these data and the data listed in Table 5.3, the threshold can be now calculated using Eq. (5.83) and the results are plotted in Figure 5.83 (continuous line). As can be seen, the approximated analysis based on the mode overlap formalism gives very close results to those obtained by the BPM method.

On the other hand, the slope efficiency s may be approximated by [55]:

$$s = \eta_P \frac{(1 - R_{2,L}) \lambda_P}{\delta \lambda_L}. \qquad (5.86)$$

The dashed line in Figure 5.83 displays the results using this approximated expression for the slope efficiency, also in good agreement with the values provided by the iterative algorithm based on BPM.

5.10 SHG Using QPM in Waveguides

Efficient generation of coherent visible radiation in solid-state devices is of considerable technological importance and one attractive approach is the frequency doubling of infrared diode laser radiation. To achieve high efficiency, it is advantageous to use optical waveguide structures that confine the light, which allows maintaining a high intensity over a long interaction path [56]. However, in general, the phase condition for efficient frequency doubling is difficult to fulfil and only certain wavelengths, specific for each material, can be phase matched in the conventional manner, using the birefringence. To address the inability of materials to phase match birefringently, quasi-phase-matching (QPM) [57] has been used for efficient second-harmonic generation (SHG) of visible radiation in LiNbO$_3$, polymers, KTP, LiTaO$_3$ or GaAs materials. QPM techniques [58] are achieved by introducing a grating consisting on periodic variation of the second-order non-linear coefficient along the propagation direction to change the relative phase difference between the fundamental and harmonic fields.

Here we present an example of the use of NL-BPM (non-linear beam propagation method) (developed in Section 4.5) for the simulation of SHG in non-linear planar waveguides using QPM techniques. In the case of waveguides, the mismatch wavenumber is defined as:

$$\Delta k = \beta_{2\omega} - 2\beta_\omega - \frac{2\pi}{\Lambda}, \qquad (5.87)$$

where β_ω and $\beta_{2\omega}$ are the propagation constants of the fundamental and the second harmonic fields, respectively, and Λ is the grating period.

The coherence length l_c is defined as the minimum length to achieve maximum conversion in a non-phase matched interaction and is given by:

$$l_c = \frac{\lambda_\omega}{4(N_{2\omega} - N_\omega)}, \qquad (5.88)$$

and the QPM period working at first order is calculated by:

$$\Lambda = 2l_c = \frac{\lambda_\omega}{2(N_{2\omega} - N_\omega)}, \qquad (5.89)$$

where λ_ω is the wavelength of the fundamental field and N_ω and $N_{2\omega}$ are the effective indices of the waveguide modes involved in the frequency conversion for the fundamental and second harmonic fields, respectively.

Theoretically, for perfect phase matching conditions ($\Delta k = 0$) the SH conversion efficiency follows the well-known dependence on the form [59]:

$$\eta \equiv \frac{P_{2\omega, OUT}}{P_{\omega, IN}} = \tanh^2(\Gamma z), \tag{5.90}$$

where Γ (with dimension of the inverse of length) is given by:

$$\Gamma = \frac{k_0 \chi_{eff}^{(2)} |\Phi_{0,\omega}|}{2\sqrt{N_\omega N_{2\omega}}}, \tag{5.91}$$

$\Phi_{0,\omega}$ being the initial effective field amplitude of the fundamental wave given by:

$$\Phi_{0,\omega} = \frac{\int \Psi_{0,\omega}^2 \Psi_{2\omega} dx}{\sqrt{\int \Psi_{0,\omega}^2 dx \int \Psi_{2\omega}^2 dx}}, \tag{5.92}$$

where $\Psi_{0,\omega}$ and $\Psi_{2\omega}$ are the fields (in V/m) of the fundamental and harmonic modes involved in the frequency conversion (usually first-order modes). On the other hand, the effective non-linear susceptibility $\chi_{eff}^{(2)}$ in Eq. (5.91) takes into account the reduction in efficiency that may arise from the use of QPM techniques. When the grating is made of regions with alternate half-periods of reversal sign of the non-linear coefficient $\chi^{(2)}$ (domain reversal), the effective non-linearity of the QPM interaction is $2/\pi$ smaller than the ordinary phase matched interaction. If the grating is a rectangular wave such that the second order non-linear coefficient $\chi^{(2)}$ has either its full value or zero in alternate half-periods of the grating (domain disordering), the effective non-linearity is reduced by $1/\pi$.

Let us now examine an example of SHG using first order QPM in LiNbO$_3$ planar waveguides [60], modelled by NL-BPM. Quasi-phase matching is accomplished through the reversal of the sign of the non-linear coefficient at odd multiples of the coherence length. The relative signs of the non-linear coefficients in LiNbO$_3$ are linked to the orientation of the ferroelectric spontaneous polarization and thus the QPM grating can be patterned by control of the spatial distribution of ferroelectric domain. The reversal of the ferroelectric domains can be induced by thermal treatment at high temperature on a patterned Ti-grating deposited onto a z-cut LiNbO$_3$ substrate. After creating the periodic ferroelectric domain pattern in the crystal, a planar waveguide is fabricated by a proton exchange process. This process gives rise to an increase on the extraordinary refractive index of the substrate and thus the waveguide support TM propagation. This waveguide configuration allows us to use the highest non-linear coefficient of the LiNbO$_3$ crystal (d_{33}) for frequency doubling, which requires that both the fundamental and second harmonic fields propagate as TM modes.

To model the device, we assume that the proton exchange process induces a step-index increase of $\Delta n = 0.01$ with respect to the index substrate, with a penetration depth of $d = 3\,\mu m$ (Figure 5.84). The extraordinary refractive indices of the LiNbO$_3$ substrate are 2.172 and 2.309 at 0.84 and 0.42 μm, respectively, $n_c = 1$ being the refractive index of the cover

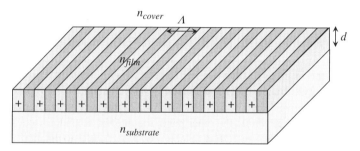

Figure 5.84 Periodically domain-inverted LiNbO$_3$-planar waveguide used for frequency doubling based on QPM in guided configuration

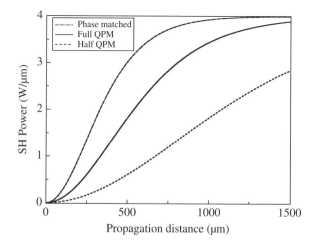

Figure 5.85 SH power in units of W/µm as a function of the propagation distance using reversal sign domains (full QPM, solid line) and alternate full value/zero non-linear coefficients (half-QPM, dashed line). For comparison purposes, the case of phase matched case is included (dashed-dotted line)

region (air). The non-linear susceptibility coefficient using this configuration is $\chi^{(2)} = 68.8$ pm/V and it is assumed that the domain inverted region depth is deeper that the waveguide core.

At these wavelengths, the effective refractive indices of the first TM-modes for the fundamental and second harmonic fields are 2.179027 and 2.318137, respectively. This results in a coherence length of $l_c = 1.5096$ µm and consequently a grating period of $\Lambda = 3.0192$ µm.

To simulate the performance of the non-linear frequency conversion device using NL-BPM, the first guided mode of the fundamental field is launched with a power per unit width of 4 W/ µm and the second harmonic field is set to zero. Figure 5.85 shows the power evolution of the SH field (in units of W/µm) as a function of the propagation distance, using first-order QMP with alternate domains (solid line). The simulation uses a transversal grid size of $\Delta x = 0.05$ µm and a propagation step of $\Delta z = 0.1$ µm. This longitudinal step is enough for obtaining accurate results and using only two iterations at each step. The evolution of the SH power obtained by NL-BPM can be accurately fitted to the theoretical relation (5.91) with $\Gamma = 607$ µm^{-1}. These results show that a device length of 1.5 mm is enough for reaching more than a 95% conversion efficiency for SHG.

For comparison purposes, the SH power for a normal phase matched waveguide ($N_{2\omega} = N_\omega$) has been simulated also by NL-BPM and the results are plotted in Figure 5.85 (dashed-dotted line). In this case, the fit gives a value of $\Gamma = 376\ \mu m^{-1}$. Using the modal profiles of the fields obtained by imaginary-distance BPM and assuming an input fundamental power of 4 W/μm, Eqs. (5.91) and (5.92) give a theoretical value of $\Gamma = 377\ \mu m^{-1}$, in very good agreement with the value obtained by NL-BPM. On the other hand, the ratio between the values of Γ provided by BPM calculations for normal and QPM SHG is 0.621, very close to the theoretical reduction factor of $2/\pi$ expected for the use of QPM with alternate domains [57].

Finally, NL-BPM calculations have been performed for QPM in which the second order non-linear coefficient $\chi^{(2)}$ has either its full value or zero in alternate half-periods (Figure 5.85, dashed line). In this case, the fitted value for Γ is 1214 μm^{-1}, as expected theoretically for a reduction of $1/\pi$ on the effective non-linear coefficient. A summary of the theoretical and BPM results is presented in Table 5.4.

Table 5.4 Comparison of the three phase-matching conditions in the SHG efficiency, evaluated by means of the Γ parameter

Type of phase matching	Γ (μm^{-1}) (theoretical)	Γ (μm^{-1}) (BPM)
Normal PM	377	376
Full QPM	598	607
Half QPM	1197	1214

Both theoretical and numerically evaluated by NL-BPM results are indicated.

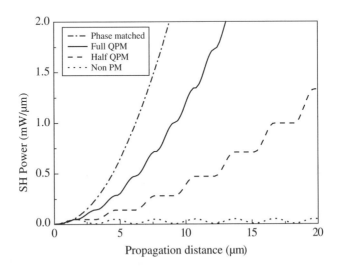

Figure 5.86 SH power evolution at short distances for normal phase matched (dashed-dotted line), domain reversal QPM (full QPM, solid line), domain disordering QPM (half QPM, dashed line) and non-phase matched (dotted line)

Figure 5.86 presents the evolution of the SH power examined at short distances in the three situations analysed previously (phase matched, full-QPM and half-QPM), where the case of non-phase matched waveguide is also included. The plots, obtained by NL-BPM, show the very different behaviour of power evolution for each case. While in the phase matched case the SH power grows in a parabolic way (dashed-dotted line), the power using full-QPM (continuous line) grows continuously, but with noticeable periodic undulations in the curve. On the other hand, the half-QPM curve (dashed line) only grows at half periods of the coherence length (1.51 μm) and during the propagation along the regions of zero value of the non-linear coefficient, the power remains unchanged. By contrast, the power evolution corresponding to the non-phase matched case (dotted line) shows a sinusoidal behaviour, where maximum conversion (with very low value) is reached at odd multiples of the coherence length.

References

[1] Deng, H., Jin, G.H., Harari, J., Vilcot, J.P. and Decoster, D. (1998) Investigation of 3-D semivectorial finite-difference beam propagation method for bent waveguides. *Journal of Lightwave Technology* **16**, 915–922.

[2] Lui, W.W., Xu, C.L., Hirono, T., Yokoyama, K. and Huang, W.P. (1998) Full vectorial wave propagation in semiconductor optical bending waveguides and equivalent straight waveguide approximation. *Journal of Lightwave Technology* **6**, 910–912.

[3] Heiblum, M. and Harris, J.H. (1975) Analysis of curved optical waveguides by conformal transformation. *IEEE Journal of Quantum Electronics* **QE-11**, 75–83.

[4] Nesterov, A. and Troppenz, U. (2003) A plane-wave boundary method for analysis of bent optical waveguides. *Journal of Lightwave Technology* **21**, 1–4.

[5] Yamauchi, J., Nito, Y. and Nakano, H. (2009) A modified semivectorial BPM retaining the effects of the longitudinal field components and its applications to the design of a spot-size converter. *Journal of Lightwave Technology* **27**, 2470–2476.

[6] Yokoyama, K., Sekino, N., Hirono, T., Tohmori, Y. and Kawaguchi, Y. (1998) Design and analysis of high coupling efficiency spot-size converter integrated laser diodes by three-dimensional BPM. *Electronics and Communications in Japan* **81**, 1–9.

[7] Sasaki, H. and Anderson, I. (1978) Theoretical and experimental studies on active Y-junctions in optical waveguides. *IEEE Journal of Quantum Electronics* **QE-14**, 883–892.

[8] Huang, W.P., Xu, C.L. and Chaudhuri, S.K. (1992) Application of the finite-difference vector beam propagation method to directional coupler devices. *IEEE Journal of Quantum Electronics* **28**, 1527–1532.

[9] Lifante, G. (2003) *Integrated Photonics: Fundamentals*. John Wiley & Sons, Inc..

[10] Bryngdahl, O. (1973) Image formation using self-imaging techniques. *Journal of the Optical Society of America* **63**, 416–419.

[11] Ulrich, R. (1975) Light-propagation and imaging in planar optical waveguides. *Nouvelle Revue d'Optique* **6**, 253–262.

[12] Chuang, R.W., Hsu, M.T. and Liao, Z.L. (2010) Integrated $SiO_2/SiON/SiO_2$ thermo-optical switch based on the multimode interference effect. *Japanese Journal of Applied Physics* **49**, 04DG21.

[13] Soldano, L.B. and Pennings, E.C.M. (1995) Optical multi-mode interference devices based on self-imaging: principles and applications. *Journal of Lightwave Technology* **13**, 615–627.

[14] Kim, J.H., Dudley, B.W. and Moyer, P.J. (2006) Experimental demonstration of replicated multimode interferometer power splitter in Zr-doped sol-gel. *Journal of Lightwave Technology* **24**, 612–616.

[15] Sheng, Z., Wang, Z., Qiu, C., Li, L., Pang, A., Wu, A., Wang, X., Zou, S. and Gan, F. (2012) A compact and low-loss MMI coupler fabricated with CMOS technology. *IEEE Photonics Journal* **4**, 2272–2277.

[16] Paiam, M.R. and MacDonald, R.I. (1997) Polarisation-insensitive 980/1550 nm wavelength (de)multiplexer using MMI couplers. *Electronics Letters* **33**, 1219–1220.

[17] Peng, S.T. and Tamir, T. (1975) Theory of periodic dielectric waveguides. *IEEE Transactions on Microwave Theory and Techniques* **MTT-23**, 123–133.

[18] Yariv, A. and Nakamura, M. (1977) Periodic structures for integrated optics. *IEEE Journal of Quantum Electronics* **QE-13**, 233–253.

[19] Zinoviev, K., Domínguez, C., Plaza, J.A. and Lechuga, L.M. (2006) Light coupling into an optical microcantilever by an embedded diffraction grating. *Applied Optics* **10**, 229–234.

[20] Hong, J., Huang, W. and Makino,T. (1992) On the transfer matrix method for distributed-feedback waveguide devices, *Journal of Lightwave Technology* **10**, 1860–1868.

[21] Smit, M.K. (1988) New focusing and dispersive component based on an optical phased-array. *Electronics Letters* **24**, 385–386.

[22] Takahashi, H., Suzuki, S., Kato, K. and Nishi, I. (1990) Arrayed waveguide grating for wavelength division multi/ demultiplexer with nanometre resolution. *Electronics Letters* **26**, 87–88.

[23] Dragone, C. (1991) An N × N optical multiplexer using a planar arrangement of two star couplers. *IEEE Photonics Technology Letters* **3**, 812–815.

[24] Smit, M.K. and van Dam, C. (1996) PHASAR-based WDM-devices: principles, design and applications. *IEEE Journal of Selected Topics in Quantum Electronics* **2**, 236–250.

[25] Amersfoort, M.R. (1998) Arrayed Waveguide Grating. Application Note A1998003, C2V, Netherlands.

[26] Chuang, R.W., Hsu, M.T., Chang, Y.C., Le, Y.J. and Chou, S.H. (2012) Integrated multimode interference coupler-based Mach–Zehnder interferometric modulator fabricated on a silicon-on-insulator substrate. *IEP Opto-electronics* **6**, 147–152.

[27] Suárez, I., Pernas, P. L. and Lifante, G. (2007) Integrated electro-optic Mach–Zehnder modulator fabricated by vapour Zn-diffusion in LiNbO$_3$. *Microwave and Optical Technology Letters* **49**, 1194–1196.

[28] Saleh, B.E.A. and Teich, M.C. (2007) *Fundamentals of Photonics*. John Wiley & Sons, Inc..

[29] Bava, G.P., Montrosset, I., Sohler, W. and Suche, H. (1987) Numerical modeling of Ti:LiNbO$_3$ integrated optical parametric oscillators. *IEEE Journal of Quantum Electronics* **QE-23**, 42–51.

[30] Zhang, D.L., Zhuang, Y.R. and Hua, P.R. (2007) Simulation of Ti diffusion into LiNbO$_3$ in Li-rich atmosphere. *Journal of Applied Physics* **101**, 013101.

[31] Alferness, R.C. (1980) Efficient waveguide electro-optic TE-TM converter/wavelength filter. *Applied Physics Letters* **36**, 513–515.

[32] Hempelmann, U. and Bersiner, L. (1993) Wave propagation in integrated acoustooptical anisotropic waveguides. *IEE Proceedings-J* **140**, 193–200.

[33] Shani, Y., Alferness, R., Koch, T., Koren, U., Oron, M., Miller, B.I. and Young, M.G. (1991) Polarization rotation in asymmetric periodic loaded rib waveguides. *Applied Physics Letters* **59**, 1278–1280.

[34] Nakayama, K., Shoji, Y. and Mizumoto, T. (2012) Single trench SiON waveguide TE-TM mode converter. *IEEE Photonics Technology Letters* **24**, 1310–1312.

[35] Yariv, A. (1973) Coupled-mode theory for guided-wave optics. *IEEE Journal of Quantum Electronics* **QE-9**, 919–933.

[36] Kogelnik, H. (1975) *Integrated Optics*, T. Tamir Springer, Heidelberg.

[37] Sosnowski, T.P. and Boyd, G.D. (1974) The efficiency of thin-film optical-waveguide modulators using electro-optic films or substrates. *IEEE Journal of Quantum Electronics* **QE-10**, 306–311.

[38] Weinert, C.M. and Heidrich, H. (1993) Vectorial simulation of passive TE/TM mode converter devices on InP. *IEEE Photonics Technology Letters* **5**, 324–326.

[39] Huang, W.P. and Mao, M.Z. (1992) Polarization rotation in periodic loaded rib waveguides. *Journal of Lightwave Technology* **10**, 1825–1831.

[40] Huang, W.P. and Xu, C.L. (1993) Simulation of three-dimensional optical waveguides by a full-vector beam propagation method. *IEEE Journal of Quantum Electronics* **29**, 2639–2649.

[41] Ando, T., Murata, T., Nakayama, H., Yamauchi, J. and Nakano, H. (2002) Analysis and measurements of polarization conversion in a periodically loaded dielectric waveguide. *IEEE Photonics Technology Letters* **14**, 1288–1290.

[42] Sheperd, D.P., Hettrick, S.J., Li, C., Mackenzie, J.I., Beach, R.J., Mitchell, S.C. and Meissner, H.E. (2001) High-power planar dielectric waveguide lasers. *Journal of Physics D: Applied Physics* **34**, 2420–2432.

[43] Mackenzie, J.I. (2007) Dielectric solid-state planar waveguide lasers: a review. *IEEE Journal of Selected Topics in Quantum Electronics* **13**, 626–637.

[44] Bolaños, W., Carvajal, J.J., Mateos, X., Cantelar, E., Lifante, G., Griebner, U. et al. (2011), Continuous-wave and Q-switched Tm-doped KY(WO$_4$)$_2$ planar waveguide laser at 1.84 µm. *Optics Express* **19**, 1449–1454.

[45] Pollnau, M. and Romanyuk, Y.E. (2007) Optical waveguide in laser crystals. *Comptes Rendus Physique* **8**, 123–137.

[46] Cantelar, E., Jaque, D. and Lifante, G. (2012) Waveguide lasers based on dielectric materials. *Optical Materials* **34**, 555–571.

[47] Lee, C.T. and Sheu, L.G. (1996) Analysis Nd:MgO:Ti:LiNbO$_3$ waveguide lasers with nonuniform concentration distributions. *Journal of Lightwave Technology* **14**, 2268–2276.

[48] Cantelar, E., Lifante, G. and Cussó, F. (2006) Modelling of Tm^{3+}-doped LiNbO$_3$ waveguide lasers. *Optical and Quantum Electronics* **38**, 111–122.

[49] Agrawal, G.P. (1984) Fast-Fourier-transform based beam-propagation model for stripe-geometry semiconductor lasers: inclusion of axial effects. *Journal of Applied Physics* **56**, 3100–3109.

[50] Lee, C.T. and Sheu, L.G. (1997) Analysis of end-pumped Nd:Ti:LiNbO$_3$ microchip waveguide Fabry–Perot lasers. *Journal of Lightwave Technology* **15**, 2147–2153.

[51] Brinkmann, R., Sohler, W., Suche, H. and Wersig, C., (1992) Fluorescence and laser operation in single-mode Ti-diffused Nd:MgO:LiNbO$_3$ waveguide structures. *IEEE Journal of Quantum Electronics* **28**, 466–470.

[52] Hempstead, M., Wilkinson, J.S. and Reeki, L. (1992) Waveguide lasers operating at 1084 nm in neodymium-diffused lithium niobate. *IEEE Photonics Technology Letters* **4**, 852–855.

[53] Jaque, D., Capmany, J., Sanz-García, J.A., Brenier, A., Boulon, G. and García-Solé, J. (1999) Nd^{3+} ion based self frequency doubling solid state lasers. *Optical Materials* **13**, 147–157.

[54] Digonnet, M.J.F. and Gaeta, C.J. (1985) Theoretical analysis of optical fiber laser amplifiers and oscillators. *Applied Optics* **24**, 333–342.

[55] Lallier, E., Pocholle, J.P., Papuchon, M. de Micheli, M.P., Li, M.J., He, Q. (1991) et al., Nd:MgO:LiNbO$_3$ channel waveguide laser devices. *IEEE Journal of Quantum Electronics* **27**, 618–625.

[56] Richter, T., Nouroozi, R., Suche, H., Sohler, W. and Schubert, C. (2013) PPLN-waveguide based tunable wavelength conversion of QAM data within the C-band. *IEEE Photonics Technology Letters* **25**, 2085–2088.

[57] Somekh, S. and Yariv, A. (1972) Phase matching by periodic modulation of the non-linear optical properties. *Optics Communications* **6**, 301–304.

[58] Houé, M. and Townsend, P.D. (1995) An introduction to methods of periodic poling for second-harmonic generation. *Journal of Physics D: Applied Physics* **28**, 1747–1763.

[59] Masoudi, H.M. and Arnold, J.M. (1995) Modeling second-order non-linear effects in optical waveguides using a parallel-processing beam propagation method. *IEEE Journal of Quantum Electronics* **31**, 2107–2113.

[60] Lim, E.J., Fejer, M.M. and Byer, R.L. (1989) Second-harmonic generation of green light in periodically poled planar lithium niobate waveguide. *Electronics Letters* **25**, 174–175.

Appendix A

Finite Difference Approximations of Derivatives

The beam propagation method based on the finite difference formulation consists of substituting the derivatives in the partial differential wave equation by the appropriate finite difference approaches, built by linear combinations of the field values at the grid points. Here we present the approximations of first order finite differences. Then, the second-order derivatives are developed, including the finite difference (FD) approaches for variable coefficients and mixed derivatives.

A.1 FD-Approximations of First-Order Derivatives

We assume that the function $f(x)$ is represented by its values at the discrete set of points:

$$x_i = x_1 + i\Delta x \quad i = 0, 1, \ldots, N, \tag{A.1}$$

Δx being the grid spacing, and we write f_i for $f(x_i)$.

Finite difference of $\dfrac{df(x)}{dx}$. The finite difference approximation of the first order derivative of the function $f(x)$ at the grid point j can be expressed as:

$$\left[\frac{df(x)}{dx}\right]_i \approx \frac{1}{\Delta x}(f_{i+1} - f_i) + \mathcal{O}(\Delta x) \quad \text{Forward difference;} \tag{A.2a}$$

$$\left[\frac{df(x)}{dx}\right]_i \approx \frac{1}{\Delta x}(f_i - f_{i-1}) + \mathcal{O}(\Delta x) \quad \text{Backward difference.} \tag{A.2b}$$

Beam Propagation Method for Design of Optical Waveguide Devices, First Edition. Ginés Lifante Pedrola.
© 2016 John Wiley & Sons, Ltd. Published 2016 by John Wiley & Sons, Ltd.

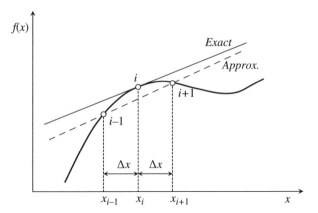

Figure A.1   Geometric interpretation of the first-order derivative of the function $f(x)$ using a central difference scheme

The error of these two finite difference representations is proportional to (Δx) and the approximation is referred as a first-order approximation.

The derivative of the function $f(x)$ at the grid point x_i can alternatively calculated by:

$$\left[\frac{df(x)}{dx}\right]_i \approx \frac{1}{2\Delta x}(f_{i+1}-f_{i-1}) + \mathcal{O}(\Delta x)^2 \quad \text{Central difference.} \tag{A.3}$$

The error using this formula in the derivative is proportional to $(\Delta x)^2$ and the approximation is referred as a second-order approximation. Figure A.1 shows the geometric interpretation of formula (A.3).

Also, this expression is often written in terms of values of the function at fictitious intermediate grid points:

$$\left[\frac{df(x)}{dx}\right]_i \approx \frac{1}{\Delta x}(f_{i+1/2}-f_{i-1/2}) + \mathcal{O}(\Delta x)^2, \tag{A.4}$$

which provides also a second-order approximation to the derivative.

A.2   FD-Approximation of Second-Order Derivatives

Now, the finite difference approximations for second order derivatives are presented, including the case of variable coefficients and the case of mixed derivatives.

Finite difference of $\dfrac{d^2 f(x)}{dx^2}$. The central finite difference scheme for the second order derivative is obtained as follows:

$$\left[\frac{d^2 f(x)}{dx^2}\right]_i = \frac{d}{dx}\left[\left(\frac{df}{dx}\right)\right]_i \approx \frac{1}{\Delta x}\left[\left(\frac{df}{dx}\right)_{i+1/2} - \left(\frac{df}{dx}\right)_{i-1/2}\right] \approx \frac{1}{\Delta x}\left[\left(\frac{f_{i+1}-f_i}{\Delta x}\right) - \left(\frac{f_i-f_{i-1}}{\Delta x}\right)\right]$$

$$= \frac{1}{(\Delta x)^2}\left(f_{i+1} - 2f_i + f_{i-1}\right) + \mathcal{O}(\Delta x)^2$$

(A.5)

where central difference schemes have been used to approximate the derivatives, making use of fictitious intermediate grid points. This expression provides a second-order approximation.

Finite difference approach of second-order derivatives involving variable coefficients $\dfrac{d}{dx}\left[g(x)\left(\dfrac{df(x)}{dx}\right)\right]$. Here we assume that both $f(x)$ and $g(x)$ functions are defined at discrete grid points. In a very similar manner as proceed before, using fictitious intermediate points and central difference schemes:

$$\frac{d}{dx}\left[g(x)\left(\frac{df(x)}{dx}\right)\right]_i \approx \frac{1}{\Delta x}\left[g_{i+1/2}\left(\frac{df}{dx}\right)_{i+1/2} - g_{i-1/2}\left(\frac{df}{dx}\right)_{i-1/2}\right]$$

$$\approx \frac{1}{\Delta x}\left[g_{i+1/2}\left(\frac{f_{i+1}-f_i}{\Delta x}\right) - g_{i-1/2}\left(\frac{f_i-f_{i-1}}{\Delta x}\right)\right]$$

(A.6)

$$= \frac{1}{(\Delta x)^2}\left(g_{i+1/2}f_{i+1} - (g_{i+1/2} + g_{i-1/2})f_i + g_{i-1/2}f_{i-1}\right) + \mathcal{O}(\Delta x)^2$$

which also provides a second-order approximation to the derivative.

Finite difference of mixed derivatives $\dfrac{\partial^2 f(x,y)}{\partial x \partial y}$. In this case, we assume that the function $f(x,y)$ is represented by its values at the discrete set of points:

$$x_i = x_1 + (i-1)\Delta x \quad i = 1,\ldots,N,$$

(A.7a)

$$y_j = y_1 + (j-1)\Delta y \quad j = 1,\ldots,M,$$

(A.7b)

being Δx and Δy the grid spacing along x and y directions, and we denote $f_{i,j}$ for $f(x_i,y_j)$. To proceed with the finite difference approach of the mixed derivative of the function $f(x,y)$, first a central difference scheme is used to perform the derivative respect to x:

$$\left[\frac{\partial^2 f(x,y)}{\partial x \partial y}\right]_{i,j} = \frac{1}{2\Delta x}\left[\left(\frac{\partial f}{\partial y}\right)_{i+1,j} - \left(\frac{\partial f}{\partial y}\right)_{i-1,j}\right] + \mathcal{O}(\Delta x)^2.$$

(A.8)

Then, using also central schemes, the inner derivatives with respect to the y coordinate are given by:

$$\left[\frac{\partial f(x,y)}{\partial y}\right]_{i+1,j} = \frac{1}{2\Delta y}\left(f_{i+1,j+1}-f_{i+1,j-1}\right)+\mathcal{O}(\Delta y)^2; \qquad (A.9a)$$

$$\left[\frac{\partial f(x,y)}{\partial y}\right]_{i-1,j} = \frac{1}{2\Delta y}\left(f_{i-1,j+1}-f_{i-1,j-1}\right)+\mathcal{O}(\Delta y)^2. \qquad (A.9b)$$

Substituting these two equations in expression (A.8) gives finally:

$$\left[\frac{\partial^2 f(x,y)}{\partial x \partial y}\right]_{i,j} \approx \frac{1}{4\Delta x \Delta y}\left(f_{i+1,j+1}-f_{i-1,j+1}-f_{i+1,j-1}+f_{i-1,j-1}\right)+\mathcal{O}\left[(\Delta x)^2,(\Delta y)^2\right], \qquad (A.10)$$

which is second-order accurate in both spatial increments.

Appendix B

Tridiagonal System: The Thomas Method Algorithm

A tridiagonal matrix M is an $n \times n$ matrix that has nonzero elements only on the main diagonal, the first diagonal below this (lower diagonal) and the first diagonal above the main diagonal (upper diagonal). This matrix M can be expressed in a generic form as:

$$M = \begin{pmatrix}
b_1 & c_1 & 0 & 0 & . & . & 0 & 0 & 0 & 0 \\
a_2 & b_2 & c_2 & 0 & . & . & 0 & 0 & 0 & 0 \\
0 & a_3 & b_3 & c_3 & . & . & 0 & 0 & 0 & 0 \\
0 & 0 & a_4 & b_4 & . & . & 0 & 0 & 0 & 0 \\
0 & 0 & 0 & a_5 & . & . & 0 & 0 & 0 & 0 \\
. & . & . & . & . & . & . & . & . & . \\
. & . & . & . & . & . & . & . & . & . \\
0 & 0 & 0 & 0 & . & . & a_{n-2} & b_{n-2} & c_{n-2} & 0 \\
0 & 0 & 0 & 0 & . & . & 0 & a_{n-1} & b_{n-1} & c_{n-1} \\
0 & 0 & 0 & 0 & . & . & 0 & 0 & a_n & b_n
\end{pmatrix} \tag{B.1}$$

with $a_1 = 0$ and $c_n = 0$. The tridiagonal matrix can be stored efficiently and usually it is stored in three one-dimensional arrays of length n containing the diagonal, lower-diagonal and upper-diagonal elements.

Following this definition, a tridiagonal system is a linear system of the form:

$$Mu = r \tag{B.2}$$

Beam Propagation Method for Design of Optical Waveguide Devices, First Edition. Ginés Lifante Pedrola.
© 2016 John Wiley & Sons, Ltd. Published 2016 by John Wiley & Sons, Ltd.

where u is a row vector of n unknowns, r is a row vector of n elements, and M is an $n \times n$ tri-diagonal matrix. This linear system can be expanded as:

$$
\begin{pmatrix}
b_1 & c_1 & 0 & 0 & . & . & 0 & 0 & 0 & 0 \\
a_2 & b_2 & c_2 & 0 & . & . & 0 & 0 & 0 & 0 \\
0 & a_3 & b_3 & c_3 & . & . & 0 & 0 & 0 & 0 \\
0 & 0 & a_4 & b_4 & . & . & 0 & 0 & 0 & 0 \\
0 & 0 & 0 & a_5 & . & . & 0 & 0 & 0 & 0 \\
. & . & . & . & . & . & . & & . & \\
. & . & . & . & . & . & . & & . & \\
0 & 0 & 0 & 0 & . & . & a_{n-2} & b_{n-2} & c_{n-2} & 0 \\
0 & 0 & 0 & 0 & . & . & 0 & a_{n-1} & b_{n-1} & c_{n-1} \\
0 & 0 & 0 & 0 & . & . & 0 & 0 & a_n & b_n
\end{pmatrix}
\begin{pmatrix}
u_1 \\ u_2 \\ u_3 \\ u_4 \\ u_5 \\ . \\ . \\ u_{n-2} \\ u_{n-1} \\ u_n
\end{pmatrix}
=
\begin{pmatrix}
r_1 \\ r_2 \\ r_3 \\ r_4 \\ r_5 \\ . \\ . \\ r_{n-2} \\ r_{n-1} \\ r_n
\end{pmatrix}
\tag{B.3}
$$

This linear system can be also expressed in a compact form by:

$$
a_j u_{j-1} + b_j u_j + c_j u_{j+1} = r_j, \quad \text{for } j = 1 \text{ to } n
\tag{B.4}
$$

For such systems, the solution can be obtained in $\mathcal{O}(n)$ operations, instead the $\mathcal{O}(n^3)$ operations required by the standard method of Gaussian elimination. One such efficient algorithm is the Thomas method, which is shown here:

1. Set: $\beta = b_1$, $u_1 = r_1/\beta$.
2. Evaluate for $j = 2$ to n:

$$
\gamma_j = \frac{c_{j-1}}{\beta}
$$

$$
\beta = b_j - a_j \gamma_j
$$

$$
u_j = \frac{r_j - a_j u_{j-1}}{\beta}
$$

3. Find for $j = 1$ to $n - 1$:

$$
k = n - j
$$

$$
u_k = u_k - \gamma_{k+1} u_{k+1}.
$$

The following routine is the implementation of the Thomas algorithm, adapted from [1]. The lower, main and upper diagonal of the tridiagonal matrix M are, respectively, given by the (complex) vectors a, b and c, and r is the right-hand side vector, n being the size of the linear system. The unknown is placed in the n-dimensional (complex) vector u. The vector gamma(j)

and *beta* are intermediate (complex) variables. Also, *j* and *k* are intermediate (integer) variables. Lets us notice that a_1 and c_n are undefined and are not referenced by the routine.

```
IF b(1) = 0 THEN STOP
beta = b(1)
u(1) = r(1) / beta
FOR j = 2 TO n
        gamma(j) = c(j-1) / beta
        beta = b(j) - a(j) * gamma(j)
        IF beta = 0 THEN STOP
        u(j) = (r(j) - a(j) * u(j-1)) / beta
NEXT j
FOR j = 1 TO n-1
        k = n - j
        u(k) = u(k) - gamma(k+1) * u(k+1)
NEXT
```

Reference

[1] Press, W.H., Teukolsky, S.A., Vetterling, W.T. and Flannery, B.P. (1996) *Numerical Recipes in Fortran 77: The Art of Scientific Computing*, Chapter 2. Cambridge University Press, New York.

Appendix C

Correlation and Relative Power between Optical Fields

C.1 Correlation between Two Optical Fields

The correlation Γ of two optical fields gives a measure of how similar the fields are between them, regardless of their amplitudes. Mathematically, the correlation between two arbitrary scalar fields $u(x,y)$ and $v(x,y)$ is calculated by means of the overlap integral between the normalized fields as [1]:

$$\Gamma \equiv \frac{\left| \iint u(x,y)v^*(x,y)dxdy \right|^2}{\iint |u(x,y)|^2 dxdy \iint |v(x,y)|^2 dxdy}. \tag{C.1}$$

If the two fields are very similar, the correlation will give a value close to 1. By contrast, if the two fields are very different, the correlation should be low and close to zero.

C.2 Power Contribution of a Waveguide Mode

Here, the power contribution of a normalized (scalar) eigenmode $\psi_\mu(x,y)$ to the total power carried by an arbitrary (scalar) optical field $u(x,y)$ in a z-invariant waveguide is calculated.

It is known that any arbitrary optical field $u(x,y)$ can be expressed as a linear combination of all modes supported by the z-invariant structure, including both guided modes (discretized) and radiation modes (continuum). Nevertheless, in practice only the eigenmodes (guided modes)

Beam Propagation Method for Design of Optical Waveguide Devices, First Edition. Ginés Lifante Pedrola.
© 2016 John Wiley & Sons, Ltd. Published 2016 by John Wiley & Sons, Ltd.

need to be considered. Assuming that fact, the arbitrary field $u(x,y)$ can be expressed as an expansion of the eigenmodes as:

$$u(x,y) = \sum_{\nu} a_{\nu} \psi_{\nu}(x,y), \tag{C.2}$$

where a_{ν} is the modal weight corresponding to the mode $\psi_{\nu}(x,y)$. The modal weights can be easily calculated by multiplying Eq. (C.2) by ψ_{μ}^* and then making an integration over the cross section of the z-invariant waveguide structure:

$$\int\int u(x,y)\psi_{\mu}^*(x,y)dxdy = \int\int \sum_{\nu} a_{\nu}\psi_{\nu}(x,y)\psi_{\mu}^*(x,y)dxdy = a_{\mu}\int\int |\psi_{\mu}(x,y)|^2 dxdy, \tag{C.3}$$

where the orthogonality between eigenmodes have been applied in the result. The modal weights are then given by:

$$a_{\mu} = \frac{\int\int u(x,y)\psi_{\mu}^*(x,y)dxdy}{\int\int |\psi_{\nu}(x,y)|^2 dxdy}, \tag{C.4}$$

On the other hand, the total power P carried by the optical field $u(x,y)$ can be calculated by:

$$P = \int\int |u(x,y)|^2 dxdy = \int\int u(x,y)u^*(x,y)dxdy = \int\int \left(\sum a_{\nu}\psi_{\nu}(x,y)\right)\left(\sum a_{\mu}^*\psi_{\mu}^*(x,y)\right)dxdy$$

$$= \sum_{\nu,\mu} a_{\nu}a_{\mu}^* \int\int \psi_{\nu}(x,y)\psi_{\mu}^*(x,y)dxdy = \sum_{\nu} |a_{\nu}|^2 \int\int |\psi_{\nu}(x,y)|^2 dxdy$$

$$\tag{C.5}$$

The fraction of the power carried by the scalar field $u(x,y)$ associated to the eigenmode μ is given by:

$$\Gamma_{\mu} \equiv \frac{Power_{\mu}}{Total\ power} = \frac{|a_{\mu}|^2 \int\int |\psi_{\mu}(x,y)|^2 dxdy}{\int\int |u(x,y)|^2 dxdy} = \frac{\left|\int\int u(x,y)\psi_{\mu}^*(x,y)dxdy\right|^2}{\int\int |u(x,y)|^2 dxdy \int\int |\psi_{\mu}(x,y)|^2 dxdy} \tag{C.6}$$

As an example, the reflectivity of the fundamental mode $\psi_0(x,y)$ at a waveguide discontinuity is calculated. First the fundamental waveguide mode is obtained by the imaginary distance beam propagation method (BPM) and is chosen as the input field. Then, the reflected field $u(x,y)$ at the discontinuity is computed, by using for instance the bidirectional beam propagation method (Bi-BPM) algorithm presented in Section 4.2. The contribution of the fundamental guided mode in the calculated reflected field is now obtained by evaluating the overlap integral between both fields. Using these results, the reflectivity R of the fundamental mode is given by:

$$R \equiv \frac{Power_0}{Launch\ power} = \frac{|a_0|^2 \int\int |\psi_0(x,y)|^2 dxdy}{\int\int |\psi_0(x,y)|^2 dxdy} = \frac{\left|\int\int u(x,y)\psi_0^*(x,y)dxdy\right|^2}{\left(\int\int |\psi_0(x,y)|^2 dxdy\right)^2} \tag{C.7}$$

This formula is in general use and measures both the dissipation in power (due to absorption) as well as the loss of shape of the propagation mode if the original z-invariant structure is modified in some section of the waveguide (gratings, bends, imperfections, discontinuities etc.) [2].

References

[1] Kunz, A., Zimulinda, F. and Heinlein, W.E. (1993) Fast three-dimensional split-step algorithm for vectorial wave propagation in integrated optics. *IEEE Photonics Technology Letters* **5**, 1073–1076.

[2] Sharma, A. and Agrawal, A. (2004) New method for nonparaxial beam propagation. *Journal of the Optical Society of America A* **21**, 1082–1087.

Appendix D

Poynting Vector Associated to an Electromagnetic Wave Using the SVE Fields

Here, we develop expressions for the z-component of the Poynting vector for electromagnetic (EM) waves using the slowly varying fields for monochromatic light for both 2D and 3D propagation applied to slowly varying structures [1]. Also, finite-difference expressions for second-order derivatives that appear in 3D propagation are derived.

D.1 Poynting Vector in 2D-Structures

D.1.1 TE Propagation in Two-Dimensional Structures

In the case of 2D structures, for transverse electric field (TE) propagation the electric and magnetic fields have the following non-null components:

$$\mathcal{E} = \left[0, \mathcal{E}_y, 0\right] ; \tag{D.1a}$$

$$\mathcal{H} = \left[\mathcal{H}_x, 0, \mathcal{H}_z\right] . \tag{D.1b}$$

Using complex notation:

$$\mathcal{E}(r,t) = \mathrm{Re}\left[E(r)e^{i\omega t}\right]; \tag{D.2a}$$

$$\mathcal{H}(r,t) = \mathrm{Re}\left[H(r)e^{i\omega t}\right], \tag{D.2b}$$

Beam Propagation Method for Design of Optical Waveguide Devices, First Edition. Ginés Lifante Pedrola.
© 2016 John Wiley & Sons, Ltd. Published 2016 by John Wiley & Sons, Ltd.

the third Maxwell equation $(\nabla \times \mathcal{E}) = -\mu_0 \dfrac{\partial \mathcal{H}}{\partial t}$ results in:

$$(\nabla \times E(r)e^{i\omega t}) = \begin{vmatrix} u_x & u_y & u_z \\ \dfrac{\partial}{\partial x} & \dfrac{\partial}{\partial y} & \dfrac{\partial}{\partial z} \\ 0 & E_y(r) & 0 \end{vmatrix} e^{i\omega t} = \left[-\dfrac{\partial E_y(r)}{\partial z} u_x + \dfrac{\partial E_y(r)}{\partial x} u_z \right] e^{i\omega t}$$

$$= -\mu_0 \dfrac{\partial}{\partial t} \left(H(r)e^{i\omega t} \right) = -i\mu_0 \omega (H_x(r)u_x + H_z(r)u_z)e^{i\omega t}$$

(D.3)

which provides a relation between the non-null components of the complex amplitudes of the electric and magnetic fields.

Equalling the x and z components on this equation, we obtain:

$$\frac{\partial E_y}{\partial z} = i\mu_0 \omega H_x;$$

(D.4a)

$$\frac{\partial E_y}{\partial x} = -i\mu_0 \omega H_z.$$

(D.4b)

Therefore, the Poynting vector (in units of W/m^2) associated to a monochromatic wave under TE propagation in an inhomogeneous medium is reduced to:

$$\langle S \rangle = \mathrm{Re}\{S\} = \frac{1}{2}\mathrm{Re}(E \times H^*) = \frac{1}{2}\mathrm{Re} \begin{vmatrix} u_x & u_y & u_z \\ 0 & E_y & 0 \\ H_x^* & 0 & H_z^* \end{vmatrix} = \frac{1}{2}\mathrm{Re}\left(E_y H_z^* u_x - E_y H_x^* u_z\right)$$

$$= \frac{1}{2}\mathrm{Re}\left(\frac{-i}{\omega\mu_0} E_y \frac{\partial E_y^*}{\partial x} u_x - \frac{i}{\omega\mu_0} E_y \frac{\partial E_y^*}{\partial z} u_z \right)$$

(D.5)

If the field is expressed as a slowly varying envelope (SVE): $E_y(x,z) = u(x,z)e^{-i\beta z}$, then the z component of the Poynting vector is given by:

$$\langle S_z \rangle = \frac{-1}{2\omega\mu_0} \mathrm{Re}\left\{ iu(x,z)e^{-i\beta z}\left(\frac{\partial u^*(x,z)}{\partial z}e^{+i\beta z} + i\beta u^*(x,z)e^{+i\beta z} \right) \right\}$$

$$= \frac{-1}{2\omega\mu_0} \mathrm{Re}\left\{ iu(x,z)\left(\frac{\partial u^*(x,z)}{\partial z} + i\beta u^*(x,z) \right) \right\} = \frac{1}{2\omega\mu_0} \mathrm{Re}\left\{ \beta|u(x,z)|^2 - iu(x,z)\frac{\partial u^*(x,z)}{\partial z} \right\}$$

(D.6)

Moreover, if the envelope amplitude $u(x,z)$ slowly varies in a wavelength:

$$\left| \frac{\partial u(x,z)}{\partial z} \right| \leq \beta|u(x,z)|,$$

(D.7)

the Poynting vector z component reduces to:

$$\langle S_z \rangle \approx \frac{\beta}{2\omega\mu_0} |u(x,z)|^2. \tag{D.8}$$

This last expression is exact if the propagating field corresponds to a waveguide eigenmode, because any propagating mode supported by a waveguide can be expressed as:

$$E_{\nu y}(x,z) = u_\nu(x)e^{-i\beta_\nu z}. \tag{D.9}$$

Taking into account formula (D.6), the power carried by a monochromatic TE-wave is therefore:

$$\begin{aligned}
P_z &= \frac{1}{2\omega\mu_0} \iint_A \mathrm{Re}\left\{ \beta |u(x,z)|^2 - iu(x,z)\frac{\partial u^*(x,z)}{\partial z} \right\} dx dy \\
&= \frac{L_y}{2\omega\mu_0} \int \mathrm{Re}\left\{ \beta |u(x,z)|^2 - iu(x,z)\frac{\partial u^*(x,z)}{\partial z} \right\} dx
\end{aligned} \tag{D.10}$$

If the slowly envelope approximation holds, the power per unit length along the y-axis for a TE-propagating field is:

$$\frac{P_z}{L_y} \approx \frac{\beta}{2\omega\mu_0} \int |u(x,z)|^2 dx = \frac{N_{eff}}{2\eta_0} \int |u(x,z)|^2 dx, \tag{D.11}$$

where the propagation constant of the mode has been expressed as function of the effective refractive index $N_{eff} = \dfrac{\beta}{k_0}$ and it has been introduced into free-space impedance defined by

$$\eta_0 \equiv \sqrt{\frac{\mu_0}{\varepsilon_0}}. \tag{D.12}$$

D.1.2   TM Propagation in Two-Dimensional Structures

In the case of 2D structures, for transverse magnetic field (TM) propagation the electric and magnetic fields have the following non-null components:

$$\mathcal{E} = [\mathcal{E}_x, 0, \mathcal{E}_z]; \tag{D.13a}$$

$$\mathcal{H} = [0, \mathcal{H}_y, 0]. \tag{D.13b}$$

The fourth Maxwell equation $(\nabla \times \mathcal{H}) = \varepsilon \dfrac{\partial \mathcal{E}}{\partial t}$ provides a relation between these non-null components:

$$(\nabla \times \mathbf{H}(r)e^{i\omega t}) = \begin{vmatrix} \mathbf{u}_x & \mathbf{u}_y & \mathbf{u}_z \\ \dfrac{\partial}{\partial x} & \dfrac{\partial}{\partial y} & \dfrac{\partial}{\partial z} \\ 0 & H_y(r) & 0 \end{vmatrix} e^{i\omega t} = \left[-\dfrac{\partial H_y(r)}{\partial z}\mathbf{u}_x + \dfrac{\partial H_y(r)}{\partial x}\mathbf{u}_z \right] e^{i\omega t}$$

$$= \varepsilon \dfrac{\partial}{\partial t}\left(\mathbf{E}(r)e^{i\omega t} \right) = i\varepsilon\omega(E_x(r)\mathbf{u}_x + E_z(r)\mathbf{u}_z)e^{i\omega t} \qquad (D.14)$$

Equalling the x and z components in this equation, we obtain:

$$\dfrac{\partial H_y}{\partial z} = -i\varepsilon\omega E_x; \qquad (D.15a)$$

$$\dfrac{\partial H_y}{\partial x} = i\varepsilon\omega E_z. \qquad (D.15b)$$

Therefore, the Poynting vector (in units of W/m^2) associated to a monochromatic wave under TM propagation in an inhomogeneous media is reduced to:

$$\langle \mathcal{S} \rangle = \dfrac{1}{2}\mathrm{Re}(E \times H^*) = \dfrac{1}{2}\mathrm{Re}\begin{vmatrix} \mathbf{u}_x & \mathbf{u}_y & \mathbf{u}_z \\ E_x & 0 & E_z \\ 0 & H_y^* & 0 \end{vmatrix} = \dfrac{1}{2}\mathrm{Re}\left(-E_z H_y^* \mathbf{u}_x + E_x H_y^* \mathbf{u}_z \right)$$

$$= \dfrac{1}{2}\mathrm{Re}\left(\dfrac{i}{\omega\varepsilon}H_y^*\dfrac{\partial H_y}{\partial x}\mathbf{u}_x + \dfrac{i}{\omega\varepsilon}H_y^*\dfrac{\partial H_y}{\partial z}\mathbf{u}_z \right) \qquad (D.16)$$

If the field is expressed as: $H_y(x,z) = u(x,z)e^{-i\beta z}$, then the intensity of the z component of the Poynting vector can be expressed as:

$$\langle S_z \rangle = \dfrac{1}{2\omega\varepsilon}\mathrm{Re}\left\{ iu^*(x,z)e^{+i\beta z}\left(\dfrac{\partial u(x,z)}{\partial z}e^{-i\beta z} - i\beta u(x,z)e^{-i\beta z} \right) \right\}$$

$$= \dfrac{1}{2\omega\varepsilon}\mathrm{Re}\left\{ iu^*(x,z)\left(\dfrac{\partial u(x,z)}{\partial z} - i\beta u(x,z) \right) \right\} = \dfrac{1}{2\omega\varepsilon}\mathrm{Re}\left\{ \beta|u(x,z)|^2 + iu^*(x,z)\dfrac{\partial u(x,z)}{\partial z} \right\}$$

$$(D.17)$$

Moreover, if the envelope amplitude $u(x,z)$ slowly varies in a wavelength distance:

$$\left| \dfrac{\partial u(x,z)}{\partial z} \right| \leq \beta|u(x,z)|, \qquad (D.18)$$

the Poynting vector z component is reduced to:

$$\langle S_z \rangle \approx \frac{\beta}{2\omega\varepsilon} |u(x,z)|^2 . \tag{D.19}$$

This last expression is exact if the propagating field corresponds to a waveguide eigenmode, because any propagating mode supported by a waveguide can be expressed as:

$$H_{\nu y}(x,z) = u_\nu(x) e^{-i\beta_\nu z}. \tag{D.20}$$

The power carried by a monochromatic TM-wave is therefore:

$$
\begin{aligned}
P_z &= \frac{1}{2\omega} \iint_A \mathrm{Re}\left\{ \frac{\beta}{\varepsilon}|u(x,z)|^2 + \frac{i}{\varepsilon} u^*(x,z) \frac{\partial u(x,z)}{\partial z} \right\} dx dy \\
&= \frac{L_y}{2\omega} \int \mathrm{Re}\left\{ \frac{\beta}{\varepsilon}|u(x,z)|^2 + \frac{i}{\varepsilon} u^*(x,z) \frac{\partial u(x,z)}{\partial z} \right\} dx
\end{aligned}
\tag{D.21}
$$

If the slowly envelope approximation holds, the power per unit length along the y-axis for a TM-propagating field is:

$$\frac{P_z}{L_y} \approx \frac{\beta}{2\omega} \int \frac{|u(x,z)|^2}{\varepsilon} dx = \frac{N_{eff}\eta_0}{2} \int \frac{|u(x,z)|^2}{n^2(x)} dx. \tag{D.22}$$

D.2 Poynting Vector in 3D-Structures

In the case of 3D structures, the electric and magnetic fields have all three components:

$$\mathcal{E} = \left[\mathcal{E}_x, \mathcal{E}_y, \mathcal{E}_z \right]; \tag{D.23a}$$

$$\mathcal{H} = \left[\mathcal{H}_x, \mathcal{H}_y, \mathcal{H}_z \right]. \tag{D.23b}$$

Using complex notation:

$$\mathcal{E}(r,t) = \mathrm{Re}\left[E(r) e^{i\omega t} \right]; \tag{D.24a}$$

$$\mathcal{H}(r,t) = \mathrm{Re}\left[H(r) e^{i\omega t} \right], \tag{D.24b}$$

the Poynting vector (in units of W/m^2) associated to an EM field is given by:

$$\langle \mathcal{S} \rangle \equiv \langle \mathcal{E} \times \mathcal{H} \rangle = \mathrm{Re}\{S\} = \frac{1}{2}\mathrm{Re}(E \times H^*) = \frac{1}{2}\mathrm{Re} \begin{vmatrix} u_x & u_y & u_z \\ E_x & E_y & E_z \\ H_x^* & H_y^* & H_z^* \end{vmatrix}. \tag{D.25}$$

If the electric and magnetic fields are expressed as function of SVE fields:

$$E(x,y,z) = \boldsymbol{\Psi}(x,y,z)e^{-iKz}; \tag{D.26a}$$

$$H(x,y,z) = \boldsymbol{\Phi}(x,y,z)e^{-iKz}, \tag{D.26b}$$

then the z component of the Poynting vector yields:

$$\langle S_z \rangle = \frac{1}{2}\mathrm{Re}\left\{ E_x H_y^* - E_y H_x^* \right\} = \frac{1}{2}\mathrm{Re}\left\{ \Psi_x \Phi_y^* - \Psi_y \Phi_x^* \right\}. \tag{D.27}$$

D.2.1   Expression as a Function of the Transverse Electric Field

Now, we want to express this formula in terms of the transverse SVE electric field components. In order to do that, the magnetic components have to be put as functions of the electric field components.

The third Maxwell equation $(\nabla \times \boldsymbol{\mathcal{E}}) = -\mu_0 \dfrac{\partial \boldsymbol{\mathcal{H}}}{\partial t}$ results in:

$$\nabla \times [E(r)e^{i\omega t}] = \begin{vmatrix} u_x & u_y & u_z \\ \dfrac{\partial}{\partial x} & \dfrac{\partial}{\partial y} & \dfrac{\partial}{\partial z} \\ E_x(r) & E_y(r) & E_z(r) \end{vmatrix} e^{i\omega t}$$

$$= \left[\left(\frac{\partial E_z}{\partial y} - \frac{\partial E_y}{\partial z} \right) u_x + \left(\frac{\partial E_x}{\partial z} - \frac{\partial E_z}{\partial x} \right) u_y + \left(\frac{\partial E_y}{\partial x} - \frac{\partial E_x}{\partial y} \right) u_z \right] e^{i\omega t} \tag{D.28}$$

$$= -\mu_0 \frac{\partial}{\partial t} \left[H(r)e^{i\omega t} \right] = -i\mu_0 \omega \left(H_x u_x + H_y u_y + H_z u_z \right) e^{i\omega t}$$

which provides a relation between the components of the complex amplitudes of the electric and magnetic fields.

Equalling the x, y and z components on this vectorial equation, we obtain:

$$\frac{\partial E_z}{\partial y} - \frac{\partial E_y}{\partial z} = -i\mu_0 \omega H_x; \tag{D.29a}$$

$$\frac{\partial E_x}{\partial z} - \frac{\partial E_z}{\partial x} = -i\mu_0 \omega H_y; \tag{D.29b}$$

$$\frac{\partial E_y}{\partial x} - \frac{\partial E_x}{\partial y} = -i\mu_0 \omega H_z. \tag{D.29c}$$

Now expressing the first and second equations in terms of the SVE fields, we get:

$$\frac{\partial \left(\Psi_z e^{-iKz} \right)}{\partial y} - \frac{\partial \left(\Psi_y e^{-iKz} \right)}{\partial z} = -i\mu_0 \omega \Phi_x e^{-iKz}; \tag{D.30a}$$

$$\frac{\partial(\Psi_z e^{-iKz})}{\partial x} - \frac{\partial(\Psi_x e^{-iKz})}{\partial z} = i\mu_0\omega\Phi_y e^{-iKz}. \tag{D.30b}$$

Expanding the derivative with respect to the z coordinate, it yields:

$$\frac{\partial\Psi_z}{\partial y}e^{-iKz} - \frac{\partial\Psi_y}{\partial z}e^{-iKz} + iK\Psi_y e^{-iKz} = -i\mu_0\omega\Phi_x e^{-iKz}; \tag{D.31a}$$

$$\frac{\partial\Psi_z}{\partial x}e^{-iKz} - \frac{\partial\Psi_x}{\partial z}e^{-iKz} + iK\Psi_x e^{-iKz} = i\mu_0\omega\Phi_y e^{-iKz}. \tag{D.31b}$$

Now applying the slow varying approximation, $\left|\frac{\partial\Psi_{x,y}}{\partial z}\right| \ll \left|K\Psi_{x,y}\right|$, it simplifies to:

$$\frac{\partial\Psi_z}{\partial y} + iK\Psi_y = -i\mu_0\omega\Phi_x; \tag{D.32a}$$

$$\frac{\partial\Psi_z}{\partial x} + iK\Psi_x = i\mu_0\omega\Phi_y. \tag{D.32b}$$

And finally, the expressions for the magnetic field components are given by:

$$\Phi_x = \frac{i}{\mu_0\omega}\left(\frac{\partial\Psi_z}{\partial y} + iK\Psi_y\right) = \frac{1}{\mu_0\omega}\left(i\frac{\partial\Psi_z}{\partial y} - K\Psi_y\right); \tag{D.33a}$$

$$\Phi_y = \frac{-i}{\mu_0\omega}\left(\frac{\partial\Psi_z}{\partial x} + iK\Psi_x\right) = \frac{-1}{\mu_0\omega}\left(i\frac{\partial\Psi_z}{\partial x} - K\Psi_x\right). \tag{D.33b}$$

Substituting these magnetic field components into the Poynting vector z component (Eq. (D.27)) it yields:

$$\langle S_z\rangle = \frac{1}{2}\mathrm{Re}\left\{\Psi_x\Phi_y^* - \Psi_y\Phi_x^*\right\} = \frac{1}{2\mu_0\omega}\mathrm{Re}\left\{\Psi_x\left(i\frac{\partial\Psi_z^*}{\partial x} + K\Psi_x^*\right) + \Psi_y\left(i\frac{\partial\Psi_z^*}{\partial y} + K\Psi_y^*\right)\right\}. \tag{D.34}$$

Now, the z component of the SVE electric field has to be put as function of the transverse components. To do that, we make use of the first Maxwell equation:

$$\nabla\cdot(\varepsilon\mathcal{E}) = 0 \quad \Rightarrow \quad \nabla\cdot\left(n^2 E\right) = \frac{\partial(n^2 E_x)}{\partial x} + \frac{\partial(n^2 E_y)}{\partial y} + \frac{\partial(n^2 E_z)}{\partial z} = 0. \tag{D.35}$$

If the refractive index varies slowly in the z direction:

$$\frac{\partial n^2}{\partial z} \approx 0, \tag{D.36}$$

then Eq. (D.35) simplifies to:

$$\frac{\partial E_z}{\partial z} \approx -\frac{1}{n^2}\left(\frac{\partial(n^2 E_x)}{\partial x} + \frac{\partial(n^2 E_y)}{\partial y}\right), \tag{D.37}$$

or in terms of the SVE fields:

$$\frac{\partial(\Psi_z e^{-iKz})}{\partial z} \approx -\frac{1}{n^2}\left(\frac{\partial(n^2 \Psi_x e^{-iKz})}{\partial x} + \frac{\partial(n^2 \Psi_y e^{-iKz})}{\partial y}\right). \tag{D.38}$$

Expanding the left hand side of the equation, it reads:

$$\frac{\partial \Psi_z}{\partial z} - iK\Psi_z \approx -\frac{1}{n^2}\left(\frac{\partial(n^2 \Psi_x)}{\partial x} + \frac{\partial(n^2 \Psi_y)}{\partial y}\right). \tag{D.39}$$

Now applying the slow varying approximation, $\left|\dfrac{\partial \Psi_z}{\partial z}\right| \ll |K\Psi_z|$, it results in:

$$\Psi_z \approx -\frac{i}{Kn^2}\left(\frac{\partial(n^2 \Psi_x)}{\partial x} + \frac{\partial(n^2 \Psi_y)}{\partial y}\right). \tag{D.40}$$

Substituting this expression in formula (D.34), it gives:

$$\langle S_z \rangle = \frac{1}{2\mu_0 \omega}\text{Re}\left\{ \begin{array}{l} \Psi_x\left(K\Psi_x^* - \dfrac{\partial}{\partial x}\left[\dfrac{1}{Kn^2}\left(\dfrac{\partial(n^2 \Psi_x^*)}{\partial x} + \dfrac{\partial\left(n^2 \Psi_y^*\right)}{\partial y}\right)\right]\right) + \\[2em] \Psi_y\left(K\Psi_y^* - \dfrac{\partial}{\partial y}\left[\dfrac{1}{Kn^2}\left(\dfrac{\partial(n^2 \Psi_x^*)}{\partial x} + \dfrac{\partial\left(n^2 \Psi_y^*\right)}{\partial y}\right)\right]\right) \end{array} \right\}. \tag{D.41}$$

Rewriting this expression, it finally yields:

$$\langle S_z \rangle = \frac{K}{2\mu_0 \omega}\text{Re}\left\{ \begin{array}{l} |\Psi_x|^2 - \dfrac{\Psi_x}{K^2}\dfrac{\partial}{\partial x}\left[\dfrac{1}{n^2}\left(\dfrac{\partial(n^2 \Psi_x^*)}{\partial x} + \dfrac{\partial\left(n^2 \Psi_y^*\right)}{\partial y}\right)\right] + \\[2em] |\Psi_y|^2 - \dfrac{\Psi_y}{K^2}\dfrac{\partial}{\partial y}\left[\dfrac{1}{n^2}\left(\dfrac{\partial(n^2 \Psi_x^*)}{\partial x} + \dfrac{\partial\left(n^2 \Psi_y^*\right)}{\partial y}\right)\right] \end{array} \right\}. \tag{D.42}$$

Let us now to obtain the finite-difference approximation of this formula using the results provided in Appendix A. For the sake of clarity, we will denote u and v as the x and y components of the SVE electric field, respectively.

Finite difference expression of $\dfrac{\partial}{\partial x}\left[\dfrac{1}{n^2}\left(\dfrac{\partial\left(n^2\Psi_x^*\right)}{\partial x}\right)\right]$

$$\frac{\partial}{\partial x}\left\{\left[\frac{1}{n^2}\left(\frac{\partial(n^2 u^*)}{\partial x}\right)\right]_{i,j}\right\}\approx\frac{1}{\Delta x}\left\{\left[\frac{1}{n^2}\left(\frac{\partial(n^2 u^*)}{\partial x}\right)\right]_{i+1/2,j}-\left[\frac{1}{n^2}\left(\frac{\partial(n^2 u^*)}{\partial x}\right)\right]_{i-1/2,j}\right\}$$

$$\approx\frac{1}{\Delta x}\left[\frac{2}{n_{i+1,j}^2+n_{i,j}^2}\left(\frac{n_{i+1,j}^2 u_{i+1,j}^*-n_{i,j}^2 u_{i,j}^*}{\Delta x}\right)-\frac{2}{n_{i,j}^2+n_{i-1,j}^2}\left(\frac{n_{i,j}^2 u_{i,j}^*-n_{i-1,j}^2 u_{i-1,j}^*}{\Delta x}\right)\right]$$

(D.43)

$$=\frac{2}{(\Delta x)^2}\left[\frac{n_{i+1,j}^2}{n_{i+1,j}^2+n_{i,j}^2}u_{i+1,j}^*-\frac{n_{i,j}^2}{n_{i,j}^2+n_{i-1,j}^2}u_{i,j}^*-\frac{n_{i,j}^2}{n_{i+1,j}^2+n_{i,j}^2}u_{i,j}^*+\frac{n_{i-1,j}^2}{n_{i,j}^2+n_{i-1,j}^2}u_{i-1,j}^*\right]$$

$$=\frac{2}{(\Delta x)^2}\left[\frac{n_{i+1,j}^2}{n_{i+1,j}^2+n_{i,j}^2}u_{i+1,j}^*-\left(\frac{n_{i,j}^2}{n_{i,j}^2+n_{i-1,j}^2}+\frac{n_{i,j}^2}{n_{i+1,j}^2+n_{i,j}^2}\right)u_{i,j}^*+\frac{n_{i-1,j}^2}{n_{i,j}^2+n_{i-1,j}^2}u_{i-1,j}^*\right]$$

Finite difference expression of $\dfrac{\partial}{\partial y}\left[\dfrac{1}{n^2}\left(\dfrac{\partial\left(n^2\Psi_y^*\right)}{\partial y}\right)\right]$

$$\frac{\partial}{\partial y}\left[\frac{1}{n^2}\left(\frac{\partial(n^2 v^*)}{\partial y}\right)\right]\approx$$

$$\frac{2}{(\Delta y)^2}\left[\frac{n_{i,j+1}^2}{n_{i,j+1}^2+n_{i,j}^2}v_{i,j+1}^*-\left(\frac{n_{i,j}^2}{n_{i,j}^2+n_{i,j-1}^2}+\frac{n_{i,j}^2}{n_{i,j+1}^2+n_{i,j}^2}\right)v_{i,j}^*+\frac{n_{i,j-1}^2}{n_{i,j}^2+n_{i,j-1}^2}v_{i,j-1}^*\right]$$

(D.44)

Finite difference expression of $\dfrac{\partial}{\partial x}\left[\dfrac{1}{n^2}\left(\dfrac{\partial\left(n^2\Psi_y^*\right)}{\partial y}\right)\right]$

$$\frac{\partial}{\partial x}\left[\frac{1}{n^2}\left(\frac{\partial(n^2 v^*)}{\partial y}\right)\right]\approx\frac{\partial}{\partial x}\left[\frac{1}{n_{i,j}^2}\left(\frac{n_{i,j+1}^2 v_{i,j+1}^*-n_{i,j-1}^2 v_{i,j-1}^*}{2\Delta y}\right)\right]$$

$$=\frac{1}{2\Delta y}\frac{\partial}{\partial x}\left[\frac{n_{i,j+1}^2}{n_{i,j}^2}v_{i,j+1}^*-\frac{n_{i,j-1}^2}{n_{i,j}^2}v_{i,j-1}^*\right]$$

(D.45)

$$\approx\frac{1}{4\Delta x\Delta y}\left[\frac{n_{i+1,j+1}^2}{n_{i+1,j}^2}v_{i+1,j+1}^*-\frac{n_{i-1,j+1}^2}{n_{i-1,j}^2}v_{i-1,j+1}^*-\frac{n_{i+1,j-1}^2}{n_{i+1,j}^2}v_{i+1,j-1}^*+\frac{n_{i-1,j-1}^2}{n_{i-1,j}^2}v_{i-1,j-1}^*\right]$$

Finite difference expression of $\dfrac{\partial}{\partial y}\left[\dfrac{1}{n^2}\left(\dfrac{\partial\left(n^2\Psi_x^*\right)}{\partial x}\right)\right]$

$$\frac{\partial}{\partial y}\left[\frac{1}{n^2}\left(\frac{\partial(n^2 u^*)}{\partial x}\right)\right]\approx\frac{\partial}{\partial y}\left[\frac{1}{n_{i,j}^2}\left(\frac{n_{i+1,j}^2 u_{i+1,j}^* - n_{i-1,j}^2 u_{i-1,j}^*}{2\Delta x}\right)\right]$$

$$=\frac{1}{2\Delta x\partial y}\frac{\partial}{\partial y}\left[\frac{n_{i+1,j}^2}{n_{i,j}^2}u_{i+1,j}^* - \frac{n_{i-1,j}^2}{n_{i,j}^2}u_{i-1,j}^*\right] \tag{D.46}$$

$$\approx\frac{1}{4\Delta x\Delta y}\left[\frac{n_{i+1,j+1}^2}{n_{i,j+1}^2}u_{i+1,j+1}^* - \frac{n_{i+1,j-1}^2}{n_{i,j-1}^2}u_{i+1,j-1}^* - \frac{n_{i-1,j+1}^2}{n_{i,j+1}^2}u_{i-1,j+1}^* + \frac{n_{i-1,j-1}^2}{n_{i,j-1}^2}u_{i-1,j-1}^*\right]$$

D.2.2   Expression as Function of the Transverse Magnetic Field

Now, we want to express the z component of the Poynting vector (Eq. (D.27)) in terms of the transverse SVE magnetic field components and thus the electric components have to be put as functions of the magnetic field components.

The fourth Maxwell equation $(\nabla\times\mathcal{H})=\varepsilon\dfrac{\partial\mathcal{E}}{\partial t}$ results in:

$$\nabla\times\left[\boldsymbol{H}(r)e^{i\omega t}\right]=\begin{vmatrix}\boldsymbol{u}_x & \boldsymbol{u}_y & \boldsymbol{u}_z \\[4pt] \dfrac{\partial}{\partial x} & \dfrac{\partial}{\partial y} & \dfrac{\partial}{\partial z} \\[4pt] H_x(r) & H_y(r) & H_z(r)\end{vmatrix}e^{i\omega t}$$

$$=\left[\left(\frac{\partial H_z}{\partial y}-\frac{\partial H_y}{\partial z}\right)\boldsymbol{u}_x+\left(\frac{\partial H_x}{\partial z}-\frac{\partial H_z}{\partial x}\right)\boldsymbol{u}_y+\left(\frac{\partial H_y}{\partial x}-\frac{\partial H_x}{\partial y}\right)\boldsymbol{u}_z\right]e^{i\omega t} \tag{D.47}$$

$$=\varepsilon\frac{\partial}{\partial t}\left[\boldsymbol{E}(r)e^{i\omega t}\right]=i\varepsilon\omega\left(E_x\boldsymbol{u}_x+E_y\boldsymbol{u}_y+E_z\boldsymbol{u}_z\right)e^{i\omega t}$$

which provides a relation between the components of the complex amplitudes of the electric and magnetic fields.

Equalling the x, y and z components on this equation, we obtain:

$$\frac{\partial H_z}{\partial y}-\frac{\partial H_y}{\partial z}=i\varepsilon\omega E_x; \tag{D.48a}$$

$$\frac{\partial H_x}{\partial z}-\frac{\partial H_z}{\partial x}=i\varepsilon\omega E_y; \tag{D.48b}$$

$$\frac{\partial H_y}{\partial x}-\frac{\partial H_x}{\partial y}=i\varepsilon\omega E_z. \tag{D.48c}$$

Now expressing the first and second equations in terms of the SVE fields, it gives:

$$\frac{\partial\left(\Phi_z e^{-iKz}\right)}{\partial y} - \frac{\partial\left(\Phi_y e^{-iKz}\right)}{\partial z} = i\varepsilon\omega\Psi_x e^{-iKz}; \tag{D.49a}$$

$$\frac{\partial\left(\Phi_z e^{-iKz}\right)}{\partial x} - \frac{\partial\left(\Phi_x e^{-iKz}\right)}{\partial z} = -i\varepsilon\omega\Psi_y e^{-iKz}. \tag{D.49b}$$

Expanding the derivatives with respect to the z coordinate:

$$\frac{\partial\Phi_z}{\partial y}e^{-iKz} - \frac{\partial\Phi_y}{\partial z}e^{-iKz} + iK\Phi_y e^{-iKz} = i\varepsilon\omega\Psi_x e^{-iKz}; \tag{D.50a}$$

$$\frac{\partial\Phi_z}{\partial x}e^{-iKz} - \frac{\partial\Phi_x}{\partial z}e^{-iKz} + iK\Phi_x e^{-iKz} = -i\varepsilon\omega\Psi_y e^{-iKz}. \tag{D.50b}$$

Now applying the slow varying approximation, $\left|\frac{\partial\Phi_{x,y}}{\partial z}\right| \ll \left|K\Phi_{x,y}\right|$, it results in:

$$\frac{\partial\Phi_z}{\partial y} + iK\Phi_y = i\varepsilon\omega\Psi_x; \tag{D.51a}$$

$$\frac{\partial\Phi_z}{\partial x} + iK\Phi_x = -i\varepsilon\omega\Psi_y. \tag{D.51b}$$

And finally, the expressions for the electric field components are given by:

$$\Psi_x = \frac{-i}{\varepsilon\omega}\left(\frac{\partial\Phi_z}{\partial y} + iK\Phi_y\right) = \frac{-1}{\varepsilon\omega}\left(i\frac{\partial\Phi_z}{\partial y} - K\Phi_y\right); \tag{D.52a}$$

$$\Psi_y = \frac{i}{\varepsilon\omega}\left(\frac{\partial\Phi_z}{\partial x} + iK\Phi_x\right) = \frac{1}{\varepsilon\omega}\left(i\frac{\partial\Phi_z}{\partial x} - K\Phi_x\right). \tag{D.52b}$$

Substituting in the Poynting vector z components (Eq. (D.27)) it yields:

$$\langle S_z\rangle = \frac{1}{2}\mathrm{Re}\left\{\Psi_x\Phi_y^* - \Psi_y\Phi_x^*\right\} = \frac{1}{2\varepsilon\omega}\mathrm{Re}\left\{\left(-i\frac{\partial\Phi_z}{\partial y} + K\Phi_y\right)\Phi_y^* + \left(-i\frac{\partial\Phi_z}{\partial x} + K\Phi_x\right)\Phi_x^*\right\}. \tag{D.53}$$

Now, the z component of the SVE electric field has to be put as function of the transverse components. To do that, we start with the second Maxwell equation:

$$\nabla\cdot\mathcal{H} = 0 \quad\Rightarrow\quad \nabla\cdot H = \frac{\partial H_x}{\partial x} + \frac{\partial H_y}{\partial y} + \frac{\partial H_z}{\partial z} = 0. \tag{D.54}$$

Last equation gives:

$$\frac{\partial H_z}{\partial z} \approx -\left(\frac{\partial H_x}{\partial x} + \frac{\partial H_y}{\partial y}\right), \tag{D.55}$$

or in terms of the SVE fields:

$$\frac{\partial(\Phi_z e^{-iKz})}{\partial z} \approx -\left(\frac{\partial(\Phi_x e^{-iKz})}{\partial x} + \frac{\partial(\Phi_y e^{-iKz})}{\partial y}\right). \tag{D.56}$$

Expanding the derivatives with respect to the z coordinate we get:

$$\frac{\partial \Phi_z}{\partial z} - iK\Phi_z \approx -\left(\frac{\partial \Phi_x}{\partial x} + \frac{\partial \Phi_y}{\partial y}\right). \tag{D.57}$$

Now applying the slow varying approximation, $\left|\frac{\partial \Phi_z}{\partial z}\right| \ll |K\Phi_z|$, it results in:

$$\Phi_z \approx -\frac{i}{K}\left(\frac{\partial \Phi_x}{\partial x} + \frac{\partial \Phi_y}{\partial y}\right). \tag{D.58}$$

Substituting this expression in formula (D.53), it gives:

$$\langle S_z \rangle = \frac{1}{2\varepsilon\omega}\mathrm{Re}\left\{\left(K\Phi_y - \frac{\partial}{\partial y}\left[\frac{1}{K}\left(\frac{\partial \Phi_x}{\partial x} + \frac{\partial \Phi_y}{\partial y}\right)\right]\right)\Phi_y^* + \left(K\Phi_x - \frac{\partial}{\partial x}\left[\frac{1}{K}\left(\frac{\partial \Phi_x}{\partial x} + \frac{\partial \Phi_y}{\partial y}\right)\right]\right)\Phi_x^*\right\}. \tag{D.59}$$

Rewriting this expression, it finally yields:

$$\langle S_z \rangle = \frac{K}{2\varepsilon\omega}\mathrm{Re}\left\{|\Phi_y|^2 - \frac{\Phi_y^*}{K^2}\frac{\partial}{\partial y}\left(\frac{\partial \Phi_x}{\partial x} + \frac{\partial \Phi_y}{\partial y}\right) + |\Phi_x|^2 - \frac{\Phi_x^*}{K^2}\frac{\partial}{\partial x}\left(\frac{\partial \Phi_x}{\partial x} + \frac{\partial \Phi_y}{\partial y}\right)\right\}. \tag{D.60}$$

Let us now obtain the finite-difference approximation of the second-order derivatives in this formula. For the sake of clarity, we will denote u and v as the x and y components of the SVE magnetic field, respectively.

Finite difference expression of $\dfrac{\partial}{\partial y}\left(\dfrac{\partial \Phi_x}{\partial x}\right)$

$$\frac{\partial}{\partial y}\left(\frac{\partial u}{\partial x}\right) \approx \frac{\partial}{\partial y}\left[\frac{u_{i+1,j} - u_{i-1,j}}{2\Delta x}\right] \approx \frac{1}{4\Delta x \Delta y}\left[u_{i+1,j+1} - u_{i+1,j-1} - u_{i-1,j+1} + u_{i-1,j-1}\right]. \tag{D.61}$$

Finite difference expression of $\dfrac{\partial}{\partial y}\left(\dfrac{\partial \Phi_y}{\partial y}\right)$

$$\frac{\partial}{\partial y}\left(\frac{\partial v}{\partial y}\right) \approx \frac{1}{\Delta y}\left[\left(\frac{\partial v}{\partial y}\right)_{i,j+1/2} - \left(\frac{\partial v}{\partial y}\right)_{i,j-1/2}\right] \approx \frac{1}{\Delta y}\left[\frac{\left(v_{i,j+1}-v_{i,j}\right)}{\Delta y} - \frac{\left(v_{i,j}-v_{i,j-1}\right)}{\Delta y}\right]$$

$$\approx \frac{1}{(\Delta y)^2}\left(v_{i,j+1}-2v_{i,j}+v_{i,j-1}\right) \tag{D.62}$$

Finite difference expression of $\dfrac{\partial}{\partial x}\left(\dfrac{\partial \Phi_x}{\partial x}\right)$

$$\frac{\partial}{\partial x}\left(\frac{\partial u}{\partial x}\right) \approx \frac{1}{\Delta x}\left[\left(\frac{\partial u}{\partial x}\right)_{i+1/2,j} - \left(\frac{\partial u}{\partial x}\right)_{i-1/2,j}\right] \approx \frac{1}{\Delta x}\left[\frac{\left(u_{i+1,j}-u_{i,j}\right)}{\Delta x} - \frac{\left(u_{i,j}-u_{i-1,j}\right)}{\Delta x}\right]$$

$$\approx \frac{1}{(\Delta x)^2}\left(u_{i+1,j}-2u_{i,j}+u_{i-1,j}\right) \tag{D.63}$$

Finite difference expression of $\dfrac{\partial}{\partial x}\left(\dfrac{\partial \Phi_y}{\partial y}\right)$

$$\frac{\partial}{\partial x}\left(\frac{\partial v}{\partial y}\right) \approx \frac{\partial}{\partial x}\left(\frac{v_{i,j+1}-v_{i,j-1}}{2\Delta y}\right) \approx \frac{1}{4\Delta x \Delta y}\left(v_{i+1,j+1}-v_{i-1,j+1}-v_{i+1,j-1}+v_{i-1,j-1}\right). \tag{D.64}$$

Reference

[1] Yamauchi, J., Nito, Y. and Nakano, H. (2009) A modified semivectorial BPM retaining the effects of the longitudinal field component and its application to the design of a spot-size converter. *Journal of Lightwave Technology* **27**, 2470–2476.

Appendix E

Finite Difference FV-BPM Based on the Electric Field Using the Scheme Parameter Control

In order to implement the alternating direction implicit (ADI) method for full vectorial beam-propagation method (BPM) based on the electric field using the scheme parameter control, we proceed as was shown in Section 3.2.3. The matrix operator in Eq. (3.37) acting on the slowly varying envelope (SVE) of the transverse electric field is split as the sum of two matrix operators as follows:

$$
2in_0k_0\frac{\partial}{\partial z}\begin{bmatrix}\Psi_x\\\Psi_y\end{bmatrix}=\begin{bmatrix}A_x+A_y & C\\ D & B_x+B_y\end{bmatrix}\begin{bmatrix}\Psi_x\\\Psi_y\end{bmatrix}=\left(\begin{bmatrix}A_x & C\\ 0 & B_x\end{bmatrix}+\begin{bmatrix}A_y & 0\\ D & B_y\end{bmatrix}\right)\begin{bmatrix}\Psi_x\\\Psi_y\end{bmatrix}
\tag{E.1}
$$

After discretizing this differential equation by the Crank–Nicolson scheme using the scheme parameter α, we obtain:

$$
(1-\alpha)\frac{2in_0k_0}{\Delta z}\left(\begin{bmatrix}u\\v\end{bmatrix}^{m+1}-\begin{bmatrix}u\\v\end{bmatrix}^{m}\right)=(1-\alpha)\left(\begin{bmatrix}A_x & C\\ 0 & B_x\end{bmatrix}+\begin{bmatrix}A_y & 0\\ D & B_y\end{bmatrix}\right)\begin{bmatrix}u\\v\end{bmatrix}^{m}
\tag{E.2a}
$$

$$
\alpha\frac{2in_0k_0}{\Delta z}\left(\begin{bmatrix}u\\v\end{bmatrix}^{m+1}-\begin{bmatrix}u\\v\end{bmatrix}^{m}\right)=\alpha\left(\begin{bmatrix}A_x & C\\ 0 & B_x\end{bmatrix}+\begin{bmatrix}A_y & 0\\ D & B_y\end{bmatrix}\right)\begin{bmatrix}u\\v\end{bmatrix}^{m+1}
\tag{E.2b}
$$

Adding these two equations, one obtains:

$$
\frac{2in_0k_0}{\Delta z}\left(\begin{bmatrix}u\\v\end{bmatrix}^{m+1}-\begin{bmatrix}u\\v\end{bmatrix}^{m}\right)=\left(\begin{bmatrix}A_x & C\\ 0 & B_x\end{bmatrix}+\begin{bmatrix}A_y & 0\\ D & B_y\end{bmatrix}\right)\left(\alpha\begin{bmatrix}u\\v\end{bmatrix}^{m+1}+(1-\alpha)\begin{bmatrix}u\\v\end{bmatrix}^{m}\right)
\tag{E.3}
$$

Beam Propagation Method for Design of Optical Waveguide Devices, First Edition. Ginés Lifante Pedrola.
© 2016 John Wiley & Sons, Ltd. Published 2016 by John Wiley & Sons, Ltd.

And rearranging terms, it yields:

$$
\left\{ 1 + \frac{i\alpha\Delta z}{2n_0 k_0} \left(\begin{bmatrix} A_x & C \\ 0 & B_x \end{bmatrix} + \begin{bmatrix} A_y & 0 \\ D & B_y \end{bmatrix} \right) \right\} \begin{bmatrix} u \\ v \end{bmatrix}^{m+1}
$$
$$
= \left\{ 1 - \frac{i(1-\alpha)\Delta z}{2n_0 k_0} \left(\begin{bmatrix} A_x & C \\ 0 & B_x \end{bmatrix} + \begin{bmatrix} A_y & 0 \\ D & B_y \end{bmatrix} \right) \right\} \begin{bmatrix} u \\ v \end{bmatrix}^{m}
$$

(E.4)

where we have denoted u and v as the x and y components of the electric field envelope, respectively. Adopting the ADI method, introducing second-order error terms in Δz, the last equation can be written as:

$$
\begin{bmatrix} u \\ v \end{bmatrix}^{m+1} = \frac{\left(1 - \dfrac{i(1-\alpha)\Delta z}{2n_0 k_0} \begin{bmatrix} A_x & C \\ 0 & B_x \end{bmatrix} \right)}{\left(1 + \dfrac{i\alpha\Delta z}{2n_0 k_0} \begin{bmatrix} A_x & C \\ 0 & B_x \end{bmatrix} \right)} \cdot \frac{\left(1 - \dfrac{i(1-\alpha)\Delta z}{2n_0 k_0} \begin{bmatrix} A_y & 0 \\ D & B_y \end{bmatrix} \right)}{\left(1 + \dfrac{i\alpha\Delta z}{2n_0 k_0} \begin{bmatrix} A_y & 0 \\ D & B_y \end{bmatrix} \right)} \cdot \begin{bmatrix} u \\ v \end{bmatrix}^{m}
$$

(E.5)

In this form, this equation can be split in two artificial sub-steps to perform a single propagation step of length Δz, according to the sequence:

$$
\begin{bmatrix} u \\ v \end{bmatrix}^{m+1/2} = \frac{\left(1 - \dfrac{i(1-\alpha)\Delta z}{2n_0 k_0} \begin{bmatrix} A_y & 0 \\ D & B_y \end{bmatrix} \right)}{\left(1 + \dfrac{i\alpha\Delta z}{2n_0 k_0} \begin{bmatrix} A_y & 0 \\ D & B_y \end{bmatrix} \right)} \cdot \begin{bmatrix} u \\ v \end{bmatrix}^{m}
$$

(E.6a)

$$
\begin{bmatrix} u \\ v \end{bmatrix}^{m+1} = \frac{\left(1 - \dfrac{i(1-\alpha)\Delta z}{2n_0 k_0} \begin{bmatrix} A_x & C \\ 0 & B_x \end{bmatrix} \right)}{\left(1 + \dfrac{i\alpha\Delta z}{2n_0 k_0} \begin{bmatrix} A_x & C \\ 0 & B_x \end{bmatrix} \right)} \cdot \begin{bmatrix} u \\ v \end{bmatrix}^{m+1/2}
$$

(E.6b)

Equation (E.6a) is indeed the formal expression of the following equation:

$$
\left(1 + \frac{i\alpha\Delta z}{2n_0 k_0} \begin{bmatrix} A_y & 0 \\ D & B_y \end{bmatrix} \right) \cdot \begin{bmatrix} u \\ v \end{bmatrix}^{m+1/2} = \left(1 - \frac{i(1-\alpha)\Delta z}{2n_0 k_0} \begin{bmatrix} A_y & 0 \\ D & B_y \end{bmatrix} \right) \cdot \begin{bmatrix} u \\ v \end{bmatrix}^{m}
$$

(E.7)

Now, this vectorial equation is separated into its two components. The first component is expressed as:

$$
\left(1 + \frac{i\alpha\Delta z}{2n_0 k_0} A_y \right) u^{m+1/2} = \left(1 - \frac{i(1-\alpha)\Delta z}{2n_0 k_0} A_y \right) u^{m}
$$

(E.8)

The second component is:

$$\left(1+\frac{i\alpha\Delta z}{2n_0k_0}B_y\right)v^{m+1/2}+\frac{i\alpha\Delta z}{2n_0k_0}Du^{m+1/2}=\left(1-\frac{i(1-\alpha)\Delta z}{2n_0k_0}B_y\right)v^m-\frac{i(1-\alpha)\Delta z}{2n_0k_0}Du^m \quad (E.9)$$

Equation (E.8) allows us to obtain the x component of the intermediate field $u^{m+1/2}$ from the known field u^m. Then, Eq. (E.9) is solved to find the y component of the intermediate field $v^{m+1/2}$ from the known fields u^m, $u^{m+1/2}$ and v^m. This concatenated sequence allows us to perform the first sub-step from m to $(m+1/2)$.

To complete the propagation along a distance Δz, it is necessary to carry out a second step. For that purpose, let us consider Eq. (E.6b), which can be expressed in the form:

$$\left(1+\frac{i\alpha\Delta z}{2n_0k_0}\begin{bmatrix}A_x & C\\0 & B_x\end{bmatrix}\right)\cdot\begin{bmatrix}u\\v\end{bmatrix}^{m+1}=\left(1-\frac{i(1-\alpha)\Delta z}{2n_0k_0}\begin{bmatrix}A_x & C\\0 & B_x\end{bmatrix}\right)\cdot\begin{bmatrix}u\\v\end{bmatrix}^{m+1/2} \quad (E.10)$$

Once again, this matrix equation is decomposed into two components, resulting in:

$$\left(1+\frac{i\alpha\Delta z}{2n_0k_0}A_x\right)u^{m+1}+\frac{i\alpha\Delta z}{2n_0k_0}Cv^{m+1}=\left(1-\frac{i(1-\alpha)\Delta z}{2n_0k_0}A_x\right)u^{m+1/2}-\frac{i(1-\alpha)\Delta z}{2n_0k_0}Cv^{m+1/2}$$

$$(E.11a)$$

$$\left(1+\frac{i\alpha\Delta z}{2n_0k_0}B_x\right)v^{m+1}=\left(1-\frac{i(1-\alpha)\Delta z}{2n_0k_0}B_x\right)v^{m+1/2} \quad (E.11b)$$

Using Eq. (E.11b) allows us to calculate the field v^{m+1} from the known field $v^{m+1/2}$. Once this field is obtained, Eq. (E.11a) is used to find the field u^{m+1} from the known fields $u^{m+1/2}$, $v^{m+1/2}$ and v^{m+1}.

Adopting this sequential ADI scheme the stability is assured, and the four equations in finite differences derived form them are all tridiagonal (see Appendix B), thus being a quite efficient way of solving the vectorial propagation of light in three-dimensional structures. Les us now see how to proceed with each of these differential equations in terms of their finite difference versions.

E.1   First Component of the First Step

Regarding the first sub-step, the finite-difference wave equation (E.8), following the Crank–Nicolson scheme, can be rewritten in a more convenient form as:

$$\left(\frac{2in_0k_0}{\Delta z}-\alpha A_y\right)u^{m+1/2}=\left(\frac{2in_0k_0}{\Delta z}+(1-\alpha)A_y\right)u^m \quad (E.12)$$

The finite difference form of $A_y u$ has been shown in Eq. (3.43a). Using this finite difference expression, Eq. (E.12) is finally given by:

$$\frac{2in_0k_0}{\Delta z}u_{i,j}^{m+1/2} - \alpha\frac{u_{i,j-1}^{m+1/2} - 2u_{i,j}^{m+1/2} + u_{i,j-1}^{m+1/2}}{(\Delta y)^2} - \frac{\alpha}{2}k_0^2\left[\left(n_{i,j}^m\right)^2 - n_0^2\right]u_{i,j}^{m+1/2} = \frac{2in_0k_0}{\Delta z}u_{i,j}^m$$

$$+ (1-\alpha)\frac{u_{i,j-1}^m - 2u_{i,j}^m + u_{i,j-1}^m}{(\Delta y)^2} + \frac{(1-\alpha)}{2}k_0^2\left[\left(n_{i,j}^m\right)^2 - n_0^2\right]u_{i,j}^m$$

(E.13)

The coefficients of this tridiagonal system are:

$$a_j = -\frac{\alpha}{(\Delta y)^2} \tag{E.14a}$$

$$b_j = \frac{2in_0k_0}{\Delta z} + \frac{2\alpha}{(\Delta y)^2} - \frac{\alpha}{2}k_0^2\left[\left(n_{i,j}^m\right)^2 - n_0^2\right] \tag{E.14b}$$

$$c_j = -\frac{\alpha}{(\Delta y)^2} \tag{E.14c}$$

$$r_j = \frac{(1-\alpha)}{(\Delta y)^2}\left[u_{i,j-1}^m + u_{i,j+1}^m\right] + \left\{\frac{2in_0k_0}{\Delta z} - \frac{2(1-\alpha)}{(\Delta y)^2} + \frac{(1-\alpha)}{2}k_0^2\left[\left(n_{i,j}^m\right)^2 - n_0^2\right]\right\}u_{i,j}^m \tag{E.14d}$$

E.2   Second Component of the First Step

To establish the finite difference approximation of Eq. (E.9), we put it in the form:

$$\left(\frac{2in_0k_0}{\Delta z} - \alpha B_y\right)v^{m+1/2} - \alpha Du^{m+1/2} = \left(\frac{2in_0k_0}{\Delta z} + (1-\alpha)B_y\right)v^m + (1-\alpha)Du^m \tag{E.15}$$

and we use the finite difference approaches of the operators $B_y v$ and Du as defined in Eqs. (3.43c) and (3.45b). Using these finite-difference expressions, Eq. (E.15) is finally given by:

$$\frac{2in_0k_0}{\Delta z}v_{i,j}^{m+1/2} - \alpha\frac{T_{i,j-1}^m v_{i,j-1}^{m+1/2} - 2R_{i,j}^m v_{i,j}^{m+1/2} + T_{i,j+1}^m v_{i,j+1}^{m+1/2}}{(\Delta y)^2} - \frac{\alpha}{2}k_0^2\left[\left(n_{i,j}^m\right)^2 - n_0^2\right]v_{i,j}^{m+1/2}$$

$$= \frac{2in_0k_0}{\Delta z}v_{i,j}^m + (1-\alpha)\frac{T_{i,j-1}^m v_{i,j-1}^m - 2R_{i,j}^m v_{i,j}^m + T_{i,j+1}^m v_{i,j+1}^m}{(\Delta y)^2} + \frac{(1-\alpha)}{2}k_0^2\left[\left(n_{i,j}^m\right)^2 - n_0^2\right]v_{i,j}^m$$

$$+ \frac{(1-\alpha)}{4\Delta x\Delta y}\left(Z1_{i,j}^m u_{i+1,j+1}^m - Z2_{i,j}^m u_{i+1,j-1}^m - Z3_{i,j}^m u_{i-1,j+1}^m + Z4_{i,j}^m u_{i-1,j-1}^m\right)$$

$$+ \frac{\alpha}{4\Delta x\Delta y}\left(Z1_{i,j}^m u_{i+1,j+1}^{m+1/2} - Z2_{i,j}^m u_{i+1,j-1}^{m+1/2} - Z3_{i,j}^m u_{i-1,j+1}^{m+1/2} + Z4_{i,j}^m u_{i-1,j-1}^{m+1/2}\right)$$

(E.16)

the corresponding coefficients being:

$$a_j = -\frac{\alpha}{(\Delta y)^2} T_{i,j-1}^m \tag{E.17a}$$

$$b_j = \frac{2in_0k_0}{\Delta z} + \frac{2\alpha}{(\Delta y)^2} R_{i,j}^m - \frac{\alpha}{2}k_0^2 \left[\left(n_{i,j}^m \right)^2 - n_0^2 \right] \tag{E.17b}$$

$$c_j = -\frac{\alpha}{(\Delta y)^2} T_{i,j+1}^m \tag{E.17c}$$

$$r_j = \frac{(1-\alpha)}{(\Delta x)^2} \left[T_{i,j-1}^m v_{i,j-1}^m + T_{i,j+1}^m v_{i,j+1}^m \right] + \left\{ \frac{2in_0k_0}{\Delta z} - \frac{2(1-\alpha)R_{i,j}^m}{(\Delta x)^2} + \frac{(1-\alpha)}{2}k_0^2 \left[\left(n_{i,j}^m \right)^2 - n_0^2 \right] \right\} v_{i,j}^m$$

$$+ \frac{(1-\alpha)}{4\Delta x \Delta y} \left(Z1_{i,j}^m u_{i+1,j+1}^m - Z2_{i,j}^m u_{i+1,j-1}^m - Z3_{i,j}^m u_{i-1,j+1}^m + Z4_{i,j}^m u_{i-1,j-1}^m \right)$$

$$+ \frac{\alpha}{4\Delta x \Delta y} \left(Z1_{i,j}^m u_{i+1,j+1}^{m+1/2} - Z2_{i,j}^m u_{i+1,j-1}^{m+1/2} - Z3_{i,j}^m u_{i-1,j+1}^{m+1/2} + Z4_{i,j}^m u_{i-1,j-1}^{m+1/2} \right)$$

$$\tag{E.17d}$$

The coefficients T, R and $Z1$–$Z4$ are those defined in Eqs. (3.44a), (3.44b) and (3.46e)–(3.46h).

E.3   Second Component of the Second Step

Using Eq. (E.11b) in the form:

$$\left(\frac{2in_0k_0}{\Delta z} - \alpha B_x \right) v^{m+1} = \left(\frac{2in_0k_0}{\Delta z} + (1-\alpha)B_x \right) v^{m+1/2} \tag{E.18}$$

Its finite-difference approach is given by:

$$\frac{2in_0k_0}{\Delta z} v_{i,j}^{m+1} - \alpha \frac{v_{i-1,j}^{m+1} - 2v_{i,j}^{m+1} + v_{i+1,j}^{m+1}}{(\Delta x)^2} - \frac{\alpha}{2}k_0^2 \left[\left(n_{i,j}^{m+1} \right)^2 - n_0^2 \right] v_{i,j}^{m+1} = \frac{2in_0k_0}{\Delta z} v_{i,j}^{m+1/2}$$

$$+ (1-\alpha) \frac{v_{i-1,j}^{m+1/2} - 2v_{i,j}^{m+1/2} + v_{i+1,j}^{m+1/2}}{(\Delta x)^2} - \frac{(1-\alpha)}{2}k_0^2 \left[\left(n_{i,j}^m \right)^2 - n_0^2 \right] v_{i,j}^{m+1/2} \tag{E.19}$$

And the coefficients of the tridiagonal system take the expressions:

$$a_i = -\frac{\alpha}{(\Delta x)^2} \tag{E.20a}$$

$$b_i = \frac{2in_0k_0}{\Delta z} + \frac{2\alpha}{(\Delta x)^2} - \frac{\alpha}{2}k_0^2 \left[\left(n_{i,j}^{m+1} \right)^2 - n_0^2 \right] \tag{E.20b}$$

$$c_i = -\frac{\alpha}{(\Delta x)^2} \tag{E.20c}$$

$$r_i = \frac{(1-\alpha)}{(\Delta x)^2}\left[v_{i-1,j}^{m+1/2} + v_{i+1,j}^{m+1/2} \right] + \left\{ \frac{2in_0k_0}{\Delta z} - \frac{2(1-\alpha)}{(\Delta x)^2} + \frac{(1-\alpha)}{2}k_0^2\left[\left(n_{i,j}^m\right)^2 - n_0^2 \right] \right\} v_{i,j}^{m+1/2} \tag{E.20d}$$

E.4   First Component of the Second Step

The last step to complete the propagation is to solve Eq. (E.11a), which is written in the form:

$$\left(\frac{2in_0k_0}{\Delta z} - \alpha A_x \right)u^{m+1} - \alpha C v^{m+1} = \left(\frac{2in_0k_0}{\Delta z} + (1-\alpha)A_x \right)u^{m+1/2} + (1-\alpha)Cv^{m+1/2} \tag{E.21}$$

And its finite-difference expressions are given by:

$$\frac{2in_0k_0}{\Delta z}u_{i,j}^{m+1} - \alpha\frac{T_{i-1,j}^{m+1}u_{i-1,j}^{m+1} - 2R_{i,j}^{m+1}u_{i,j}^{m+1} + T_{i+1,j}^{m+1}u_{i+1,j}^{m+1}}{(\Delta x)^2} - \frac{\alpha}{2}k_0^2\left[\left(n_{i,j}^{m+1}\right)^2 - n_0^2 \right]u_{i,j}^{m+1}$$

$$= \frac{2in_0k_0}{\Delta z}u_{i,j}^{m+1/2} + (1-\alpha)\frac{T_{i-1,j}^m u_{i-1,j}^{m+1/2} - 2R_{i,j}^m u_{i,j}^{m+1/2} + T_{i+1,j}^m u_{i+1,j}^{m+1/2}}{(\Delta x)^2} + \frac{(1-\alpha)}{2}k_0^2\left[\left(n_{i,j}^m\right)^2 - n_0^2 \right]u_{i,j}^{m+1/2}$$

$$+ \frac{(1-\alpha)}{4\Delta x\Delta y}\left(S1_{i,j}^m v_{i+1,j+1}^{m+1/2} - S2_{i,j}^m v_{i+1,j-1}^{m+1/2} - S3_{i,j}^m v_{i-1,j+1}^{m+1/2} + S4_{i,j}^m v_{i-1,j-1}^{m+1/2} \right)$$

$$+ \frac{\alpha}{4\Delta x\Delta y}\left(S1_{i,j}^{m+1} v_{i+1,j+1}^{m+1} - S2_{i,j}^{m+1} v_{i+1,j-1}^{m+1} - S3_{i,j}^{m+1} v_{i-1,j+1}^{m+1} + S4_{i,j}^{m+1} v_{i-1,j-1}^{m+1} \right) \tag{E.22}$$

The coefficients T, R and $S1$–$S4$ are those defined in Eqs. (3.41) and (3.46a)–(3.46d). Finally, the coefficients of this tridiagonal system are given by:

$$a_i = -\frac{\alpha}{(\Delta x)^2}T_{i-1,j}^{m+1} \tag{E.23a}$$

$$b_i = \frac{2in_0k_0}{\Delta z} + \frac{2\alpha}{(\Delta x)^2}R_{i,j}^{m+1} - \frac{\alpha}{2}k_0^2\left[\left(n_{i,j}^{m+1}\right)^2 - n_0^2 \right] \tag{E.23b}$$

$$c_i = -\frac{\alpha}{(\Delta x)^2}T_{i+1,j}^{m+1} \tag{E.23c}$$

$$r_i = \frac{(1-\alpha)}{(\Delta x)^2} \left[T_{i-1,j}^m u_{i-1,j}^{m+1/2} + T_{i+1,j}^m u_{i+1,j}^{m+1/2} \right]$$

$$+ \left\{ \frac{2in_0 k_0}{\Delta z} - \frac{2(1-\alpha)R_{i,j}^m}{(\Delta x)^2} + \frac{(1-\alpha)}{2} k_0^2 \left[\left(n_{i,j}^m \right)^2 - n_0^2 \right] \right\} u_{i,j}^{m+1/2}$$

$$+ \frac{(1-\alpha)}{4\Delta x \Delta y} \left(S1_{i,j}^m v_{i+1,j+1}^{m+1/2} - S2_{i,j}^m v_{i+1,j-1}^{m+1/2} - S3_{i,j}^m v_{i-1,j+1}^{m+1/2} + S4_{i,j}^m v_{i-1,j-1}^{m+1/2} \right)$$

$$+ \frac{\alpha}{4\Delta x \Delta y} \left(S1_{i,j}^{m+1} v_{i+1,j+1}^{m+1} - S2_{i,j}^{m+1} v_{i+1,j-1}^{m+1} - S3_{i,j}^{m+1} v_{i-1,j+1}^{m+1} + S4_{i,j}^{m+1} v_{i-1,j-1}^{m+1} \right)$$

$$(E.23d)$$

Appendix F

Linear Electro-Optic Effect

A crystal is said to be electro-optic if the application of an external electrical field modifies its index ellipsoid. If the coefficients η_i of the impermeability matrix, or $(1/n^2)_i$, which defines the optical indicatrix, change linearly with the external field, the material exhibits linear electro-optic effect, also called the Pockels effect. The modification of the index ellipsoid by the linear electro-optic effect can be described by the electro-optic matrix r_{ij} through the relation [1]:

$$\Delta\eta_i = \sum_{j=1}^{3} r_{ij}E_j, \quad \text{with} \quad i = 1, 2, \ldots, 6,$$

(F.1)

where E_1, E_2 and E_3 stand for E_x, E_y and E_z, respectively. In its expanded form, this relation is expressed as:

$$\begin{pmatrix} \Delta\eta_1 \\ \Delta\eta_2 \\ \Delta\eta_3 \\ \Delta\eta_4 \\ \Delta\eta_5 \\ \Delta\eta_6 \end{pmatrix} = \begin{pmatrix} r_{11} & r_{12} & r_{13} \\ r_{21} & r_{22} & r_{23} \\ r_{31} & r_{32} & r_{33} \\ r_{41} & r_{42} & r_{43} \\ r_{51} & r_{52} & r_{53} \\ r_{61} & r_{62} & r_{63} \end{pmatrix} \begin{pmatrix} E_x \\ E_y \\ E_z \end{pmatrix}.$$

(F.2)

Consequently, the application of the external electric field modifies the index ellipsoid according to:

$$(\eta_1 + \Delta\eta_1)x^2 + (\eta_2 + \Delta\eta_2)y^2 + (\eta_3 + \Delta\eta_3)z^2 + 2(\eta_4 + \Delta\eta_4)zy + 2(\eta_5 + \Delta\eta_5)xz$$
$$+ 2(\eta_6 + \Delta\eta_6)yx = 1$$

(F.3)

Beam Propagation Method for Design of Optical Waveguide Devices, First Edition. Ginés Lifante Pedrola.
© 2016 John Wiley & Sons, Ltd. Published 2016 by John Wiley & Sons, Ltd.

Taking into account the expression of this index ellipsoid, is clear that, in general, the linear electro-optic effect causes both deformation and rotation of the index ellipsoid.

If the original axes were principal axes (which is the most common situation in practice), the application of an external field gives rise not only to changes of the diagonal elements of the permittivity tensor, but also causes the apparition of off-diagonal elements:

$$
\boldsymbol{\varepsilon} = \begin{pmatrix} \varepsilon_x & 0 & 0 \\ 0 & \varepsilon_y & 0 \\ 0 & 0 & \varepsilon_z \end{pmatrix} \xrightarrow{\textit{External field}} \boldsymbol{\varepsilon} = \begin{pmatrix} \varepsilon_{xx} & \varepsilon_{xy} & \varepsilon_{xz} \\ \varepsilon_{yx} & \varepsilon_{yy} & \varepsilon_{yz} \\ \varepsilon_{zx} & \varepsilon_{zy} & \varepsilon_{zz} \end{pmatrix},
\tag{F.4}
$$

$$
\frac{x^2}{n_x^2} + \frac{y^2}{n_y^2} + \frac{z^2}{n_z^2} = 1 \xrightarrow{\textit{External field}}
$$

$$
\left[\frac{1}{n_x^2} + \Delta \left(\frac{1}{n_1^2} \right) \right] x^2 + \left[\frac{1}{n_y^2} + \Delta \left(\frac{1}{n_2^2} \right) \right] y^2 + \left[\frac{1}{n_z^2} + \Delta \left(\frac{1}{n_3^2} \right) \right] z^2 + 2 \left[\Delta \left(\frac{1}{n_4^2} \right) \right] yz
\tag{F.5}
$$

$$
+ 2 \left[\Delta \left(\frac{1}{n_5^2} \right) \right] xz + 2 \left[\Delta \left(\frac{1}{n_6^2} \right) \right] xy = 1.
$$

Under these circumstances, it is always possible to find a new reference system $X'Y'Z'$ in which the permittivity tensor becomes again diagonal and corresponds to the principal axes of the crystal after the application of the external field:

$$
\xrightarrow{\textit{New axes}} \boldsymbol{\varepsilon}' = \begin{pmatrix} \varepsilon_{x'} & 0 & 0 \\ 0 & \varepsilon_{y'} & 0 \\ 0 & 0 & \varepsilon_{z'} \end{pmatrix}.
\tag{F.6}
$$

When referred to this coordinate system, the index ellipsoid takes the reduced form of:

$$
\xrightarrow{\textit{New axes}} \frac{x'^2}{n_{x'}^2} + \frac{y'^2}{n_{y'}^2} + \frac{z'^2}{n_{z'}^2} = 1,
\tag{F.7}
$$

and the refractive indices referred to this coordinate system are given by:

$$
n_{x'} = \sqrt{\frac{\varepsilon_{x'}}{\varepsilon_0}}, n_{y'} = \sqrt{\frac{\varepsilon_{y'}}{\varepsilon_0}}, n_{z'} = \sqrt{\frac{\varepsilon_{z'}}{\varepsilon_0}}.
\tag{F.8}
$$

In order to find the new principal axis and the new principal refractive indexes, we remind that for light polarized along any of the principal axis, the vectors E and D are parallel, and thus in this case it is fulfilled that:

$$
\varepsilon_0 E = \eta D = \frac{1}{n^2} D,
\tag{F.9}
$$

where n represent the refractive index associated for polarization and which is in general different for each principal axis. Therefore, the vector v that defines the direction of one principal axis must satisfy:

$$
\begin{pmatrix} \eta_1 & \eta_6 & \eta_5 \\ \eta_6 & \eta_2 & \eta_4 \\ \eta_5 & \eta_4 & \eta_3 \end{pmatrix} \begin{pmatrix} v_x \\ v_y \\ v_z \end{pmatrix} = \lambda \begin{pmatrix} v_x \\ v_y \\ v_z \end{pmatrix},
\tag{F.10}
$$

which forms an eigenvalue problem. The eigenvalues λ are the inverse of the squared principal refractive indices and the eigenvectors are the direction of the new principal axes.

Equation (F.10) can be transformed in the following matricial equation:

$$
\begin{pmatrix} \eta_1 - \lambda & \eta_6 & \eta_5 \\ \eta_6 & \eta_2 - \lambda & \eta_4 \\ \eta_5 & \eta_4 & \eta_3 - \lambda \end{pmatrix} \begin{pmatrix} v_x \\ v_y \\ v_z \end{pmatrix} = \begin{pmatrix} 0 \\ 0 \\ 0 \end{pmatrix}.
\tag{F.11}
$$

Apart from the trivial solution, this set of homogeneous linear equation has solution only if the determinant of the matrix is zero:

$$
\begin{vmatrix} \eta_1 - \lambda & \eta_6 & \eta_5 \\ \eta_6 & \eta_2 - \lambda & \eta_4 \\ \eta_5 & \eta_4 & \eta_3 - \lambda \end{vmatrix} = 0.
\tag{F.12}
$$

This condition gives rise to a cubic equation in λ; thus, in general we have three different solutions for the principal refractive indices. For each eigenvalue, λ, relation (F.11) is used to obtain the associated eigenvector, that is, the directions of the new principal axes:

$$
\lambda_1 \rightarrow v_1 = (v_{1x}, v_{1y}, v_{1z});
$$

$$
\lambda_2 \rightarrow v_2 = (v_{2x}, v_{2y}, v_{2z});
$$

$$
\lambda_3 \rightarrow v_3 = (v_{3x}, v_{3y}, v_{3z}).
$$

Reference

[1] Davis, C.C., (1996) *Lasers and Electro-Optics: Fundamentals and Engineering*. Cambridge University Press.

Appendix G

Electro-Optic Effect in GaAs Crystal

GaAs is a cubic crystal belonging to the $\bar{4}3m$ symmetry group, with a refractive index of 3.60 at $\lambda = 1\ \mu m$ and it exhibits a linear electro-optic effect. The electro-optic matrix of crystals with this symmetry is given by [1]:

$$r_{ij} = \begin{pmatrix} 0 & 0 & 0 \\ 0 & 0 & 0 \\ 0 & 0 & 0 \\ r_{41} & 0 & 0 \\ 0 & r_{52} & 0 \\ 0 & 0 & r_{63} \end{pmatrix}, \tag{G.1}$$

with $r_{41} = r_{52} = r_{62}$ being the only non-vanishing electro-optic coefficient for point group $\bar{4}3m$. In the case of GaAs at $\lambda = 1\ \mu m$ the value of this electro-optic coefficient is $r_{41} = 1.1 \times 10^{-12}\ m/V$ [2].

We are interested in finding the new principal axis and the principal refractive indices of a GaAs crystal after the application of an external electric field E parallel to the [110] crystallographic direction (Figure G.1) [3].

In the absence of any external electrical field the GaAs crystal is isotropic and its permittivity tensor, impermeability tensor and index ellipsoid are given, respectively, by:

$$\boldsymbol{\varepsilon} = \varepsilon_0 \begin{pmatrix} n^2 & 0 & 0 \\ 0 & n^2 & 0 \\ 0 & 0 & n^2 \end{pmatrix}, \boldsymbol{\eta} = \begin{pmatrix} \dfrac{1}{n^2} & 0 & 0 \\ 0 & \dfrac{1}{n^2} & 0 \\ 0 & 0 & \dfrac{1}{n^2} \end{pmatrix}, \dfrac{x^2}{n^2} + \dfrac{y^2}{n^2} + \dfrac{z^2}{n^2} = 1. \tag{G.2}$$

Beam Propagation Method for Design of Optical Waveguide Devices, First Edition. Ginés Lifante Pedrola.
© 2016 John Wiley & Sons, Ltd. Published 2016 by John Wiley & Sons, Ltd.

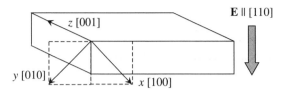

Figure G.1 External electric field applied along the [110] crystallographic direction of a GaAs electro-optic crystal

Equation (G.1) is used to obtain the variation of the impermeability matrix elements after the application of an external electric field E parallel to the [110] crystallographic direction $(E = E(1/\sqrt{2}, 1/\sqrt{2}, 0))$, yielding:

$$\Delta\eta_1 = \Delta\eta_2 = \Delta\eta_3 = \Delta\eta_6 = 0; \tag{G.3a}$$

$$\Delta\eta_4 = \Delta\eta_5 = \frac{rE}{\sqrt{2}}, \tag{G.3b}$$

where we have denoted r to be the electro-optic coefficient r_{41} for the sake of clarity.

Thus, the impermeability matrix of the perturbed crystal takes the form:

$$\eta = \begin{pmatrix} \dfrac{1}{n^2} & 0 & \dfrac{rE}{\sqrt{2}} \\[2mm] 0 & \dfrac{1}{n^2} & \dfrac{rE}{\sqrt{2}} \\[2mm] \dfrac{rE}{\sqrt{2}} & \dfrac{rE}{\sqrt{2}} & \dfrac{1}{n^2} \end{pmatrix}, \tag{G.4}$$

and its permittivity tensor is readily obtained by using relation 1.49, resulting in:

$$\varepsilon = \frac{\varepsilon_0 n^2}{1-(n^2 rE)^2} \begin{pmatrix} 1 - \dfrac{n^4 r^2 E^2}{2} & \dfrac{n^4 r^2 E^2}{2} & -\dfrac{n^2 rE}{\sqrt{2}} \\[3mm] \dfrac{n^4 r^2 E^2}{2} & 1 - \dfrac{n^4 r^2 E^2}{2} & -\dfrac{n^2 rE}{\sqrt{2}} \\[3mm] -\dfrac{n^2 rE}{\sqrt{2}} & -\dfrac{n^2 rE}{\sqrt{2}} & 1 \end{pmatrix}. \tag{G.5}$$

Let us observe the apparition of off-diagonal elements in these matrices, which indicates that the original axes are no longer the principal crystal directions after the application of the external electric field.

To obtain the principal refractive indices and the principal axes of the perturbed crystal, we must solve the next secular equation for the parameter λ:

$$
\begin{vmatrix}
\dfrac{1}{n^2} - \lambda & 0 & \dfrac{rE}{\sqrt{2}} \\[2mm]
0 & \dfrac{1}{n^2} - \lambda & \dfrac{rE}{\sqrt{2}} \\[2mm]
\dfrac{rE}{\sqrt{2}} & \dfrac{rE}{\sqrt{2}} & \dfrac{1}{n^2} - \lambda
\end{vmatrix} = 0.
\tag{G.6}
$$

The expansion of the determinant gives the following cubic equation:

$$
\left(\frac{1}{n^2} - \lambda \right)^3 - r^2 E^2 \left(\frac{1}{n^2} - \lambda \right) = 0.
\tag{G.7}
$$

The three solutions of this equation for the eigenvalue λ are:

$$
\lambda_1 = \frac{1}{n^2};
\tag{G.8a}
$$

$$
\lambda_2 = \frac{1}{n^2} - rE;
\tag{G.8b}
$$

$$
\lambda_3 = \frac{1}{n^2} + rE.
\tag{G.8c}
$$

Once the eigenvalues λ_1, λ_2 and λ_3 have been determined, the associated eigenvectors (which define the principal direction of the perturbed crystal) are obtained by solving the following vectorial equation:

$$
\begin{pmatrix}
\dfrac{1}{n^2} - \lambda_i & 0 & \dfrac{rE}{\sqrt{2}} \\[2mm]
0 & \dfrac{1}{n^2} - \lambda_i & \dfrac{rE}{\sqrt{2}} \\[2mm]
\dfrac{rE}{\sqrt{2}} & \dfrac{rE}{\sqrt{2}} & \dfrac{1}{n^2} - \lambda_i
\end{pmatrix}
\begin{pmatrix} v_x \\ v_y \\ v_z \end{pmatrix}
=
\begin{pmatrix} 0 \\ 0 \\ 0 \end{pmatrix}.
\tag{G.9}
$$

For the eigenvalue λ_1 this vectorial equation yields:

$$
\begin{pmatrix}
0 & 0 & \dfrac{rE}{\sqrt{2}} \\[2mm]
0 & 0 & \dfrac{rE}{\sqrt{2}} \\[2mm]
\dfrac{rE}{\sqrt{2}} & \dfrac{rE}{\sqrt{2}} & 0
\end{pmatrix}
\begin{pmatrix} v_x \\ v_y \\ v_z \end{pmatrix}
=
\begin{pmatrix} 0 \\ 0 \\ 0 \end{pmatrix}.
\tag{G.10}
$$

After separating this vectorial equation into its components, it gives rise to the following equations:

$$0v_x + 0v_y + \frac{rE}{\sqrt{2}}v_z = 0; \quad \rightarrow v_z = 0, \tag{G.11a}$$

$$0v_x + 0v_y + \frac{rE}{\sqrt{2}}v_z = 0; \quad \rightarrow v_z = 0, \tag{G.11b}$$

$$\frac{rE}{\sqrt{2}}v_x + \frac{rE}{\sqrt{2}}v_y + 0v_z = 0; \quad \rightarrow v_x = -v_y. \tag{G.11c}$$

Therefore, the normalized eigenvector associated to the first eigenvalue is:

$$v_1 = \frac{1}{\sqrt{2}}\begin{pmatrix} 1 \\ -1 \\ 0 \end{pmatrix}. \tag{G.12}$$

The vectorial equation (G.9) for the eigenvalue λ_2 is:

$$\begin{pmatrix} rE & 0 & \dfrac{rE}{\sqrt{2}} \\ 0 & rE & \dfrac{rE}{\sqrt{2}} \\ \dfrac{rE}{\sqrt{2}} & \dfrac{rE}{\sqrt{2}} & rE \end{pmatrix}\begin{pmatrix} v_x \\ v_y \\ v_z \end{pmatrix} = \begin{pmatrix} 0 \\ 0 \\ 0 \end{pmatrix}, \tag{G.13}$$

which results in:

$$rEv_x + 0v_y + \frac{rE}{\sqrt{2}}v_z = 0; \quad \rightarrow v_z = -\sqrt{2}v_x, \tag{G.14a}$$

$$0v_x + rEv_y + \frac{rE}{\sqrt{2}}v_z = 0; \quad \rightarrow v_z = -\sqrt{2}v_y, \tag{G.14b}$$

$$\frac{rE}{\sqrt{2}}v_x + \frac{rE}{\sqrt{2}}v_y + rEv_z = 0; \quad \rightarrow v_z = -\frac{(v_x + v_y)}{\sqrt{2}} \tag{G.14c}$$

The normalized eigenvector associated to the eigenvalue λ_2 is given by:

$$v_2 = \frac{1}{2}\begin{pmatrix} 1 \\ 1 \\ -\sqrt{2} \end{pmatrix}. \tag{G.15}$$

Proceeding in a similar way for the vectorial equation (G.9) using the eigenvalue λ_3, it results in:

$$
\begin{pmatrix}
-rE & 0 & \dfrac{rE}{\sqrt{2}} \\[2mm]
0 & -rE & \dfrac{rE}{\sqrt{2}} \\[2mm]
\dfrac{rE}{\sqrt{2}} & \dfrac{rE}{\sqrt{2}} & -rE
\end{pmatrix}
\begin{pmatrix} v_x \\ v_y \\ v_z \end{pmatrix}
=
\begin{pmatrix} 0 \\ 0 \\ 0 \end{pmatrix}.
\tag{G.16}
$$

Now separating it into its components gives:

$$
-rEv_x + 0v_y + \frac{rE}{\sqrt{2}}v_z = 0; \quad \rightarrow v_z = \sqrt{2}v_x,
\tag{G.17a}
$$

$$
0v_x - rEv_y + \frac{rE}{\sqrt{2}}v_z = 0; \quad \rightarrow v_z = \sqrt{2}v_y,
\tag{G.17b}
$$

$$
\frac{rE}{\sqrt{2}}v_x + \frac{rE}{\sqrt{2}}v_y - rEv_z = 0; \quad \rightarrow v_z = \frac{(v_x + v_y)}{\sqrt{2}}.
\tag{G.17c}
$$

The normalized eigenvector associated to the eigenvalue λ_3 is thus given by:

$$
v_3 = \frac{1}{2}\begin{pmatrix} 1 \\ 1 \\ \sqrt{2} \end{pmatrix}.
\tag{G.18}
$$

Let us note that the three vectors v_1, v_2 and v_3 are orthogonal between them. Referring to the geometry and crystal orientation depicted in Figure E.1, the new principal axes are drawn in Figure G.2. The principal refractive indices associated with the directions defined by the vectors v_1, v_2 and v_3 are deduced from Eq. (G.8a–c) ($\lambda_i = \dfrac{1}{n_i^2}$), yielding:

$$
n_1 = n;
\tag{G.19a}
$$

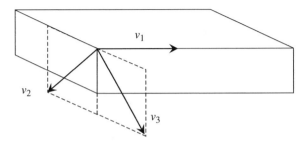

Figure G.2 Directions of the crystal principal axes after the application of an external electric field along the vertical direction ([110] crystallographic direction)

$$n_2 = \frac{n}{\sqrt{1-n^2 rE}};$$ (G.19b)

$$n_3 = \frac{n}{\sqrt{1+n^2 rE}}.$$ (G.19c)

For low electric field values ($n^2 rE \ll 1$), which is fulfilled in most practical situations, the following approximation hold:

$$n_1 = n;$$ (G.20a)

$$n_2 \approx n + \frac{1}{2} n^3 rE;$$ (G.20b)

$$n_3 \approx n - \frac{1}{2} n^3 rE.$$ (G.20c)

Referring now to the coordinate system normally used to define the optical waveguides (Figure G.3), is clear that this reference system $X'Y'Z'$ does not coincide with the principal axis indicated in Figure G.2. Thus, the permittivity matrix in the coordinate system defined in Figure G.3 should have non-null off-diagonal elements. The matrix transformation T between the coordinate system defined by the vectors v_1, v_2 and v_3 and the reference system $X'Y'Z'$ used in Figure G.3 is given by:

$$T = \frac{1}{\sqrt{2}} \begin{pmatrix} 1 & 1 & 0 \\ 0 & 0 & \sqrt{2} \\ 1 & -1 & 0 \end{pmatrix}.$$ (G.21)

Using the matrix transformation, the permeability tensor is calculated by:

$$\varepsilon' = T \varepsilon T^{-1}.$$ (G.22)

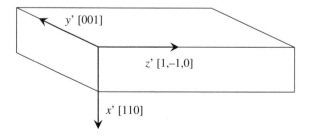

Figure G.3 Coordinate system used to define the optical waveguides and their correspondent crystallographic directions of the GaAs crystal

Taking into account Eq. (G.22) and the expressions (G.5) and (G.22), it yields:

$$
\varepsilon' = \varepsilon_0 \begin{pmatrix} \dfrac{n^2}{1-\left(n^2 rE\right)^2} & \dfrac{-n^4 rE}{1-\left(n^2 rE\right)^2} & 0 \\[4mm] \dfrac{-n^4 rE}{1-\left(n^2 rE\right)^2} & \dfrac{n^2}{1-\left(n^2 rE\right)^2} & 0 \\[4mm] 0 & 0 & n^2 \end{pmatrix}, \tag{G.23}
$$

which represents the permeability matrix of the perturbed crystal expressed in the coordinate system $X'Y'Z'$. These axes conform the usual reference system to describe and analyse the light propagation in waveguides.

Now particularizing for $n = 3.60$, $E = 50$ V/μm and $r = 1.1$ pm/V, results in:

$$
\varepsilon' = \varepsilon_0 \begin{pmatrix} 12.96 & -0.0092 & 0 \\ -0.0092 & 12.96 & 0 \\ 0 & 0 & 12.96 \end{pmatrix}, \tag{G.24}
$$

which is the permeability tensor used in Section 4.6.1 to study the light propagation in anisotropic waveguides.

References

[1] Davis, C.C. (1996) *Lasers and Electro-Optics: Fundamentals and Engineering*, Cambridge University Press.
[2] Namba, S. (1961) Electro-optical effect of zincblende. *Journal of the Optical Society of America* **51**, 76–79.
[3] McKenna, J. and Reinhart, F.K. (1976) Double-heterostructure GaAs-Al$_x$Ga$_{1-x}$As [110] p-n-junction-diode modulator. *Journal of Applied Physics* **47**, 2069–2078.

Appendix H

Electro-Optic Effect in LiNbO$_3$ Crystal

LiNbO$_3$ crystal is an anisotropic uniaxial medium, with $n_o = 2.2865$ and $n_e = 2.2030$ at $\lambda = 633$ nm, which exhibits a linear electro-optic effect. Taking into account that the LiNbO$_3$ crystal belongs to the $3m$ symmetry class, its electro-optic matrix is given by [1]:

$$r_{ij} = \begin{pmatrix} 0 & -r_{22} & r_{13} \\ 0 & r_{22} & r_{13} \\ 0 & 0 & r_{33} \\ 0 & r_{51} & 0 \\ r_{51} & 0 & 0 \\ -r_{22} & 0 & 0 \end{pmatrix}, \tag{H.1}$$

where the electro-optic coefficients at $\lambda = 633$ nm have the following values [2]:

$$r_{13} = 8.6 \, \text{pm/V}, r_{22} = 3.4 \, \text{pm/V}, r_{33} = 30.8 \, \text{pm/V}, r_{51} = 28 \, \text{pm/V}.$$

Here, we are interested in finding the new principal axis and the principal refractive indices when an external electric field E is applied along the x direction (Figure H.1), assuming that $r_{22} \ll r_{51}$.

If the crystal is oriented along their principal axes (Figure H.1), the permittivity tensor, the impermeability tensor and the index ellipsoid are given, respectively, by:

Beam Propagation Method for Design of Optical Waveguide Devices, First Edition. Ginés Lifante Pedrola.
© 2016 John Wiley & Sons, Ltd. Published 2016 by John Wiley & Sons, Ltd.

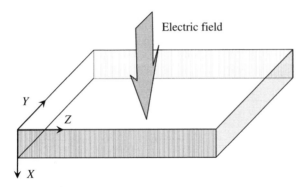

Figure H.1 External electric field applied along the *x*-axis to an electro-optic crystal

$$\boldsymbol{\varepsilon} = \varepsilon_0 \begin{pmatrix} n_o^2 & 0 & 0 \\ 0 & n_o^2 & 0 \\ 0 & 0 & n_e^2 \end{pmatrix}, \boldsymbol{\eta} = \begin{pmatrix} \dfrac{1}{n_0^2} & 0 & 0 \\ 0 & \dfrac{1}{n_0^2} & 0 \\ 0 & 0 & \dfrac{1}{n_e^2} \end{pmatrix}, \dfrac{x^2}{n_o^2} + \dfrac{y^2}{n_o^2} + \dfrac{z^2}{n_e^2} = 1. \qquad (H.2)$$

From Eq. (E.2), and using Eq. (H.1), one obtains the variation of the impermeability matrix after the application of an external electric field parallel to the *x* direction:

$$\Delta\eta_1 = \Delta\eta_2 = \Delta\eta_3 = \Delta\eta_4 = 0; \qquad (H.3a)$$

$$\Delta\eta_5 = r_{51}E; \qquad (H.3b)$$

$$\Delta\eta_6 = -r_{22}E. \qquad (H.3c)$$

In what follows, we will assume that $\Delta\eta_6 \approx 0$ from the approximation $r_{22} \ll r_{51}$.
The impermeability matrix of the perturbed crystal thus takes the form:

$$\boldsymbol{\eta} = \begin{pmatrix} \dfrac{1}{n_o^2} & 0 & r_{51}E \\ 0 & \dfrac{1}{n_o^2} & 0 \\ r_{51}E & 0 & \dfrac{1}{n_e^2} \end{pmatrix}. \qquad (H.4)$$

Now, the permittivity tensor is obtained by using relation 1.49, giving:

$$\varepsilon = \frac{\varepsilon_0 n_o^2 n_e^2}{1 - (r_{51} E n_o n_e)^2} \begin{pmatrix} \dfrac{1}{n_e^2} & 0 & -r_{51}E \\[2mm] 0 & \dfrac{1}{n_e^2} - (r_{51}En_o)^2 & 0 \\[2mm] -r_{51}E & 0 & \dfrac{1}{n_o^2} \end{pmatrix}. \tag{H.5}$$

Let us note that both tensors are symmetric and let us observe the apparition of off-diagonal elements in the matrices.

To obtain the principal refractive indices of the perturbed crystal, we must solve Eq. (F.12):

$$\begin{vmatrix} \dfrac{1}{n_0^2} - \lambda & 0 & r_{51}E \\[2mm] 0 & \dfrac{1}{n_0^2} - \lambda & 0 \\[2mm] r_{51}E & 0 & \dfrac{1}{n_e^2} - \lambda \end{vmatrix} = 0. \tag{H.6}$$

This secular equation results in:

$$\left(\frac{1}{n_o^2} - \lambda \right) \left[\left(\frac{1}{n_o^2} - \lambda \right) \left(\frac{1}{n_e^2} - \lambda \right) - r_{51}^2 E^2 \right] = 0. \tag{H.7}$$

A first solution of this equation can be obtained straightforwardly:

$$\lambda_1 = \frac{1}{n_o^2}. \tag{H.8}$$

Dividing Eq. (H.7) by $\left(\dfrac{1}{n_o^2} - \lambda \right)$ one obtains:

$$\left(\frac{1}{n_o^2} - \lambda \right) \left(\frac{1}{n_e^2} - \lambda \right) = r_{51}^2 E^2. \tag{H.9}$$

Now assuming the approximation:

$$|r_{51}E| << \left| \frac{1}{n_o^2} - \frac{1}{n_e^2} \right|, \tag{H.10}$$

we obtain the two remaining solutions for the eigenvalue, λ:

$$\lambda_2 = \frac{1}{n_o^2} + r_{51}^2 E^2 \left(\frac{1}{n_o^2} - \frac{1}{n_e^2}\right)^{-1}; \tag{H.11a}$$

$$\lambda_3 = \frac{1}{n_o^2} - r_{51}^2 E^2 \left(\frac{1}{n_o^2} - \frac{1}{n_e^2}\right)^{-1}. \tag{H.11b}$$

Once the eigenvalues λ_1, λ_2 and λ_3 have been obtained, we now proceed to calculate the associated eigenvectors.

For the eigenvalue λ_1, the resulting vectorial equation from (H.11a and b) is:

$$\begin{pmatrix} 0 & 0 & r_{51}E \\ 0 & 0 & 0 \\ r_{51}E & 0 & \frac{1}{n_e^2} - \frac{1}{n_o^2} \end{pmatrix} \begin{pmatrix} v_x \\ v_y \\ v_z \end{pmatrix} = \begin{pmatrix} 0 \\ 0 \\ 0 \end{pmatrix}, \tag{H.12}$$

giving rise to the following equations:

$$0v_x + 0v_y + r_{51}Ev_z = 0; \quad \rightarrow v_z = 0, \tag{H.13a}$$

$$0v_x + 0v_y + 0v_z = 0; \tag{H.13b}$$

$$r_{51}Ev_x + 0v_y + \left(\frac{1}{n_e^2} - \frac{1}{n_o^2}\right)v_z = 0; \quad \rightarrow v_x = 0. \tag{H.13c}$$

Therefore, the eigenvector associated to the first eigenvalue is:

$$\mathbf{v}_1 = \begin{pmatrix} 0 \\ 1 \\ 0 \end{pmatrix}. \tag{H.14}$$

The vectorial equation (E.11a and b) for the eigenvalue λ_2 is:

$$\begin{pmatrix} -\dfrac{r_{51}^2 E^2}{\left(\dfrac{1}{n_o^2} - \dfrac{1}{n_e^2}\right)} & 0 & r_{51}E \\ 0 & -\dfrac{r_{51}^2 E^2}{\left(\dfrac{1}{n_o^2} - \dfrac{1}{n_e^2}\right)} & 0 \\ r_{51}E & 0 & \left(\dfrac{1}{n_e^2} - \dfrac{1}{n_o^2}\right) - \dfrac{r_{51}^2 E^2}{\left(\dfrac{1}{n_o^2} - \dfrac{1}{n_e^2}\right)} \end{pmatrix} \begin{pmatrix} v_x \\ v_y \\ v_z \end{pmatrix} = \begin{pmatrix} 0 \\ 0 \\ 0 \end{pmatrix}. \tag{H.15}$$

This vectorial equation results in:

$$-\frac{r_{51}^2 E^2}{\left(\frac{1}{n_o^2}-\frac{1}{n_e^2}\right)}v_x+0v_y+r_{51}Ev_z=0; \qquad \rightarrow v_z=\frac{r_{51}E}{\left(\frac{1}{n_o^2}-\frac{1}{n_e^2}\right)}v_x; \qquad (\text{H.16a})$$

$$0v_x-\frac{r_{51}^2 E^2}{\left(\frac{1}{n_o^2}-\frac{1}{n_e^2}\right)}v_y+0v_z=0; \qquad \rightarrow v_y=0; \qquad (\text{H.16b})$$

$$r_{51}Ev_x+0v_y+\left[\left(\frac{1}{n_e^2}-\frac{1}{n_o^2}\right)-\frac{r_{51}^2 E^2}{\left(\frac{1}{n_o^2}-\frac{1}{n_e^2}\right)}\right]v_z=0; \qquad \rightarrow v_z\approx\frac{r_{51}E}{\left(\frac{1}{n_o^2}-\frac{1}{n_e^2}\right)}v_x, \qquad (\text{H.16c})$$

where we have made use of the approximation Eq. (H.10).

The normalized eigenvector associated to the eigenvalue λ_2 is given by:

$$v_2\approx\frac{1}{\sqrt{1+\frac{r_{51}^2 E^2}{\left(\frac{1}{n_o^2}-\frac{1}{n_e^2}\right)^2}}}\begin{pmatrix}1\\0\\\frac{r_{51}E}{\left(\frac{1}{n_o^2}-\frac{1}{n_e^2}\right)}\end{pmatrix}. \qquad (\text{H.17})$$

Finally, proceeding in a similar way for the vectorial equation (E.11a and b) using eigenvalue λ_3:

$$\begin{pmatrix}\left(\frac{1}{n_o^2}-\frac{1}{n_e^2}\right)+\frac{r_{51}^2 E^2}{\left(\frac{1}{n_o^2}-\frac{1}{n_e^2}\right)} & 0 & r_{51}E\\[2em] 0 & \left(\frac{1}{n_o^2}-\frac{1}{n_e^2}\right)+\frac{r_{51}^2 E^2}{\left(\frac{1}{n_o^2}-\frac{1}{n_e^2}\right)} & 0\\[2em] r_{51}E & 0 & \frac{r_{51}^2 E^2}{\left(\frac{1}{n_o^2}-\frac{1}{n_e^2}\right)}\end{pmatrix}\begin{pmatrix}v_x\\v_y\\v_z\end{pmatrix}=\begin{pmatrix}0\\0\\0\end{pmatrix}, \qquad (\text{H.18})$$

we obtain:

$$\left(\left(\frac{1}{n_o^2}-\frac{1}{n_e^2}\right)+\frac{r_{51}^2 E^2}{\left(\frac{1}{n_o^2}-\frac{1}{n_e^2}\right)}\right)v_x+0v_y+r_{51}Ev_z=0 \qquad \rightarrow v_x\approx-\frac{r_{51}E}{\left(\frac{1}{n_o^2}-\frac{1}{n_e^2}\right)}v_z; \qquad (\text{H.19a})$$

$$0v_x+\left(\left(\frac{1}{n_o^2}-\frac{1}{n_e^2}\right)+\frac{r_{51}^2 E^2}{\left(\frac{1}{n_o^2}-\frac{1}{n_e^2}\right)}\right)v_y+0v_z=0; \qquad v_y=0; \qquad (\text{H.19b})$$

$$r_{51}Ev_x + 0v_y + \frac{r_{51}^2 E^2}{\left(\dfrac{1}{n_o^2} - \dfrac{1}{n_e^2}\right)}v_z = 0; \quad \rightarrow v_x = -\frac{r_{51}E}{\left(\dfrac{1}{n_o^2} - \dfrac{1}{n_e^2}\right)}v_z, \quad \text{(H.19c)}$$

and thus the associated normalized eigenvector is given by:

$$v_3 \approx \frac{1}{\sqrt{1 + \dfrac{r_{51}^2 E^2}{\left(\dfrac{1}{n_o^2} - \dfrac{1}{n_e^2}\right)^2}}}\begin{pmatrix} \dfrac{-r_{51}E}{\left(\dfrac{1}{n_o^2} - \dfrac{1}{n_e^2}\right)} \\ 0 \\ 1 \end{pmatrix}. \quad \text{(H.20)}$$

Let us note that the three vectors v_1, v_2 and v_3 are orthogonal between them. Particularizing for $E = 1$ V/μm, the perturbed permittivity tensor is:

$$\boldsymbol{\varepsilon} = \varepsilon_0 \begin{pmatrix} 5.2280823 & 0 & -0.0007105 \\ 0 & 5.2280822 & 0 \\ -0.0007105 & 0 & 4.853209 \end{pmatrix}, \quad \text{(H.21)}$$

and the new principal refractive indices and the associated principal axes are:

$$n_1 = 2.286500000; \quad v_1 = (0, 1, 0), \quad \text{(H.22a)}$$

$$n_2 = 2.286500317; \quad v_2 = (0.9999982, 0, 0.0018951), \quad \text{(H.22b)}$$

$$n_3 = 2.202999716; \quad v_3 = (-0.0018952, 0, 0.9999982). \quad \text{(H.22c)}$$

Note: if we do not neglect the contribution of the r_{22} EO coefficient (EO, electro-optic), the resulting perturbed permittivity tensor is:

$$\boldsymbol{\varepsilon} = \varepsilon_0 \begin{pmatrix} 5.2280823 & 0.0000929 & -0.0007105 \\ 0.0000929 & 5.2280822 & -0.0000001 \\ -0.0007105 & -0.0000001 & 4.853209 \end{pmatrix}. \quad \text{(H.23)}$$

Now comparing this with the tensor previously obtained, it is apparent that the approximation $r_{22} \ll r_{51}$ is valid for practical purposes.

References

[1] Davis, C.C. (1996) *Lasers and Electro-Optics: Fundamentals and Engineering.* Cambridge University Press.
[2] Ghatak, A.K. and Thyagarajan, K. (1989) *Optical Electronics.* Cambridge University Press.

Appendix I

Padé Polynomials for Wide-Band TD-BPM

To determine the Padé polynomials, let us write the wave equation (4.116) in terms of the temporal envelope scalar field $u(\mathbf{r},t)$ and the centred frequency ω_0:

$$\frac{\partial^2 u}{\partial t^2} + 2i\omega_0 \frac{\partial u}{\partial t} - \frac{c^2}{n^2}\nabla^2 u - \omega_0^2 u = 0. \tag{I.1}$$

This equation can be rewritten in the following formal form:

$$\frac{\partial u}{\partial t} = -i\omega_0 \frac{\frac{\mathcal{P}}{2}}{1 - \frac{i}{2\omega_0}\frac{\partial}{\partial t}} u, \tag{I.2}$$

thus providing a recursive way to include accurately the effect of the temporal derivative:

$$\left.\frac{\partial}{\partial t}\right|_n = -i\omega_0 \frac{\frac{\mathcal{P}}{2}}{1 - \frac{i}{2\omega_0}\left.\frac{\partial}{\partial t}\right|_{n-1}}. \tag{I.3}$$

From this expression, the first Padé approximants can be easily obtained, assuming that the zero-order partial t-derivative is zero:

$$\left.\frac{\partial}{\partial t}\right|_0 = 0; \tag{I.4}$$

Beam Propagation Method for Design of Optical Waveguide Devices, First Edition. Ginés Lifante Pedrola.
© 2016 John Wiley & Sons, Ltd. Published 2016 by John Wiley & Sons, Ltd.

$$\frac{\partial u}{\partial t} = -i\omega_0\frac{\mathcal{P}}{2}u, \tag{I.5}$$

and therefore:

$$\left.\frac{\partial}{\partial t}\right|_1 = -i\omega_0\frac{\dfrac{\mathcal{P}}{2}}{1} = -i\omega_0\frac{M_1}{N_0}, \tag{I.6}$$

which is the Padé (1,0) approximant and constitutes the basic approximation for narrow-band time-domain beam propagation method (TD-BPM). The next-order approximation is obtained by substituting the last derivative in Eq. (I.3):

$$\left.\frac{\partial}{\partial t}\right|_2 = -i\omega_0\frac{\dfrac{\mathcal{P}}{2}}{1 - \dfrac{i}{2\omega_0}\left(-\dfrac{i\omega_0\mathcal{P}}{2}\right)} = -i\omega_0\frac{\dfrac{\mathcal{P}}{2}}{1 - \dfrac{\mathcal{P}}{4}}. \tag{I.7}$$

This corresponds to the Padé (1,1) approximant. The next order approaches are:

$$\left.\frac{\partial}{\partial t}\right|_3 = -i\omega_0\frac{\dfrac{\mathcal{P}}{2}}{1 - \dfrac{i}{2\omega_0}\left(-i\omega_0\dfrac{\dfrac{\mathcal{P}}{2}}{1 - \dfrac{\mathcal{P}}{4}}\right)} = -i\omega_0\frac{\dfrac{\mathcal{P}}{2} - \dfrac{\mathcal{P}^2}{8}}{1 - \dfrac{\mathcal{P}}{2}}; \tag{I.8}$$

$$\left.\frac{\partial}{\partial t}\right|_4 = -i\omega_0\frac{\dfrac{\mathcal{P}}{2}}{1 - \dfrac{i}{2\omega_0}\left(-i\omega_0\dfrac{\dfrac{\mathcal{P}}{2} - \dfrac{\mathcal{P}^2}{8}}{1 - \dfrac{\mathcal{P}}{2}}\right)} = -i\omega_0\frac{\dfrac{\mathcal{P}}{2} - \dfrac{\mathcal{P}^2}{4}}{1 - \dfrac{3\mathcal{P}}{4} + \dfrac{\mathcal{P}^2}{16}}. \tag{I.9}$$

This last one is, in fact, the Padé (2,2) approximant. Finally, proceeding in the same way, we obtain the Padé (3,3) approximants as follows:

$$\left.\frac{\partial}{\partial t}\right|_5 = -i\omega_0\frac{\dfrac{\mathcal{P}}{2}}{1 - \dfrac{i}{2\omega_0}\left(-i\omega_0\dfrac{\dfrac{\mathcal{P}}{2} - \dfrac{\mathcal{P}^2}{4}}{1 - \dfrac{3\mathcal{P}}{4} + \dfrac{\mathcal{P}^2}{16}}\right)} = -i\omega_0\frac{\dfrac{\mathcal{P}}{2} - \dfrac{3\mathcal{P}^2}{8} + \dfrac{\mathcal{P}^3}{32}}{1 - \mathcal{P} + \dfrac{3\mathcal{P}^2}{16}}; \tag{I.10}$$

$$\frac{\partial}{\partial t}\bigg|_6 = -i\omega_0 \cfrac{\dfrac{\mathcal{P}}{2}}{1 - \dfrac{i}{2\omega_0}\left(-i\omega_0\cfrac{\dfrac{\mathcal{P}}{2} - \dfrac{3\mathcal{P}^2}{8} + \dfrac{\mathcal{P}^3}{32}}{1 - \mathcal{P} + \dfrac{3\mathcal{P}^2}{16}}\right)} = -i\omega_0 \cfrac{\dfrac{\mathcal{P}}{2} - \dfrac{\mathcal{P}^2}{2} + \dfrac{3\mathcal{P}^3}{32}}{1 - \dfrac{5\mathcal{P}}{4} + \dfrac{3\mathcal{P}^2}{8} - \dfrac{\mathcal{P}^3}{64}}. \tag{I.11}$$

These results are summarized in Table 4.3.

Appendix J

Obtaining the Dispersion Relation for a Monomode Waveguide Using FDTD

The longitudinal propagation constant β of a monomode waveguide can be computed by means of FDTD (finite-difference time domain) simulation. To do that, a pulse is launched into the waveguide, and the light propagation is performed by FDTD. By taking the ratio of the discrete Fourier transforms of the pulse at two points along the waveguide longitudinal axis, the propagation constant β of the guided mode can be computed over the pulse's full bandwidth [1].

The E_z field component is recorded as the pulsed waveguide mode propagates past two observation points located along the centre of the core at the opposite end of the waveguide. Then, the frequency spectrum $w(z,\omega)$ is obtained by FFT of the temporal evolution of the field at both locations:

$$w(z,\omega) = \frac{1}{\sqrt{2\pi}} \int\limits_{-\infty}^{+\infty} E_z(z,t)e^{i\omega t}dt. \tag{J.1}$$

Now, it is assumed that the frequency spectra at two separated points z and $(z+\Delta z)$ are related through:

$$w(z+\Delta z,\omega) = w(z,\omega)e^{-i\beta\Delta z}, \tag{J.2}$$

Beam Propagation Method for Design of Optical Waveguide Devices, First Edition. Ginés Lifante Pedrola.
© 2016 John Wiley & Sons, Ltd. Published 2016 by John Wiley & Sons, Ltd.

which can be used to calculate the relation between the propagation constant of the waveguide mode β and the angular frequency ω:

$$\beta(\omega) = \frac{i}{\Delta z} \ln \left[\frac{w(z + \Delta z, \omega)}{w(z, \omega)} \right], \tag{J.3}$$

which gives the relation dispersion $\omega(\beta)$ for the waveguide mode. From this formula the phase velocity can then be easily calculated:

$$v_{phase} \equiv \frac{\omega}{\beta(\omega)}. \tag{J.4}$$

Reference

[1] Hagness, S.C., Rafizadeh, D., Ho, S.T. and Taflove, A. (1997) FDTD microcavity simulations: design and experimental realization of waveguide-coupled single-mode ring and whispering-gallery-mode disk resonators. *Journal of Lightwave Technology* **15**, 2154–2165.

Appendix K

Electric Field Distribution in Coplanar Electrodes

Optical modulation or switching in integrated photonic devices can be accomplished through the refractive index change of the electro-optic (EO) material in the waveguide region by applying an external modulating electric field and modifying the propagation constant of the guided optical modes. The external electrical field is often applied by using metal stripes as electrodes, placed side by side on the surface of the substrate.

Here we examine the electric field distribution produced by two common electrode configurations: the symmetric coplanar strip (CS) and the symmetric complementary coplanar strip (CCS) (Figure K.1). Closed form analytical expressions are presented for these two types of electrode, derived from conformal mapping techniques [1] and they are applied to study the electric field distribution in LiNbO$_3$ crystals The optimum position of the waveguides with respect to the electrodes is then obtained by computing the overlap integral of the externally applied electric field and the optical field of the waveguide mode.

K.1 Symmetric Coplanar Strip Configuration

The symmetric CS configuration is shown in Figure K.2, which consists of two identical electrodes situated symmetrically from $x = 0$, and located at the substrate surface at the plane $y = 0$. The location of the electrodes are determined by setting the positions b and a, which are the x coordinates limits of the right electrode. The electrodes width is then $W = (a - b)$, and the gap between them is $G = 2b$.

Beam Propagation Method for Design of Optical Waveguide Devices, First Edition. Ginés Lifante Pedrola.
© 2016 John Wiley & Sons, Ltd. Published 2016 by John Wiley & Sons, Ltd.

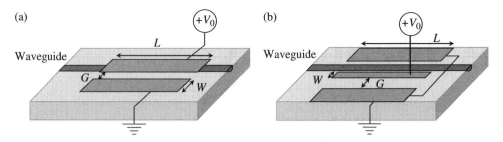

Figure K.1   (a) Coplanar electrode configuration (CS) and (b) complementary coplanar strip (CCS)

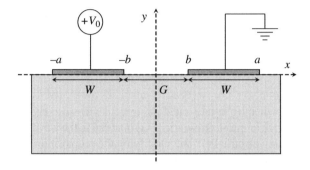

Figure K.2   Geometry and relevant parameters of a coplanar electrode configuration

In what follows, the substrate is assumed to be anisotropic, with relative permittivities ε_x and ε_y. The x and y components of the electric field can be calculated by using conformal transformations [1], and are given by:

$$E_x = -\frac{V_0}{2K(k)}\mathrm{Re}(f(z)); \tag{K.1}$$

$$E_y = \sqrt{\frac{\varepsilon_x}{\varepsilon_y}}\frac{V_0}{2K(k)}\mathrm{Im}(f(z)), \tag{K.2}$$

where the complex function $f(z)$ is:

$$f(z) = \frac{b}{\sqrt{(b^2 - k^2 z^2)(b^2 - z^2)}}, \tag{K.3}$$

with $k = b/a$, and the complex variable z is related with the x and y coordinates by:

$$z = x + i\sqrt{\frac{\varepsilon_x}{\varepsilon_y}}y. \tag{K.4}$$

On the other hand, the quantity K in the expressions of the field is the complete elliptic integral $K(k)$ of the first kind, defined by [2]:

$$K(k) = \int_0^{\pi/2} \frac{d\theta}{\sqrt{1-k^2\sin^2\theta}}. \tag{K.5}$$

The quantity k (with $k = b/a$ in our case) is called the modulus of the elliptic integral $K(k)$. If the complementary modulus k' is defined by:

$$k' = \sqrt{1-k^2}, \tag{K.6}$$

then the capacitance per unit length C/L and the characteristic impedance Z_c of the strip line of CS configuration can be expressed by [3]:

$$C/L = \varepsilon_0 \varepsilon_{eff} \frac{K(k')}{K(k)}; \tag{K.7}$$

$$Z_c = \frac{120\pi}{\sqrt{\varepsilon_{eff}}} \frac{K(k)}{K(k')}, \tag{K.8}$$

where L is the electrode length and the effective permittivity ε_{eff} is defined by:

$$\varepsilon_{eff} = \left(1 + \sqrt{\varepsilon_x \varepsilon_y}\right)/2. \tag{K.9}$$

The capacitance determines the maximum operation frequency of the modulator and the characteristic impedance should be chosen to match the impedance of the driving source, typically 50 Ω. These two magnitudes are important design parameters for determining the performance of EO-modulators [4].

For using the strongest r_{33} EO coefficient of the LiNbO$_3$ crystal for EO modulation, a c-cut oriented substrate can be used under the CS electrode configuration (the c-axis refers to the optical axis of a uniaxial crystal). Under this geometry, the responsible (external) electric field component for EO modulation using the r_{33} coefficient is the vertical component (parallel to the optical c-axis of the crystal) and the affected electrical field component of the optical mode is also its vertical component (this direction corresponds to the y-axis in Figure K.3). Therefore, a quasi-transverse magnetic field (TM) polarized waveguide mode must be used. The electric field lines induced by CS electrodes with $b = 4$ μm and $a = 16$ μm, computed by using formulas (K.1) and (K.2), are plotted in Figure K.3, for a c-cut LiNbO$_3$ substrate, where at high frequencies its relative permittivities are $\varepsilon_x = 43$ and $\varepsilon_y = 28$ [5]. According to the electric field lines, the optical waveguide should be positioned below one of the electrodes to utilize the vertical component of the external electric field. For this modulator a capacitance per unit length of 2.8 pF/cm is obtained, having a characteristic impedance of 50 Ω.

In an X-cut LiNbO$_3$ substrate, and using the largest r_{33} EO coefficient, the waveguide should be positioned between the electrodes, and the optical mode should be a Y-propagating quasi-TE-polarized mode [4].

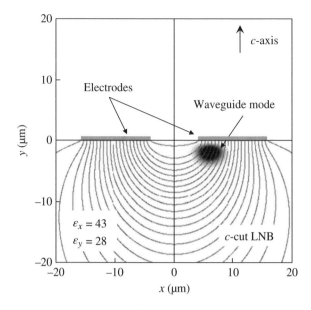

Figure K.3 Electric field lines produced by symmetric coplanar strip electrode configuration in a *c*-cut LiNbO$_3$ substrate

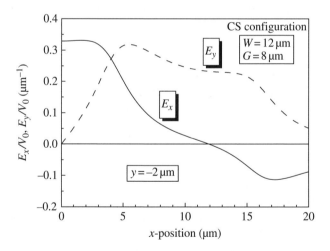

Figure K.4 Electric field components (normalized to V_0) as function of the *x*-position created by coplanar strip electrodes, at a depth of $y = -2\,\mu m$, in a *c*-cut LiNO$_3$ substrate. Parameters of the electrodes are given in the inset

Figure K.4 plots the *x* and *y* components of the electric field (normalized to the voltage V_0) as function of the *x* coordinate, calculated along the line $y = -2\,\mu m$. According with the figure, as we are interested on using the E_y component, the waveguide should be positioned around $x \sim 6\,\mu m$, where the field reaches its maximum value. Nevertheless, to determine the optimum position of the waveguide for EO modulation a closer analysis must be performed, to find the conditions under which the propagation constant of the waveguide mode is more sensitive to the external field. This analysis takes into account both the externally applied field profile and the optical profile of the waveguide mode.

The change in the propagation constant $\Delta\beta$ of a waveguide mode in response to an externally applied voltage can be derived from a perturbation analysis [1]. In particular, the $\Delta\beta$ corresponding to a TM-guided mode for the configuration described is given by:

$$\Delta\beta = \frac{2\pi}{\lambda}\Delta n = \frac{\pi}{\lambda}n^3 r_{33}\frac{V_0}{G}\Gamma, \tag{K.10}$$

where n is the extraordinary index at the wavelength λ, and Γ is the overlap integral factor defined as [6]:

$$\Gamma = \frac{G}{V_0}\frac{\iint E_{op}^2(x,y)E_{el}(x,y)dxdy}{\iint E_{op}^2(x,y)dxdy}. \tag{K.11}$$

here, E_{op} is the optical field component of the waveguide mode and E_{el} is the component of the externally applied electric field, where both components are parallel to the optical axis under the present configuration (y components refer to the axis in Figure K.3). It is clear that the overlap depends on the transversal mode distribution of the waveguide mode. Thus, to find the optimum placement of electrodes with respect to the position of channel waveguides, the optical field profile of the waveguide must be taken into account, in addition of the field profile induced by the metallic electrodes. In what follows, we assume a Hermite–Gaussian function for the fundamental mode of the channel waveguide and thus:

$$E_{op} = Ay\exp\left[-\frac{1}{2}\left(\frac{x-x_0}{w_x}\right)^2\right]\exp\left[-\frac{1}{2}\left(\frac{y}{w_y}\right)^2\right], \tag{K.12}$$

where $2w_x$ and $1.38w_y$ are the widths of $1/e$ mode intensity along the x and y directions, respectively, and x_0 denotes the shift along the x-axis.

Figure K.5 shows the overlap factor Γ for CS configuration as function of the lateral displacement x_0 of the waveguide mode, using electrodes with $b = 4\,\mu m$ and $a = 16\,\mu m$, and for a waveguide mode with $w_x = w_y = 2\,\mu m$. The optimum position of the waveguide is at $x = 6.0\,\mu m$, where the overlap factor reaches a value of $\Gamma = 0.46$. For a uniform field, the overlap factor obviously is 1.

Besides matched impedance and high operation frequency, the modulator must be designed to have a low value of operation voltage. The figure of merit which describes this performance parameter is given by the product of the necessary voltage to induce a phase shift of π (V_π) for a given modulator length (L) and can be readily obtained from Eq. (K.10):

$$\Delta\beta L = \pi \quad\Rightarrow\quad V_\pi L = \frac{G}{n^3 r_{33}}\frac{\lambda}{\Gamma}. \tag{K.13}$$

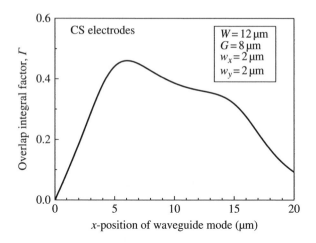

Figure K.5 Overlap integral factor for a Hermite–Gaussian fundamental mode ($w_x = w_y = 2\,\mu$m) with the field induced by CS electrodes ($W = 12\,\mu$m, $G = 8\,\mu$m) in a c-cut LiNbO$_3$ substrate, as a function of the x-position of the waveguide

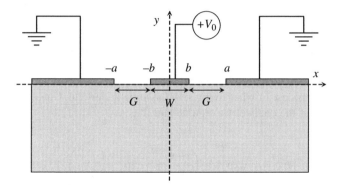

Figure K.6 Geometry and relevant parameters of the symmetric complementary coplanar electrode configuration

Using the parameters for LiNbO$_3$ ($n_e = 2.203$, $r_{33} = 30.5 \times 10^{-12}$ m/V at $\lambda = 0.633\,\mu$m) and the results of the CS configuration described previously ($G = 8\,\mu$m, $\Gamma = 0.46$), this results in a figure of merit of 3.4 V·cm.

K.2   Symmetric Complementary Coplanar Strip Configuration

The symmetric CCS configuration is shown in Figure K.6, which consists of a central electrode of width W (centred at $x = 0$) symmetrically located between two semi-infinite electrodes separated at a distance G. The position of the electrodes are determined by setting the values b and a, which are the x coordinates of the right gap. The central electrode width is then $W = 2b$ and the gap between the electrodes is $G = a - b$. While the central electrode is connected to a voltage V_0, the lateral electrodes are grounded.

The x and y components of the electric field generated by the metallic electrodes for the CCS configuration are expressed as [1]:

$$E_x = -\frac{U}{2K'(k)}\,\mathrm{Im}(f(z));$$

(K.14)

$$E_y = -\sqrt{\frac{\varepsilon_x}{\varepsilon_y}}\,\frac{U}{2K'(k)}\,\mathrm{Re}(f(z)),$$

(K.15)

where the complementary elliptic integral $K'(k)$ of modulus k is related to the elliptic integral function K by:

$$K'(k) = K(k') = K\left(\sqrt{1-k^2}\right).$$

(K.16)

The capacitance per unit length C/L and the characteristic impedance Z_c of the strip line of CCS configuration are given by [7]:

$$C/L = 4\varepsilon_0 \varepsilon_{eff}\,\frac{K(k)}{K(k')};$$

(K.17)

$$Z_c = \frac{30\pi}{\sqrt{\varepsilon_{eff}}}\,\frac{K(k')}{K(k)}.$$

(K.18)

As an example, the electric field lines generated by this electrode configuration in an x-cut LiNbO$_3$ substrate are drawn in FigureK.7 for a width of $W = 4\,\mu m$ and a gap of $G = 12\,\mu m$. The

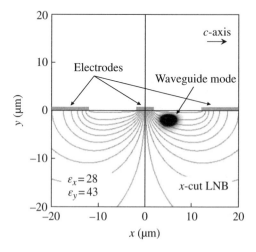

Figure K.7 Electric field lines induced by CCS electrodes in an x-cut LiNbO$_3$ substrate. The c-axis of the uniaxial crystal lies along the x-axis in the figure

crystallographic direction of the optical c-axis is taken parallel to the x-axis in the figure, thus the permittivities are now $\varepsilon_x = 28$ and $\varepsilon_y = 43$. This configuration, and the mentioned electrode parameters, corresponds to a capacitance per unit length of 3.1 pF/cm and a characteristic impedance of 45 Ω.

The CCS geometry can be used to provide strong and quite uniform electric field horizontal component [8] in the region between the electrodes, and thus the waveguide should be placed below the electrode gap. Under this geometry, the r_{33} EO coefficient of the LiNbO$_3$ crystal can be used. To do that, the sample should be an X-cut substrate (optical c-axis along the x direction

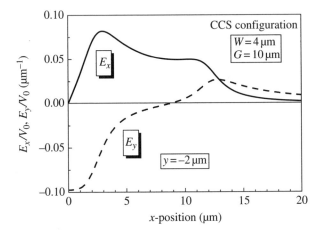

Figure K.8 Normalized electric field amplitude induced by CCS electrodes in an x-cut LiNbO$_3$ substrate (c-axis along the horizontal direction). Continuous line: horizontal component; dashed line: vertical component

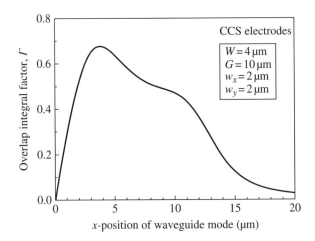

Figure K.9 Overlap integral factor between the horizontal electric field component of the CCS electrodes and the optical field of a quasi-TE mode as function of its lateral displacement

in Figure K.7) and the optical mode should be also polarized along this direction. Therefore, the light propagating along the waveguide should travel as a quasi-transverse electric field (TE) mode. Figure K.8 shows the horizontal and vertical electric field components created by the CCS electrodes, calculated by using Eqs. (K.14) and (K.15) at a depth of $y = -2\,\mu m$. The maximum of the horizontal field component is located at $x_0 \approx 3\,\mu m$.

To find the optimum placement of the waveguide with respect to the electrodes, the overlap integral is computed, taking into account the waveguide mode profile (Eq. (K.12)) at $\lambda = 0.633\,\mu m$ and the horizontal component of the externally applied field computed by Eq. (K.14). The result is presented in Figure K.9 and indicates that the waveguide should be positioned at around $x_0 = 3.9\,\mu m$. The placement of the waveguide mode at this position gives a value of the overlap integral factor of $\Gamma = 0.68$ and a figure of merit of 2.9 V·cm is obtained.

References

[1] Ramer, O.G. (1982) Integrated optic electrooptic modulator electrode analysis. *IEEE Journal of Quantum Electronics* **QE-18**, 386–392.

[2] Press, W.H., Teukolsky, S.A., Vetterling, W.T. and Flannery, B.P. (1992) *Numerical Recipes in FORTRAN 77: The Art of Scientific Computing*, Chapter 6, 2nd edn, Cambridge University Press.

[3] Kubota, K., Noda, J. and Mikami, O. (1980) Travelling wave optical modulator using a directional coupler LiNbO₃ waveguide. *IEEE Journal of Quantum Electronics* **QE-16**, 754–760.

[4] Alferness, R.C. (1982) Waveguide electrooptic modulators. *IEEE Transactions of Microwave Theory and Techniques* **MTT-30**, 1121–1137.

[5] Turner, E.H. (1966) High-frequency electro-optic coefficients of lithium niobate. *Applied Physics Letters* **8**, 303–305.

[6] Marcuse, D. (1974) *Theory of Dielectric Waveguides*, New York: Academic Press.

[7] Kim, C.M. and Ramaswamy, R.V. (1989) Overlap integral factors in integrated optic modulators and switches. *Journal of Lightwave Technology* **7**, 1063–1070.

[8] Becker, R.A. and Kincaid, B.E. (1993) Improved electrooptic efficiency in guided-wave modulators. *Journal of Lightwave Technology* **11**, 2076–2079.

Appendix L

Three-Dimensional Anisotropic BPM Based on the Electric Field Formulation

Here, we develop the numerical algorithm for 3D-light propagation in anisotropic inhomogeneous structures, following the finite-difference scheme and using the alternating direction implicit (ADI) approach and the efficient Thomas algorithm. The formulation based on the electric field is used for the derivation of the finite-difference beam propagation method (FD-BPM) equations.

We start with the full vectorial wave equations for the transversal slowly varying envelope (SVE) electric field components (Eq. (3.34)) expressed as:

$$2in_0k_0\frac{\partial \Psi_x}{\partial z} = P_{xx}\Psi_x + P_{xy}\Psi_y; \tag{L.1a}$$

$$2in_0k_0\frac{\partial \Psi_y}{\partial z} = P_{yy}\Psi_y + P_{yx}\Psi_x. \tag{L.1b}$$

If the medium is an anisotropic crystal with the permittivity tensor on the form:

$$\bar{\bar{\varepsilon}} = \begin{pmatrix} \varepsilon_{xx} & \varepsilon_{xy} & 0 \\ \varepsilon_{yx} & \varepsilon_{yy} & 0 \\ 0 & 0 & \varepsilon_{zz} \end{pmatrix}, \tag{L.2}$$

Beam Propagation Method for Design of Optical Waveguide Devices, First Edition. Ginés Lifante Pedrola.
© 2016 John Wiley & Sons, Ltd. Published 2016 by John Wiley & Sons, Ltd.

then the differential operators P_{ij} are expressed by [1]:

$$P_{xx}\Psi_x = \frac{\partial}{\partial x}\left(\frac{1}{\varepsilon_{zz}}\frac{\partial}{\partial x}(\varepsilon_{xx}\Psi_x)\right) + \frac{\partial^2\Psi_x}{\partial y^2} + k_0^2\left(\frac{\varepsilon_{xx}}{\varepsilon_0} - n_0^2\right)\Psi_x; \tag{L.3a}$$

$$P_{yy}\Psi_y = \frac{\partial^2\Psi_y}{\partial x^2} + \frac{\partial}{\partial y}\left(\frac{1}{\varepsilon_{zz}}\frac{\partial}{\partial y}(\varepsilon_{yy}\Psi_y)\right) + k_0^2\left(\frac{\varepsilon_{yy}}{\varepsilon_0} - n_0^2\right)\Psi_y; \tag{L.3b}$$

$$P_{xy}\Psi_y = \frac{\partial}{\partial x}\left(\frac{1}{\varepsilon_{zz}}\frac{\partial}{\partial y}(\varepsilon_{yy}\Psi_y)\right) - \frac{\partial^2\Psi_y}{\partial x\partial y} + \frac{\varepsilon_{xy}}{\varepsilon_0}k_0^2\Psi_y; \tag{L.3c}$$

$$P_{yx}\Psi_x = \frac{\partial}{\partial y}\left(\frac{1}{\varepsilon_{zz}}\frac{\partial}{\partial x}(\varepsilon_{xx}\Psi_x)\right) - \frac{\partial^2\Psi_x}{\partial y\partial x} + \frac{\varepsilon_{yx}}{\varepsilon_0}k_0^2\Psi_x. \tag{L.3d}$$

Equations (L.1a) and (L.1b) can be written in matricial form as:

$$2in_0k_0\frac{\partial}{\partial z}\begin{bmatrix}\Psi_x\\\Psi_y\end{bmatrix} = \begin{bmatrix}P_{xx} & P_{xy}\\P_{yx} & P_{yy}\end{bmatrix}\begin{bmatrix}\Psi_x\\\Psi_y\end{bmatrix}. \tag{L.4}$$

To achieve an algorithm of second-order accuracy in the longitudinal direction using the ADI method, the matrix operator in Eq. (L.4) is decomposed according to [2]:

$$\begin{bmatrix}P_{xx} & P_{xy}\\P_{yx} & P_{yy}\end{bmatrix} = \begin{bmatrix}A_x + A_y & C\\D & B_x + B_y\end{bmatrix}. \tag{L.5}$$

here, A_x and A_y denote the x- and y-dependent parts of the operator P_{xx}, respectively, according to:

$$A_x\Psi_x \equiv \frac{\partial}{\partial x}\left(\frac{1}{\varepsilon_{zz}}\frac{\partial}{\partial x}(\varepsilon_{xx}\Psi_x)\right) + \frac{1}{2}k_0^2\left(\frac{\varepsilon_{xx}}{\varepsilon_0} - n_0^2\right)\Psi_x; \tag{L.6a}$$

$$A_y\Psi_x \equiv \frac{\partial^2\Psi_y}{\partial y^2} + \frac{1}{2}k_0^2\left(\frac{\varepsilon_{xx}}{\varepsilon_0} - n_0^2\right)\Psi_y. \tag{L.6b}$$

Similarly, the operators B_x and B_y denote the x- and y-dependent parts of P_{yy}, respectively, defined by:

$$B_x\Psi_y \equiv \frac{\partial^2\Psi_y}{\partial x^2} + \frac{1}{2}k_0^2\left(\frac{\varepsilon_{yy}}{\varepsilon_0} - n_0^2\right)\Psi_y; \tag{L.7a}$$

$$B_y\Psi_y \equiv \frac{\partial}{\partial y}\left(\frac{1}{\varepsilon_{zz}}\frac{\partial}{\partial y}(\varepsilon_{yy}\Psi_y)\right) + \frac{1}{2}k_0^2\left(\frac{\varepsilon_{yy}}{\varepsilon_0} - n_0^2\right)\Psi_y. \tag{L.7b}$$

The differential operators C and D denote the cross-coupling terms and are just the P_{xy} and P_{yx} operators, respectively, given by:

$$C\Psi_y \equiv \frac{\partial}{\partial x}\left(\frac{1}{\varepsilon_{zz}}\frac{\partial}{\partial y}\left(\varepsilon_{yy}\Psi_y\right)\right) - \frac{\partial^2 \Psi_y}{\partial x \partial y} + \frac{\varepsilon_{xy}}{\varepsilon_0}k_0^2\Psi_y; \qquad (L.8a)$$

$$D\Psi_x \equiv \frac{\partial}{\partial y}\left(\frac{1}{\varepsilon_{zz}}\frac{\partial}{\partial x}\left(\varepsilon_{xx}\Psi_x\right)\right) - \frac{\partial^2 \Psi_x}{\partial y \partial x} + \frac{\varepsilon_{yx}}{\varepsilon_0}k_0^2\Psi_x. \qquad (L.8b)$$

For the sake of readability, we will denote u and v as the x- and y-component of the slowly varying electric field (Ψ_x and Ψ_y), respectively. The finite difference scheme of the previously defined operator $A_x u$ is implemented as:

$$A_x u = \frac{T_{i-1,j}u_{i-1,j} - 2R_{i,j}u_{i,j} + T_{i+1,j}u_{i+1,j}}{(\Delta x)^2} + \frac{1}{2}k_0^2\left[nxx_{i,j}^2 - n_0^2\right]u_{i,j}, \qquad (L.9)$$

where the coefficients $T_{i\pm1,j}$ and $R_{i,j}$ are defined by:

$$T_{i\pm1,j} \equiv \frac{2nxx_{i\pm1,j}^2}{nzz_{i,j}^2 + nzz_{i\pm1,j}^2}; \qquad (L.10a)$$

$$R_{i,j} \equiv \frac{nxx_{i,j}^2}{nzz_{i+1,j}^2 + nzz_{i,j}^2} + \frac{nxx_{i,j}^2}{nzz_{i,j}^2 + nzz_{i-1,j}^2}, \qquad (L.10b)$$

where we have denoted $nxx^2 = \varepsilon_{xx}/\varepsilon_0$ and $nzz^2 = \varepsilon_{zz}/\varepsilon_0$.

Additionally, the finite-difference expressions of the operators $A_y u$, $B_x v$ and $B_y v$ are given by:

$$A_y u = \frac{u_{i,j-1} - 2u_{i,j} + u_{i,j+1}}{(\Delta y)^2} + \frac{1}{2}k_0^2\left[nxx_{i,j}^2 - n_0^2\right]u_{i,j}; \qquad (L.11a)$$

$$B_x v = \frac{v_{i-1,j} - 2v_{i,j} + v_{i+1,j}}{(\Delta x)^2} + \frac{1}{2}k_0^2\left[nyy_{i,j}^2 - n_0^2\right]v_{i,j}; \qquad (L.11b)$$

$$B_y v = \frac{T_{i,j-1}v_{i,j-1} - 2R_{i,j}v_{i,j} + T_{i,j+1}v_{i,j+1}}{(\Delta y)^2} + \frac{1}{2}k_0^2\left[nyy_{i,j}^2 - n_0^2\right]v_{i,j}, \qquad (L.11c)$$

where the coefficients $T_{i\pm1,j}$ and $R_{i,j}$ are:

$$T_{i\pm1,j} \equiv \frac{2nyy_{i\pm1,j}^2}{nzz_{i,j}^2 + nzz_{i\pm1,j}^2}; \qquad (L.12a)$$

$$R_{i,j} \equiv \frac{nyy_{i,j}^2}{nzz_{i+1,j}^2 + nzz_{i,j}^2} + \frac{nyy_{i,j}^2}{nzz_{i,j}^2 + nzz_{i-1,j}^2}, \qquad (L.12b)$$

with $nyy^2 = \varepsilon_{yy}/\varepsilon_0$.

Finally, the finite-difference approaches of the operators Cv and Du are implemented as:

$$Cv = \frac{1}{4\Delta x \Delta y}\left(S1_{i,j}v_{i+1,j+1} - S2_{i,j}v_{i+1,j-1} - S3_{i,j}v_{i-1,j+1} + S4_{i,j}v_{i-1,j-1}\right) + S5_{i,j}v_{i,j}; \qquad (L.13a)$$

$$Du = \frac{1}{4\Delta x \Delta y}\left(Z1_{i,j}u_{i+1,j+1} - Z2_{i,j}u_{i-1,j+1} - Z3_{i,j}u_{i+1,j-1} + Z4_{i,j}u_{i-1,j-1}\right) + Z5_{i,j}u_{i,j}, \qquad (L.13b)$$

and the coefficients $S1$–$S5$ and $Z1$–$Z5$ are given by:

$$S1_{i,j} \equiv \frac{nyy_{i+1,j+1}^2}{nzz_{i+1,j}^2} - 1; \qquad (L.14a)$$

$$S2_{i,j} \equiv \frac{nyy_{i+1,j-1}^2}{nzz_{i+1,j}^2} - 1; \qquad (L.14b)$$

$$S3_{i,j} \equiv \frac{nyy_{i-1,j+1}^2}{nzz_{i-1,j}^2} - 1; \qquad (L.14c)$$

$$S4_{i,j} \equiv \frac{nyy_{i-1,j-1}^2}{nzz_{i-1,j}^2} - 1; \qquad (L.14d)$$

$$S5_{i,j} \equiv nxy_{i,j}^2 k_0^2; \qquad (L.14e)$$

$$Z1_{i,j} \equiv \frac{nxx_{i+1,j+1}^2}{nzz_{i,j+1}^2} - 1; \qquad (L.14f)$$

$$Z2_{i,j} \equiv \frac{nxx_{i-1,j+1}^2}{nzz_{i,j+1}^2} - 1; \qquad (L.14g)$$

$$Z3_{i,j} \equiv \frac{nxx_{i+1,j-1}^2}{nzz_{i,j-1}^2} - 1; \qquad (L.14h)$$

$$Z4_{i,j} \equiv \frac{nxx_{i-1,j-1}^2}{nzz_{i,j-1}^2} - 1; \qquad (L.14i)$$

$$Z5_{i,j} \equiv nyx_{i,j}^2 k_0^2, \qquad (L.14j)$$

being $nxy^2 = \varepsilon_{xy}/\varepsilon_0$ and $nyx^2 = \varepsilon_{yx}/\varepsilon_0$.

In order to implement the ADI method for 3D-anisotropic beam propagation method (BPM), we split the matrix operator in Eq. (L.5) as the sum of two matrix operators as follows:

$$2in_0 k_0 \frac{\partial}{\partial z}\begin{bmatrix}\Psi_x \\ \Psi_y\end{bmatrix} = \begin{bmatrix} A_x + A_y & C \\ D & B_x + B_y \end{bmatrix}\begin{bmatrix}\Psi_x \\ \Psi_y\end{bmatrix} = \left(\begin{bmatrix} A_x & C \\ 0 & B_x \end{bmatrix} + \begin{bmatrix} A_y & 0 \\ D & B_y \end{bmatrix}\right)\begin{bmatrix}\Psi_x \\ \Psi_y\end{bmatrix}. \qquad (L.15)$$

After discretizing this differential equation using the standard the Crank–Nicolson scheme, we obtain:

$$\frac{2in_0k_0}{\Delta z}\left(\begin{bmatrix} u \\ v \end{bmatrix}^{m+1} - \begin{bmatrix} u \\ v \end{bmatrix}^{m}\right) = \frac{1}{2}\left(\begin{bmatrix} A_x & C \\ 0 & B_x \end{bmatrix} + \begin{bmatrix} A_y & 0 \\ D & B_y \end{bmatrix}\right)\left(\begin{bmatrix} u \\ v \end{bmatrix}^{m+1} + \begin{bmatrix} u \\ v \end{bmatrix}^{m}\right), \qquad (L.16)$$

where we have denoted u and v as the x and y component of the electric field envelope, respectively. Adopting the ADI method, introducing second-order error terms in Δz, the last equation can be written as:

$$\begin{bmatrix} u \\ v \end{bmatrix}^{m+1} = \frac{\left(1 - \dfrac{i\Delta z}{4n_0k_0}\begin{bmatrix} A_x & C \\ 0 & B_x \end{bmatrix}\right)}{\left(1 + \dfrac{i\Delta z}{4n_0k_0}\begin{bmatrix} A_x & C \\ 0 & B_x \end{bmatrix}\right)} \cdot \frac{\left(1 - \dfrac{i\Delta z}{4n_0k_0}\begin{bmatrix} A_y & 0 \\ D & B_y \end{bmatrix}\right)}{\left(1 + \dfrac{i\Delta z}{4n_0k_0}\begin{bmatrix} A_y & 0 \\ D & B_y \end{bmatrix}\right)} \cdot \begin{bmatrix} u \\ v \end{bmatrix}^{m}. \qquad (L.17)$$

In this form, Eq. (L.17) can be split in two artificial sub-steps to perform a single propagation step of length Δz, according to the sequence:

$$\begin{bmatrix} u \\ v \end{bmatrix}^{m+1/2} = \frac{\left(1 - \dfrac{i\Delta z}{4n_0k_0}\begin{bmatrix} A_y & 0 \\ D & B_y \end{bmatrix}\right)}{\left(1 + \dfrac{i\Delta z}{4n_0k_0}\begin{bmatrix} A_y & 0 \\ D & B_y \end{bmatrix}\right)} \cdot \begin{bmatrix} u \\ v \end{bmatrix}^{m}; \qquad (L.18a)$$

$$\begin{bmatrix} u \\ v \end{bmatrix}^{m+1} = \frac{\left(1 - \dfrac{i\Delta z}{4n_0k_0}\begin{bmatrix} A_x & C \\ 0 & B_x \end{bmatrix}\right)}{\left(1 + \dfrac{i\Delta z}{4n_0k_0}\begin{bmatrix} A_x & C \\ 0 & B_x \end{bmatrix}\right)} \cdot \begin{bmatrix} u \\ v \end{bmatrix}^{m+1/2}. \qquad (L.18b)$$

Equation (L.18a) is indeed the formal expression of the following equation:

$$\left(1 + \frac{i\Delta z}{4n_0k_0}\begin{bmatrix} A_y & 0 \\ D & B_y \end{bmatrix}\right) \cdot \begin{bmatrix} u \\ v \end{bmatrix}^{m+1/2} = \left(1 - \frac{i\Delta z}{4n_0k_0}\begin{bmatrix} A_y & 0 \\ D & B_y \end{bmatrix}\right) \cdot \begin{bmatrix} u \\ v \end{bmatrix}^{m}. \qquad (L.19)$$

Now, this vectorial equation is separated into its two components. The first component is expressed as:

$$\left(1 + \frac{i\Delta z}{4n_0k_0}A_y\right)u^{m+1/2} = \left(1 - \frac{i\Delta z}{4n_0k_0}A_y\right)u^{m}. \qquad (L.20a)$$

The second component is:

$$\left(1 + \frac{i\Delta z}{4n_0 k_0} B_y\right) v^{m+1/2} + \frac{i\Delta z}{4n_0 k_0} Du^{m+1/2} = \left(1 - \frac{i\Delta z}{4n_0 k_0} B_y\right) v^m - \frac{i\Delta z}{4n_0 k_0} Du^m. \qquad \text{(L.20b)}$$

Equation (L.20a) allows us to obtain the x-component of the intermediate field $u^{m+1/2}$ from the known field u^m. Then, Eq. (L.20b) is solved to find the y-component of the intermediate field $v^{m+1/2}$ from the known fields u^m, $u^{m+1/2}$ and v^m. This concatenated sequence allows us to perform the first sub-step of the propagation from m to $(m + 1/2)$.

To complete the propagation along a distance Δz, it is necessary to carry out a second step. For that purpose, let us consider Eq. (L.18b), which can be expressed in the form:

$$\left(1 + \frac{i\Delta z}{4n_0 k_0} \begin{bmatrix} A_x & C \\ 0 & B_x \end{bmatrix}\right) \cdot \begin{bmatrix} u \\ v \end{bmatrix}^{m+1} = \left(1 - \frac{i\Delta z}{4n_0 k_0} \begin{bmatrix} A_x & C \\ 0 & B_x \end{bmatrix}\right) \cdot \begin{bmatrix} u \\ v \end{bmatrix}^{m+1/2}. \qquad \text{(L.21)}$$

Once again, this matricial equation is decomposed in its two components, resulting in:

$$\left(1 + \frac{i\Delta z}{4n_0 k_0} A_x\right) u^{m+1} + \frac{i\Delta z}{4n_0 k_0} Cv^{m+1} = \left(1 - \frac{i\Delta z}{4n_0 k_0} A_x\right) u^{m+1/2} - \frac{i\Delta z}{4n_0 k_0} Cv^{m+1/2}; \qquad \text{(L.22a)}$$

$$\left(1 + \frac{i\Delta z}{4n_0 k_0} B_x\right) v^{m+1} = \left(1 - \frac{i\Delta z}{4n_0 k_0} B_x\right) v^{m+1/2}. \qquad \text{(L.22b)}$$

Using Eq. (L.22b) allows us to calculate the field v^{m+1} from the known field $v^{m+1/2}$. Once this field is obtained, Eq. (L.22a) is used to find the field u^{m+1} from the known fields $u^{m+1/2}$, $v^{m+1/2}$ and v^{m+1}, completing thus the whole step from m to $m + 1$ (Figure 3.24).

L.1 Numerical Implementation

L.1.1 First Component of the First Step

Regarding the first sub-step, the finite difference wave equation (L.20a), following the standard Crank–Nicolson scheme, can be rewritten in a more convenient form as:

$$\left(\frac{4in_0 k_0}{\Delta z} - A_y\right) u^{m+1/2} = \left(\frac{4in_0 k_0}{\Delta z} + A_y\right) u^m. \qquad \text{(L.23)}$$

The finite-difference form of the operator $A_y u$ has already been shown in Eq. (L.11a). Using that finite-difference expression, Eq. (L.23) is finally given by:

$$\begin{aligned}
&\frac{4in_0 k_0}{\Delta z} u_{i,j}^{m+1/2} - \frac{u_{i,j-1}^{m+1/2} - 2u_{i,j}^{m+1/2} + u_{i,j+1}^{m+1/2}}{(\Delta y)^2} - \frac{1}{2} k_0^2 \left[\left(nxx_{i,j}^{m+1/2}\right)^2 - n_0^2\right] u_{i,j}^{m+1/2} = \frac{4in_0 k_0}{\Delta z} u_{i,j}^m \\
&+ \frac{u_{i,j-1}^m - 2u_{i,j}^m + u_{i,j+1}^m}{(\Delta y)^2} + \frac{1}{2} k_0^2 \left[\left(nxx_{i,j}^m\right)^2 - n_0^2\right] u_{i,j}^m
\end{aligned} \qquad \text{(L.24)}$$

From this equation, the coefficients of the tridiagonal system remain as:

$$a_j = -\frac{1}{(\Delta y)^2};$$ (L.25a)

$$b_j = \frac{4in_0k_0}{\Delta z} + \frac{2}{(\Delta y)^2} - \frac{1}{2}k_0^2\left[\left(nxx_{i,j}^{m+1/2}\right)^2 - n_0^2\right];$$ (L.25b)

$$c_j = -\frac{1}{(\Delta y)^2};$$ (L.25c)

$$r_j = \frac{1}{(\Delta y)^2}\left[u_{i,j-1}^m + u_{i,j+1}^m\right] + \left\{\frac{4in_0k_0}{\Delta z} - \frac{2}{(\Delta y)^2} + \frac{1}{2}k_0^2\left[\left(nxx_{i,j}^m\right)^2 - n_0^2\right]\right\}u_{i,j}^m.$$ (L.25d)

L.1.2   Second Component of the First Step

To establish the finite-difference approximation of Eq. (L.20b), we put it in the form:

$$\left(\frac{4in_0k_0}{\Delta z} - B_y\right)v^{m+1/2} - Du^{m+1/2} = \left(\frac{4in_0k_0}{\Delta z} + B_y\right)v^m + Du^m,$$ (L.26)

and we use the finite difference forms of the operators $B_y v$ and Du as they were previously defined in Eqs. (L.11c) and (L.13b). Using these expressions, the finite difference equation corresponding to Eq. (L.26) is given by:

$$\frac{4in_0k_0}{\Delta z}v_{i,j}^{m+1/2} - \frac{T_{i,j-1}^{m+1/2}v_{i,j-1}^{m+1/2} - 2R_{i,j}^{m+1/2}v_{i,j}^{m+1/2} + T_{i,j+1}^{m+1/2}v_{i,j+1}^{m+1/2}}{(\Delta y)^2}$$

$$-\frac{1}{2}k_0^2\left[\left(nyy_{i,j}^{m+1/2}\right)^2 - n_0^2\right]v_{i,j}^{m+1/2}$$

$$= \frac{4in_0k_0}{\Delta z}v_{i,j}^m + \frac{T_{i,j-1}^m v_{i,j-1}^m - 2R_{i,j}^m v_{i,j}^m + T_{i,j+1}^m v_{i,j+1}^m}{(\Delta y)^2} + \frac{1}{2}k_0^2\left[\left(nyy_{i,j}^m\right)^2 - n_0^2\right]v_{i,j}^m$$

$$+\frac{1}{4\Delta x\Delta y}\left(Z1_{i,j}^m u_{i+1,j+1}^m - Z2_{i,j}^m u_{i-1,j+1}^m - Z3_{i,j}^m u_{i+1,j-1}^m + Z4_{i,j}^m u_{i-1,j-1}^m\right) + Z5_{i,j}^m u_{i,j}^m$$

$$+\frac{1}{4\Delta x\Delta y}\left(Z1_{i,j}^{m+1/2} u_{i+1,j+1}^{m+1/2} - Z2_{i,j}^{m+1/2} u_{i-1,j+1}^{m+1/2} - Z3_{i,j}^{m+1/2} u_{i+1,j-1}^{m+1/2} + Z4_{i,j}^{m+1/2} u_{i-1,j-1}^{m+1/2}\right)$$

$$+Z5_{i,j}^{m+1/2} u_{i,j}^{m+1/2}$$ (L.27)

The coefficients being:

$$a_j = -\frac{1}{(\Delta y)^2}T_{i,j-1}^{m+1/2};$$ (L.28a)

$$b_j = \frac{4in_0k_0}{\Delta z} + \frac{2}{(\Delta y)^2}R_{i,j}^{m+1/2} - \frac{1}{2}k_0^2\left[\left(nyy_{i,j}^{m+1/2}\right)^2 - n_0^2\right]; \tag{L.28b}$$

$$c_j = -\frac{1}{(\Delta y)^2}T_{i,j+1}^{m+1/2}; \tag{L.28c}$$

$$r_j = \frac{1}{(\Delta y)^2}\left[T_{i,j-1}^m v_{i,j-1}^m + T_{i,j+1}^m v_{i,j+1}^m\right] + \left\{\frac{4in_0k_0}{\Delta z} - \frac{2R_{i,j}^m}{(\Delta y)^2} + \frac{1}{2}k_0^2\left[\left(nyy_{i,j}^m\right)^2 - n_0^2\right]\right\}v_{i,j}^m$$

$$+ \frac{1}{4\Delta x\Delta y}\left(Z1_{i,j}^m u_{i+1,j+1}^m - Z2_{i,j}^m u_{i-1,j+1}^m - Z3_{i,j}^m u_{i+1,j-1}^m + Z4_{i,j}^m u_{i-1,j-1}^m\right) + Z5_{i,j}^m u_{i,j}^m$$

$$+ \frac{1}{4\Delta x\Delta y}\left(Z1_{i,j}^{m+1/2} u_{i+1,j+1}^{m+1/2} - Z2_{i,j}^{m+1/2} u_{i-1,j+1}^{m+1/2} - Z3_{i,j}^{m+1/2} u_{i+1,j-1}^{m+1/2} + Z4_{i,j}^{m+1/2} u_{i-1,j-1}^{m+1/2}\right)$$

$$+ Z5_{i,j}^{m+1/2} u_{i,j}^{m+1/2} \tag{L.28d}$$

where the T, R and $Z1$–$Z4$ coefficients are those defined in Eqs. (L.12a), (L.12b) and (L.14f)–(L.14i).

L.1.3 Second Component of the Second Step

Using Eq. (L22b) in the form:

$$\left(\frac{4in_0k_0}{\Delta z} - B_x\right)v^{m+1} = \left(\frac{4in_0k_0}{\Delta z} + B_x\right)v^{m+1/2}, \tag{L.29}$$

its finite-difference approach is given by:

$$\frac{4in_0k_0}{\Delta z}v_{i,j}^{m+1} - \frac{v_{i-1,j}^{m+1} - 2v_{i,j}^{m+1} + v_{i+1,j}^{m+1}}{(\Delta x)^2} - \frac{1}{2}k_0^2\left[\left(nyy_{i,j}^{m+1}\right)^2 - n_0^2\right]v_{i,j}^{m+1} = \frac{4in_0k_0}{\Delta z}v_{i,j}^{m+1/2}$$

$$+ \frac{v_{i-1,j}^{m+1/2} - 2v_{i,j}^{m+1/2} + v_{i+1,j}^{m+1/2}}{(\Delta x)^2} - \frac{1}{2}k_0^2\left[\left(nyy_{i,j}^{m+1/2}\right)^2 - n_0^2\right]v_{i,j}^{m+1/2} \tag{L.30}$$

and the coefficients of the tridiagonal system take the expressions:

$$a_i = -\frac{1}{(\Delta x)^2}; \tag{L.31a}$$

$$b_i = \frac{4in_0k_0}{\Delta z} + \frac{2}{(\Delta x)^2} - \frac{1}{2}k_0^2\left[\left(nyy_{i,j}^{m+1}\right)^2 - n_0^2\right]; \tag{L.31b}$$

$$c_i = -\frac{1}{(\Delta x)^2}; \tag{L.31c}$$

$$r_i = \frac{1}{(\Delta x)^2}\left[v_{i-1,j}^{m+1/2} + v_{i+1,j}^{m+1/2}\right] + \left\{\frac{4in_0k_0}{\Delta z} - \frac{2}{(\Delta x)^2} + \frac{1}{2}k_0^2\left[\left(nyy_{i,j}^{m+1/2}\right)^2 - n_0^2\right]\right\}v_{i,j}^{m+1/2}.$$

(L.31d)

L.1.4   First Component of the Second Step

The last step to complete the propagation is to solve Eq. (L.22a), which is written in the form:

$$\left(\frac{4in_0k_0}{\Delta z} - A_x\right)u^{m+1} - Cv^{m+1} = \left(\frac{4in_0k_0}{\Delta z} + A_x\right)u^{m+1/2} + Cv^{m+1/2}.$$

(L.32)

The finite-difference approach of this equation is given by:

$$\frac{4in_0k_0}{\Delta z}u_{i,j}^{m+1} - \frac{T_{i-1,j}^{m+1}u_{i-1,j}^{m+1} - 2R_{i,j}^{m+1}u_{i,j}^{m+1} + T_{i+1,j}^{m+1}u_{i+1,j}^{m+1}}{(\Delta x)^2} - \frac{1}{2}k_0^2\left[\left(nxx_{i,j}^{m+1}\right)^2 - n_0^2\right]u_{i,j}^{m+1}$$

$$= \frac{4in_0k_0}{\Delta z}u_{i,j}^{m+1/2} + \frac{T_{i-1,j}^{m+1/2}u_{i-1,j}^{m+1/2} - 2R_{i,j}^{m+1/2}u_{i,j}^{m+1/2} + T_{i+1,j}^{m+1/2}u_{i+1,j}^{m+1/2}}{(\Delta x)^2}$$

$$+ \frac{1}{2}k_0^2\left[\left(nxx_{i,j}^{m+1/2}\right)^2 - n_0^2\right]u_{i,j}^{m+1/2}$$

$$+ \frac{1}{4\Delta x\Delta y}\left(S1_{i,j}^{m+1/2}v_{i+1,j+1}^{m+1/2} - S2_{i,j}^{m+1/2}v_{i+1,j-1}^{m+1/2} - S3_{i,j}^{m+1/2}v_{i-1,j+1}^{m+1/2} + S4_{i,j}^{m+1/2}v_{i-1,j-1}^{m+1/2}\right)$$

$$+ S5_{i,j}^{m+1/2}v_{i,j}^{m+1/2} + \frac{1}{4\Delta x\Delta y}\left(S1_{i,j}^{m+1}v_{i+1,j+1}^{m+1} - S2_{i,j}^{m+1}v_{i+1,j-1}^{m+1} - S3_{i,j}^{m+1}v_{i-1,j+1}^{m+1} + S4_{i,j}^{m+1}v_{i-1,j-1}^{m+1}\right)$$

$$+ S5_{i,j}^{m+1}v_{i,j}^{m+1}$$

(L.33)

where the T, R and $S1$–$S5$ coefficients are those defined in Eqs. (L.10a), (L.10b) and (L.14a)–(L.14e). From this finite-difference equation, the coefficients of the tridiagonal system to be solved by the Thomas method are given by:

$$a_i = -\frac{1}{(\Delta x)^2}T_{i-1,j}^{m+1};$$

(L.34a)

$$b_i = \frac{4in_0k_0}{\Delta z} + \frac{2}{(\Delta x)^2}R_{i,j}^{m+1} - \frac{1}{2}k_0^2\left[\left(nxx_{i,j}^{m+1}\right)^2 - n_0^2\right];$$

(L.34b)

$$c_i = -\frac{1}{(\Delta x)^2}T_{i+1,j}^{m+1};$$

(L.34c)

$$r_i = \frac{1}{(\Delta x)^2} \left[T_{i-1,j}^{m+1/2} u_{i-1,j}^{m+1/2} + T_{i+1,j}^{m+1/2} u_{i+1,j}^{m+1/2} \right]$$

$$+ \left\{ \frac{4in_0 k_0}{\Delta z} - \frac{2R_{i,j}^{m+1/2}}{(\Delta x)^2} + \frac{1}{2} k_0^2 \left[\left(nxx_{i,j}^{m+1/2} \right)^2 - n_0^2 \right] \right\} u_{i,j}^{m+1/2}$$

$$+ \frac{1}{4\Delta x \Delta y} \left(S1_{i,j}^{m+1/2} v_{i+1,j+1}^{m+1/2} - S2_{i,j}^{m+1/2} v_{i+1,j-1}^{m+1/2} - S3_{i,j}^{m+1/2} v_{i-1,j+1}^{m+1/2} + S4_{i,j}^{m+1/2} v_{i-1,j-1}^{m+1/2} \right)$$

$$+ S5_{i,j}^{m+1/2} v_{i,j}^{m+1/2} + \frac{1}{4\Delta x \Delta y} \left(S1_{i,j}^{m+1} v_{i+1,j+1}^{m+1} - S2_{i,j}^{m+1} v_{i+1,j-1}^{m+1} - S3_{i,j}^{m+1} v_{i-1,j+1}^{m+1} + S4_{i,j}^{m+1} v_{i-1,j-1}^{m+1} \right)$$

$$+ S5_{i,j}^{m+1} v_{i,j}^{m+1}$$

$$(L.34d)$$

References

[1] Xu, C.L., Huang, W.P., Chrostowski, J. and Chaudhuri, S.K. (1994) A full-vectorial beam propagation method for anisotropic waveguides. *Journal of Lightwave Technology* **12**, 1926–1931.
[2] Hsuch, Y.L., Yang, M.C. and Chang, H.C. (1999) Three-dimensional noniterative full-vectorial beam propagation method based on the alternating direction implicit method. *Journal of Lightwave Technology* **17**, 2389–2397.

Appendix M

Rate Equations in a Four-Level Atomic System

The performance of optical amplifiers or lasers based on rare earths can be derived from the spectroscopic properties of the active ions, where rates equations must be established from the dynamics of the relevant levels involved in the amplification mechanism [1]. Here, we present the rate equations that describe the temporal evolution of the population densities of the levels in a four-level system (Figure M.1). Then, closed expressions for the population densities are derived in the continuous case regime for an ideal four-level system, where some useful approximations can be made.

The optical amplification or laser oscillation at λ_S takes place between the levels (3) and (2), while absorption at the wavelength pump λ_P corresponds to the (1) $\rightarrow$ (4) transition. As indicated in Figure M.1, the absorption from the ground state to the level (4) is regulated by the pump rate R_{14}. The level 4 relaxes quickly by non-radiative processes to level (3), with a probability W_{43}^{NR}. In addition, level (3) may relax by spontaneous emission, (3) $\rightarrow$ (2) radiative decay, with probability A_{32}, (3) $\rightarrow$ (1) radiative decay, with probability A_{31}, or by multi-phonon decay to the level 2 with a non-radiative probability W_{32}^{NR}. The level (2) relaxes to the ground state (1) by a non-radiative transition, with probability W_{21}^{NR}. Stimulated emissions may occur at λ_P ((4) $\rightarrow$ (1)), with a rate W_{41} and at λ_S ((3) $\rightarrow$ (2)), with a rate W_{32}. Finally, there also exists the possibility of absorption at the signal wavelength, having a rate R_{23}.

Taking into consideration those processes, the rate equations describing the population dynamics of the four-level system are [1]:

$$\frac{dN_4}{dt} = R_{14}N_1 - \left(W_{43}^{NR} + W_{41}\right)N_4; \tag{M.1a}$$

$$\frac{dN_3}{dt} = W_{43}^{NR}N_4 - \left(A_{32} + A_{31} + W_{32}^{NR} + W_{32}\right)N_3 + R_{23}N_2; \tag{M.1b}$$

Beam Propagation Method for Design of Optical Waveguide Devices, First Edition. Ginés Lifante Pedrola.
© 2016 John Wiley & Sons, Ltd. Published 2016 by John Wiley & Sons, Ltd.

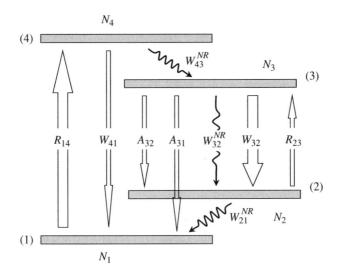

Figure M.1   Relevant processes occurring in a four-level atomic system

$$\frac{dN_2}{dt} = \left(A_{32} + W_{32}^{NR} + W_{32}\right)N_3 - W_{21}^{NR}N_2 - R_{23}N_2; \tag{M.1c}$$

$$\frac{dN_1}{dt} = W_{21}^{NR}N_2 + W_{41}N_4 + A_{31}N_3 - R_{14}N_1; \tag{M.1d}$$

$$N_T = N_1 + N_2 + N_3 + N_4, \tag{M.1e}$$

where N_i represents the population density in the i-th level ($i = 1, 2, 3, 4$) and N_T is the density of active ions in the matrix. At the steady state (continuous wave regime, CW), all the populations remain constant, meaning that all the derivatives in the above equations are equal to zero, which permit us to obtain the population of the levels.

In an ideal four-level system, some useful approximation can be made. First, the non-radiative de-excitation of the upper level is assumed to be much faster than the pumping rate $\left(W_{43}^{NR} \gg R_{14}\right)$, and then from Eq. (M.1a) the approximation $N_4 \ll N_1$ holds. More, if the non-radiative de-excitation of the level (2) is very fast $\left(W_{21}^{NR} \gg A_{32}, W_{32}, W_{32}^{NR}\right)$ then the population of the level (2) can be neglected respect to the level (3), according with Eq. (M.1c). Under these circumstances, the population of the levels can be easily calculated according to [2]:

$$N_1 \approx \frac{W_{32} + A_{32} + W_{32}^{NR}}{R_{14} + W_{32} + A_{32} + W_{32}^{NR}}N_T; \tag{M.2a}$$

$$N_2 \approx 0; \tag{M.2b}$$

$$N_3 \approx \frac{R_{14}}{R_{14} + W_{32} + A_{32} + W_{32}^{NR}}N_T; \tag{M.2c}$$

$$N_4 \approx 0; \tag{M.2d}$$

Besides the intrinsic spectroscopic parameters, the pump rate $R_{14} = R_P$ is related to the pump intensity I_P and the absorption cross-section σ_{14} of the $(1) \rightarrow (4)$ transition through:

$$R_P = \frac{\sigma_{14}}{hc/\lambda_P} I_P, \tag{M.3}$$

where h is Planck's constant and c is the speed of light in free space. Similarly, the stimulated emission rate $W_{32} = W_S$ is given by the relation:

$$W_S = \frac{\sigma_{32}}{hc/\lambda_S} I_S, \tag{M.4}$$

σ_{32} being the emission cross-section corresponding to the $(3) \rightarrow (2)$ transition and I_S is the intensity of the radiation at the signal wavelength λ_S.

Experimentally, it is difficult to measure separately the radiative and the non-radiative probability of the de-excitation $(3) \rightarrow (2)$. Indeed, the measured lifetime τ_{exp} of the level (3) decay incorporates both mechanisms, in such a way that:

$$\frac{1}{\tau_{exp}} = A_{32} + A_{31} + W_{32}^{NR}, \tag{M.5}$$

can be incorporated directly in the calculations of Eqs. (M.2a) and (M.2c).

References

[1] Ghatak, A.K. and Thyagarajan, K. (1989) Chapter 8 in *Optical Electronics*, Cambridge University Press, Cambridge.

[2] Lee, C.T. and Sheu, L.G. (1996) Analysis Nd:MgO:Ti:LiNbO$_3$ waveguide lasers with nonuniform concentration distributions. *Journal of Lightwave Technology* **14**, 2268–2276.

Appendix N

Overlap Integrals Method

The power evolution of pump and signal beams inside a z-invariant active channel waveguide can be solved by using the overlap integrals method, which considers the pump and signal beams as monochromatic waves with fixed normalized transversal distributions [1]. This algorithm can be used to simulate the performance of waveguide optical amplifiers as well as waveguide lasers [2, 3]. Here we present the basic of the method applied to the simulation of an end-pumped waveguide laser based on an atomic ideal four-level system [4].

The waveguide laser here modelled consists of a straight channel waveguide doped with active rare-earth ions, where two mirrors, M_1 and M_2, are attached to the ends of the waveguide to provide the feedback for laser oscillation (Figure N.1). The mirrors have partial reflectances at the pump and signal wavelengths, thus defining a 'double pass' cavity, along which both pump and laser beams travel forward and backward through the active waveguide.

If the active ions form an ideal four-level system (see Figure N.2, where the pump lifts the atoms from level (1)–(4) and laser oscillation takes place between levels (3) and (2)), under continuous-wave conditions at the steady-state, the rate equations which govern the temporal evolution of the population densities reduce to:

$$\frac{dN_4}{dt} \approx 0; \tag{N.1a}$$

$$\frac{dN_3}{dt} = R_{14}N_1 - (A_{3m} + W_{32})N_3; \tag{N.1b}$$

$$\frac{dN_2}{dt} \approx 0; \tag{N.1c}$$

$$N_1 = N_T - N_3, \tag{N.1d}$$

Beam Propagation Method for Design of Optical Waveguide Devices, First Edition. Ginés Lifante Pedrola.
© 2016 John Wiley & Sons, Ltd. Published 2016 by John Wiley & Sons, Ltd.

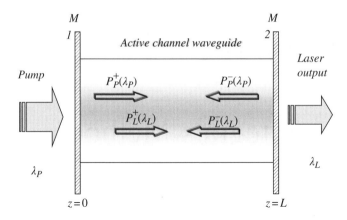

Figure N.1   Scheme of an end-pumped waveguide laser, showing the forward and backward beams

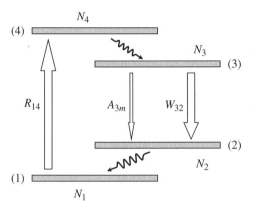

Figure N.2   Transitions occurring in an ideal four-level system

where $N_i(x,y,z)$ is the population density in the i-th level, and $N_T(x,y,z)$ is the concentration of the active ions. On the other hand, $R_{14}(x,y,z)$ and $W_{32}(x,y,z)$ are the stimulated absorption and emission rates and A_{3m} is the total de-excitation probability (radiative and non-radiative) of level (3), which is the inverse of the experimental lifetime τ_{exp} of the upper laser level (3).

The steady-state evolution of the forward and backward components of the pump beam (P_P^+ and P_P^- respectively) and those corresponding to the laser beam (P_L^+ and P_L^-, respectively) along the propagation direction z is described by:

$$\frac{dP_P^\pm(z)}{dz} = \pm\left[-\sigma_{14}\eta_1(z,\lambda_P)-\widetilde{\alpha}_P\right]P_P^\pm(z); \tag{N.2a}$$

$$\frac{dP_L^\pm(z)}{dz} = \pm\left[\sigma_{32}\eta_3(z,\lambda_L)-\widetilde{\alpha}_L\right]P_L^\pm(z), \tag{N.2b}$$

where σ_{14} represents the absorption cross section associated to the $(1){\rightarrow}(4)$ transition at the pump wavelength λ_P, σ_{32} is the emission cross section of the $(3){\rightarrow}(2)$ laser transition at the laser wavelength λ_L and $\tilde{\alpha}_{P,L}$ accounts for the passive waveguide losses at the pump or laser wavelength. On the other hand, $\eta_i(z, \lambda_{P,L})$, with $i = 1, 3$, denotes the overlapping integrals defined as:

$$\eta_i(z,\lambda_{P,L}) = \int\!\!\int_A \psi(x,y,\lambda_{P,L}) N_i(x,y,z)\,dxdy, \tag{N.3}$$

where ψ is the modal intensity distribution and N_i represents the steady-state population density of the i-th level. The integration extends to the cross section area (A) of the active channel waveguide. The modal intensity distribution ψ is normalized as:

$$\int\!\!\int_A \psi(x,y,\lambda_{P,L})\,dxdy = 1. \tag{N.4}$$

The stimulated absorption rates R_{14} depends on the longitudinal coordinate z through the power of the pump beam, and on the transversal coordinates (x and y) through the modal intensity distribution, according to:

$$R_{14}(x,y,z) = \frac{\lambda_P \sigma_{14}}{hc} \left[P_P^+(z) + P_P^-(z)\right] \psi(x,y,\lambda_P). \tag{N.5}$$

In a similar way, the stimulated emission rate W_{32} can be calculated by:

$$W_{32}(x,y,z) = \frac{\lambda_L \sigma_{32}}{hc} \left[P_L^+(z) + P_L^-(z)\right] \psi(x,y,\lambda_L). \tag{N.6}$$

After a finite discretization of the active channel waveguide along both the transversal and longitudinal dimensions, the solution of the model is obtained from the set of coupled differential equations (N.1) and (N.2), which are coupled through Eqs. (N.5) and (N.6). For doing that, at each point in the transversal plane the stimulated transition probabilities are evaluated taking into account the modal intensity distributions (Eqs. (N.5) and (N.6)). Then the rate equations are solved (Eqs. (N.1a)–(N.1c)), obtaining the different steady-state populations $N_i(x,y,z)$. When the functions $N_i(x,y,z)$ are known, the contribution of each point in the transverse plane to the overlapping integrals is evaluated by using Eq. (N.3). Once the transversal plane is solved, the pump and laser beams are propagated along the z direction according to Eqs. (N.2a) and (N.2b).

The boundary conditions that link the forward and backward beams at the input and output mirrors are given as function of the reflectances of the mirrors by:

$$P_P^+(z=0) = R_{1,P}P_P^-(z=0) + (1-R_{1,P})P_{Input,P}; \tag{N.7a}$$

$$P_L^+(z=0) = R_{1,L}P_L^-(z=0); \tag{N.7b}$$

$$P_P^-(z=L) = R_{2,P}P_P^+(z=L); \tag{N.7c}$$

$$P_L^-(z=L) = R_{2,L}P_L^+(z=L), \tag{N.7d}$$

where the subscript on the reflectances indicates mirror 1 or 2 at the pump (P) or laser (L) wavelength, and $P_{Input,P}$ is the input pump power at $z = 0$.

The algorithm starts by injecting the input pump power at $z = 0$, from where the atomic level populations and cavity powers evolve according to the dynamic equations. Initially the forward component of the laser beam at $z = 0$ is initialized at an arbitrarily low value (for instance, around 1 μW) from where it develops independently of this initial condition. The procedure iterates back and forth until the boundary conditions given by Eqs. (N.7a)–(N.7d) are fulfilled. Once the steady-state solution is reached, the waveguide laser output power is given by:

$$P_{Output,L} = (1 - R_{2,L})P_L^+ (z = L). \tag{N.8}$$

References

[1] Giles, C.R. and Desurvire, E. (1991) Modeling erbium-doped fiber amplifiers. *IEEE Journal of Lightwave Technology* **9**, 271–283.

[2] Vallés, J.A., Lázaro, J.A. and Rebolledo, M.A. (1996) Modeling of integrated erbium-doped waveguide amplifiers with overlapping factors methods. *IEEE Journal of Quantum Electronics* **32**, 1685–1694.

[3] Cantelar, E., Lifante, G. and Cussó, F. (2006) Modelling of Tm^{3+}-doped LiNbO$_3$ waveguide lasers. *Optical and Quantum Electronics* **38**, 111–122.

[4] Lee, C.T. and Sheu, L.G. (1996) Analysis Nd:MgO:Ti:LiNbO$_3$ waveguide lasers with nonuniform concentration distributions. *Journal of Lightwave Technology* **14**, 2268–2276.

Index

Beam Propagation Method for Design of Optical Waveguide Devices, First Edition. Ginés Lifante Pedrola.
© 2016 John Wiley & Sons, Ltd. Published 2016 by John Wiley & Sons, Ltd.